Advances in Planetary Science – Vol. 5

MARS ON EARTH

A STUDY OF THE QAIDAM BASIN

Advances in Planetary Science

Print ISSN: 2529-8054
Online ISSN: 2529-8062

Series Editor: Wing-Huen Ip *(National Central University, Taiwan)*

The series on Advances in Planetary Science aims to provide readers with overviews on many exciting developments in planetary research and related studies of exoplanets and their habitability. Besides a running account of the most up-to-date research results, coverage will also be given to descriptions of milestones in space exploration in the recent past by leading experts in the field.

Published

Vol. 5 *Mars on Earth: A Study of the Qaidam Basin*
edited by Long Xiao

Vol. 4 *Planetary Habitability in Binary Systems*
by Elke Pilat-Lohinger, Siegfried Eggl and Ákos Bazsó

Vol. 3 *Planetary Habitability and Stellar Activity*
by Arnold Hanslmeier

Vol. 2 *Origin and Evolution of Comets: Ten Years after the Nice Model and One Year after Rosetta*
by Hans Rickman

Vol. 1 *Nuclear Planetary Science: Planetary Science Based on Gamma-Ray, Neutron and X-Ray Spectroscopy*
by Nobuyuki Hasebe, Kyeong Ja Kim, Eido Shibamura and Kunitomo Sakurai

Advances in Planetary Science – Vol. 5

MARS ON EARTH

A STUDY OF THE QAIDAM BASIN

Editor

Long Xiao

China University of Geosciences, Wuhan, China
Macau University of Science and Technology, Macau, China

World Scientific

NEW JERSEY • LONDON • SINGAPORE • BEIJING • SHANGHAI • HONG KONG • TAIPEI • CHENNAI • TOKYO

Published by

World Scientific Publishing Co. Pte. Ltd.
5 Toh Tuck Link, Singapore 596224
USA office: 27 Warren Street, Suite 401-402, Hackensack, NJ 07601
UK office: 57 Shelton Street, Covent Garden, London WC2H 9HE

Library of Congress Cataloging-in-Publication Data
Names: Xiao, Long (Geologist), editor.
Title: Mars on Earth : a study of the Qaidam Basin / editor, Long Xiao, China University of Geosciences, Wuhan, China, Macau University of Science and Technology, Macau, China.
Description: New Jersey : World Scientific, [2020] | Series: Advances in planetary science, 2529-8054 ; Vol. 5 | Includes bibliographical references and index.
Identifiers: LCCN 2020040391 | ISBN 9789811224010 (hardcover) | ISBN 9789811224027 (ebook) | ISBN 9789811224034 (ebook other)
Subjects: LCSH: Mars (Planet)--Geomorphology. | Mars (Planet)--Geology--Earth analogs. | Tsaidam Basin (China)
Classification: LCC QB643.G46 M365 2020 | DDC 559.9/23--dc23
LC record available at https://lccn.loc.gov/2020040391

British Library Cataloguing-in-Publication Data
A catalogue record for this book is available from the British Library.

For any available supplementary material, please visit
https://www.worldscientific.com/worldscibooks/10.1142/11930#t=suppl

Desk Editors: Anthony Alexander/Amanda Yun

Typeset by Stallion Press
Email: enquiries@stallionpress.com

https://doi.org/10.1142/9789811224027_fmatter

Preface

Mars, the fourth planet and the most Earth-like body in the solar system, has been a focus of deep space exploration by many space agencies for many years. Both global orbiting and *in situ* explorations have been carried out in the past sixty years. The results indicate that there are a variety of landforms and geological features formed by long-term and multiple-type of geological activities. The current topography, geomorphologies, and surface composition of Mars are direct expressions of past endogenic and exogenic geological forces among which water-related landforms have been hotly studied because they showed evidence of ancient habitable environments comparable to Earth. Paleolake basins are the most suitable sites for the life origination and preservation on Mars, and were targeted for life detection by many *in situ* exploration missions. However, due to the limitation of exploration data, we cannot fully understand the mechanism of the formation of various geomorphological and the environmental characteristics reflected in the ancient paleolake basins on the surface of Mars. Therefore, analogy studies with similar regions on Earth are an important way to better understand the geological processes and habitability of the Martian environment.

On Earth, a couple of dried lakes have been selected for comparative studies with Mars. The results of these comparative studies not only provide indirect evidence for understanding the mechanism of the formation of Mars-like environments, but also provide ground verification targets for the implementation of Mars exploration plans. This book focuses on the Qaidam Basin in northeastern Tibetan Plateau, one of the most important analogue sites, an area very similar to the Martian environment. There are plenty of water systems and river networks, lacustrine deposits, playas and aeolian landforms similar to the Martian surface. The environment here is very extreme. Cold, dry, thin air and strong ultraviolet radiation are the main environmental characteristics. It is a demonstration area for studying the survival and preservation of microorganisms in extreme environments, and sheds light on the habitability of Mars.

In this book, we first introduce the ways used for studying Mars and present the updated surface features of Mars in general. Then we describe the general background of the Qaidam Basin, its geological settings, geographic and geomorphological features, desert landforms, and climate environments. Followed by detailed studies of the lakes, aeolian landforms, polygonal features, valleys, gullies within the Qaidam Basin and their counterparts on Mars, we discuss the habitability and astrobiological significances in the end.

The material in this book is based on our studies over the last 10 years. However, without Prof. Wing Ip's encouragement, it would not have been possible to compile all these Mars-like geomorphology and geological features together to be published by World Scientific. In addition to listed co-authors, we thank all colleagues and students who took part in our field trips and provided constructive suggestions that have improved the quality of the manuscript.

https://doi.org/10.1142/9789811224027_fmatter

About the Editor

Long Xiao is a Professor at the State Key Laboratory of Geological Processes and Mineral Resources, Planetary Science Institute, School of Earth Sciences, China University of Geosciences (CUG), China. He founded the first planetary science institute in Chinese universities. He is an expert in planetary geology and comparative planetology, with more than 160 papers published in scientific journals, three first-authored books, and three co-authored books. He also serves on the editorial board of a number of *Research in Astronomy and Astrophysics, Acta Geochimica, Journal of Earth Science, Geochimica, Science China: Physics, Mechanics & Astronomy*. He has participated in China's Chang'e lunar exploration missions (Chang'e-1, -2, -3, -4, and -5) and Mars exploration program (Tianwen 1). He has organized a number of International workshops: e.g. International Symposium on Lunar and Planetary Sciences (ISLPS), Summer School of Planetary Science and Exploration in Eastern Asia (SSPSEEA) and conferences in planetary science, and has served as a referee in many Referee for International Journals: *Nature Geosciences*, *Earth and Planetary Science Letters*, *Geophysical Research Letters*, *Journal of Geophysical Research*, *Icarus*, *Astronomy Advance*, *Lithos*, *Precambrian Research*, *Earth-Science Reviews*, etc. He is also a member of several Member of the Expert Group on Space Science and Applications for Manned Space Engineering of China; Strategic expert member of Institute of Space Science, Chinese Academy of Sciences Scientific Adviser of Lunar Exploration and Space Engineering Centre, China National Space Administration; Member of Chinese Antarctic Meteorite Expert Committee; Member of the Specialized Group on Aerospace and Deep Space of the Science and Technology Committee of CNSA.

https://doi.org/10.1142/9789811224027_fmatter

About the Contributors

Ziye Cheng is a Full Lecturer at the Institute of Geography and Tourism, Guangdong University of Finance & Economics. She received her Ph.D. from the China University of Geosciences (CUG), in 2016. Her research focuses on the microbes of extreme environments on Earth and their astrobiological implications.

Yanan Dang is a postdoctoral researcher at the State Key Laboratory of Lunar and Planetary Sciences, Macau University of Science and Technology (MUST), Macau, China. Her study focuses on the field of planetary geology, in particular, study of landform analogy of typical landforms on Earth and Mars.

Jun Huang is an Associate Professor at Planetary Science Institute, China University of Geosciences (CUG). His research focuses on the composition, physical properties and processes, and morphology of planetary surfaces, with an emphasis on Mars and the Moon.

Ting Huang is a postdoctoral researcher at the State Key Laboratory of Lunar and Planetary Sciences, Macau University of Science and Technology (MUST), Macau, China. Her research focuses on the field of comparative planetology and astrobiology, especially of the extremophiles and ancient biomarkers in Martian analogue sites.

Yutong Shi is a doctoral student at Planetary Science Institute, China University of Geosciences (CUG). Her research includes the morphology and climate implication of valley networks on Mars and soil sedimentology at the Gale crater.

Jiang Wang, Ph.D., is a Lecturer at Planetary Science Institute, China University of Geosciences (CUG), mainly engaged in the research related to comparative planetary geology, especially aeolian landforms. He has published more than 10 papers in academic journals.

Yi Xu is an assistant professor at the State Key Laboratory of Lunar and Planetary Sciences, Macau University of Science and Technology (MUST), Macau, China. Her research fields include analogy study of typical landforms on Earth and Mars, surface process and shallow subsurface structure and properties of the Moon and Mars. She has participated in Chinese Chang'e lunar exploration missions and Tianwen-1 Mars exploration program.

Mingjie Zhang is a doctoral student at the Key Laboratory of Lunar and Planetary Science, Macau University of Science and Technology (MUST), Macau, China. Her research focuses on sedimentary landforms on Mars, through the study of both analogs and simulations.

Jiannan Zhao is a postdoctoral researcher at Planetary Science Institute, China University of Geosciences (CUG). His research involves the geomorphology, mineralogy, and chronology of the Moon and Mars, focusing on Martian hydrologic activity. He has published more than 20 papers in academic journals and has participated in China's Chang'e lunar exploration missions and the Tianwen-1 Mars exploration program.

https://doi.org/10.1142/9789811224027_fmatter

Acknowledgments

We are grateful for funding from the Science and Technology Development Fund (FDCT) of Macau (Nos. 121/2017/A3, 119/2017/A3), National Natural Science Foundation of China (Nos. 41830214, 41772050, and 41773061), The Pre-research Project on Civil Aerospace Technologies of CNSA (D020101), MOST Special Fund from the State Key Laboratory of Geological Processes and Mineral Resources, China University of Geosciences (No. MSFGPMR05), and the Fundamental Research Funds for the Central Universities, China University of Geosciences (Wuhan) (Nos. CUGYCJH18-01 and CUG2017G02). We would like to thank Prof. James Head, Dr. Harald Hiesinger, Dr. Dennis Reiss, Dr. Zhiyong Xiao, Dr. Zhongping Lai, and Dr. Goro Komatsu for suggestions and constructive discussions. Dr. Le Qiao, Ms. Zhuqing Xue, Mr. Limin Zhao, Mr. Yu Wei, Mr. Ping Zhu, Mr. Bo Wang, Mr. Hansheng Liu, Mr. Jiawei Zhao, Mr. Siyuan Zhao, and Mr. Ning Zhang are thanked for fieldwork assistances.

https://doi.org/10.1142/9789811224027_fmatter

Contents

Preface v

About the Editor vii

About the Contributors ix

Acknowledgments xi

Chapter 1 Ways to Study Mars **1**
Long Xiao

1.1 Introduction 1
1.2 Global Orbiting Survey 1
1.2.1 Composition 2
1.2.2 Geomorphology and topography 5
1.2.3 Interior 7
1.3 *In Situ* Exploration 8
1.4 Sample Return 10
1.5 Analog Study 11
1.5.1 The role of analogies 11
1.5.2 Terrestrial analogies in Martian geology 14
1.6 Martian Meteorites 15
1.7 Simulation 22
1.8 Summary 24
Acknowledgments 24
References 25

Chapter 2 The Surface of Mars: An Overview **35**
Long Xiao

2.1 Introduction 35
2.2 Endogenic Geomorphology 36
  2.2.1 Tectonic features 36
  2.2.2 Volcanic features 40
2.3 Exogenic Geomorphology 45
  2.3.1 Water-related landforms 45
  2.3.2 Wind and aeolian landforms 52
  2.3.3 Ice and polar deposits 55
  2.3.4 Impact cratering formed landforms 56
  2.3.5 Mass wasting 57
2.4 Geological History 59
2.5 Summary 61
Acknowledgments 63
References 63

Chapter 3 The Qaidam Basin, NE Tibetan Plateau **71**
Long Xiao, Jiang Wang, Yanan Dang, Jiannan Zhao and Mingjie Zhang

3.1 Introduction 71
3.2 Geological Setting 73
  3.2.1 Stratigraphy 73
  3.2.2 Structure 78
  3.2.3 Geological history 81
3.3 Geographical and Geomorphological Features 82
  3.3.1 Surrounding mountain ranges 82
  3.3.2 Inside the basin 86
  3.3.3 Water system and lakes 87
  3.3.4 Playas and polygonal landforms 90
3.4 Wind Erosion and Aeolian Landform 93
  3.4.1 Macroscale wind erosion features: Yardangs 94
  3.4.2 Meso-micro-scale wind erosional features: Ventifacts and Tafoni 96
  3.4.3 Sand Dunes 101
3.5 Plants in Deserts 102
3.6 Climate Environment 105
Acknowledgments 106
References 106

Chapter 4 Lakes **111**
Jiannan Zhao and Ting Huang

4.1 Introduction 111
4.2 Paleolakes on Mars 112
4.2.1 Classification and distribution of Martian paleolakes 113
4.2.2 Martian paleolake morphology 116
4.2.3 Mineralogy of lake deposits 119
4.2.4 Age of Martian paleolakes 121
4.2.5 Origin of Martian paleolakes 122
4.3 Lakes and Playas in the Qaidam Basin 123
4.3.1 Lakes in the Qaidam Basin 124
4.3.2 Playas in the Qaidam Basin 137
4.4 Summary 145
Acknowledgments 147
References 148

Chapter 5 Aeolian Landforms **157**
Jiang Wang, Jun Huang and Long Xiao

5.1 Introduction 157
5.2 Yardangs 158
5.2.1 Yardangs on Mars 158
5.2.2 Yardangs in the Qaidam Basin 164
5.2.3 Evolution of yardangs in the Qaidam Basin 171
5.2.4 Control factors of yardang morphology 175
5.3 Dunes 183
5.3.1 Dunes on Mars 183
5.3.2 Dunes in the Qaidam Basin 188
5.4 Summary 192
Acknowledgments 193
References 193

Chapter 6 Polygons **199**
Yanan Dang, Long Xiao, and Yi Xu

6.1 Introduction 199
6.1.1 Thermal contraction polygon 200
6.1.2 Desiccation (Shrinkage) crack polygon 201
6.1.3 Lava polygon 201
6.1.4 Salt polygon 203

6.2 Polygonal Terrains on Mars 203
6.2.1 Distribution of salt polygons on Mars 206
6.2.2 Morphology of salt polygons on Mars 207
6.3 Polygonal Ground in the Qaidam Basin 212
6.3.1 Distribution of polygons in the Qaidam Basin 214
6.3.2 Morphology of salt polygons in the Qaidam Basin 214
6.3.3 Formation and evolution of salt polygons in the Qaidam Basin 217
6.4 Comparison of Salt Polygons in the Qaidam Basin and on Mars 224
6.4.1 Morphological observations 224
6.4.2 Clues for depositional environments on Mars 233
6.5 Implications for Martian Environmental and Climatic Change 234
6.5.1 Implications for climate change 234
6.5.2 Hydrology of lacustrine/playa environments on Mars and its implications for polygons formation 235
6.6 Summary 238
Acknowledgments 239
References 239

Chapter 7 Valleys **249**
Jiannan Zhao, Yutong Shi, and Long Xiao

7.1 Introduction 249
7.2 Valleys on Mars 250
7.2.1 Martian valley morphology 250
7.2.2 Spatial and temporal distribution of Martian valleys 257
7.2.3 Origin of Martian valleys 259
7.3 Valleys in the Qaidam Basin 261
7.3.1 Distribution and morphology of valleys in the Qaidam Basin 261
7.3.2 Origin of valleys in the Qaidam Basin 267
7.4 Summary 269
Acknowledgments 269
References 270

Chapter 8 Gullies **275**
Yutong Shi, Jiannan Zhao and Jiang Wang

8.1 Introduction 275

8.2 Gullies on Mars 276
 8.2.1 Global distribution of Martian gullies 276
 8.2.2 Martian gully morphology 278
 8.2.3 Origin of Martian gullies 281
8.3 Gullies in the Qaidam Basin 284
 8.3.1 Distribution and morphology of gullies in the Qaidam Basin 284
 8.3.2 Origin of gullies in the Qaidam Basin 285
8.4 Summary 289
Acknowledgments 289
References 289

Chapter 9 Habitability and Astrobiological Significances 293
Ting Huang, Ziye Cheng, and Long Xiao

9.1 Introduction 293
9.2 Water on Mars 294
9.3 Organic Compounds Found on Mars 297
 9.3.1 Organics in Martian meteorites 297
 9.3.2 Organics detected by Mars missions 298
 9.3.3 Methane in Mars atmosphere 300
9.4 Habitability Assessment of Mars Analog Sites 301
9.5 The Habitability of the Qaidam Basin 303
 9.5.1 Microorganism studies of saline lakes and evaporative sediments in the Qaidam Basin 303
 9.5.2 Lipids recovered from evaporites of the Qaidam Basin 311
9.6 The Astrobiological Implications from the Qaidam Basin 315
9.7 Summary 318
Acknowledgments 319
References 319

Index 329

https://doi.org/10.1142/9789811224027_0001

Chapter 1

Ways to Study Mars

Long Xiao

State Key Laboratory of Geological Processes and Mineral Resources, Planetary Science Institute, School of Earth Sciences, China University of Geosciences, Wuhan 430074, PR China

1.1 Introduction

The study of Mars began with telescope observations and has since expanded to the exploration with orbiters and landers, the study of Martian meteorites, simulation studies, and planned sample-return studies in the near future. Each of these methods has advantages. The orbiters can be used for large-scale and even global-scale exploration (including imaging and composition detection, etc.), and the *in situ* landers (including the rovers) can facilitate detailed investigation of specific sites of local regions. The most accurate composition data can be obtained by studying the Martian samples in the laboratory, and then the interpretation results of remote sensing data can be verified. Geological interpretation of extraterrestrial phenomena mainly relies on our understanding of similar phenomena on Earth, thus analog study is essentially important for planetary geosciences.

1.2 Global Orbiting Survey

The global-scale exploration of Mars is mainly based on remote sensing from orbital platforms. Remote sensing is usually used to obtain qualitative or quantitative information of electromagnetic radiation transmission of the target object through instruments on the orbiting spacecraft. The quality of remote sensing data is determined by spatial resolution, spectral resolution, observation time, and scientific content. Historically, remote sensing observations have been the preferred way to explore planetary objects and, for most objects in the solar system, such observations remain the only way.

Remote sensing experiments utilize different techniques, based on where and how the electromagnetic radiation is generated, to investigate the object. While passive remote sensing takes advantage of either radiation generated by the target body or by external sources such as the Sun (e.g. X-ray radiation, infrared radiation, microwaves, see Fig. 1.1) which interact with the target body, active remote sensing is based on electromagnetic radiation generated by the spacecraft experiment, which then interacts with the target body (e.g. radar, laser). Both passive and active remote sensing have been widely applied in Mars exploration.

1.2.1 *Composition*

Elemental and mineral composition can be obtained by using short wavelength/high energy (gamma ray and X-ray) and middle wavelength (visible and infrared) electromagnetic spectra, respectively (Fig. 1.1). For Mars, only one gamma ray spectrometer (GRS) has been used to detect elemental distribution on Mars, while several near infrared, short-wave infrared, and thermal infrared spectrometers have been used to acquire mineral composition.

The GRS is an instrument on the 2001 Mars Odyssey spacecraft. The spatial resolution of the instrument is about 300 km. It returns compositional data from Mars's surface, identifying certain elements (Cl, Fe, K, Si, and Th) and the location of hydrogen/water (http://grs.lpl.arizona.edu/specials/) (Bandfield, 2007; Boynton *et al*., 2002). Additionally, gamma-ray spectroscopy is uniquely suited to the task of characterizing the elemental composition of materials beneath an optically thick (>mm) layer of dust or soil. Gamma rays sample to depths of order tens of cm, making them relatively insensitive to the composition of a thin (<1 cm) surficial layer. In principle, depth-varying composition information is available in gamma-ray spectra, as most elements produce gamma-ray emissions at multiple energies, so gamma-ray measurements can in principle also be used to identify depth-dependent variations in subsurface composition (Peplowski *et al*., 2018).

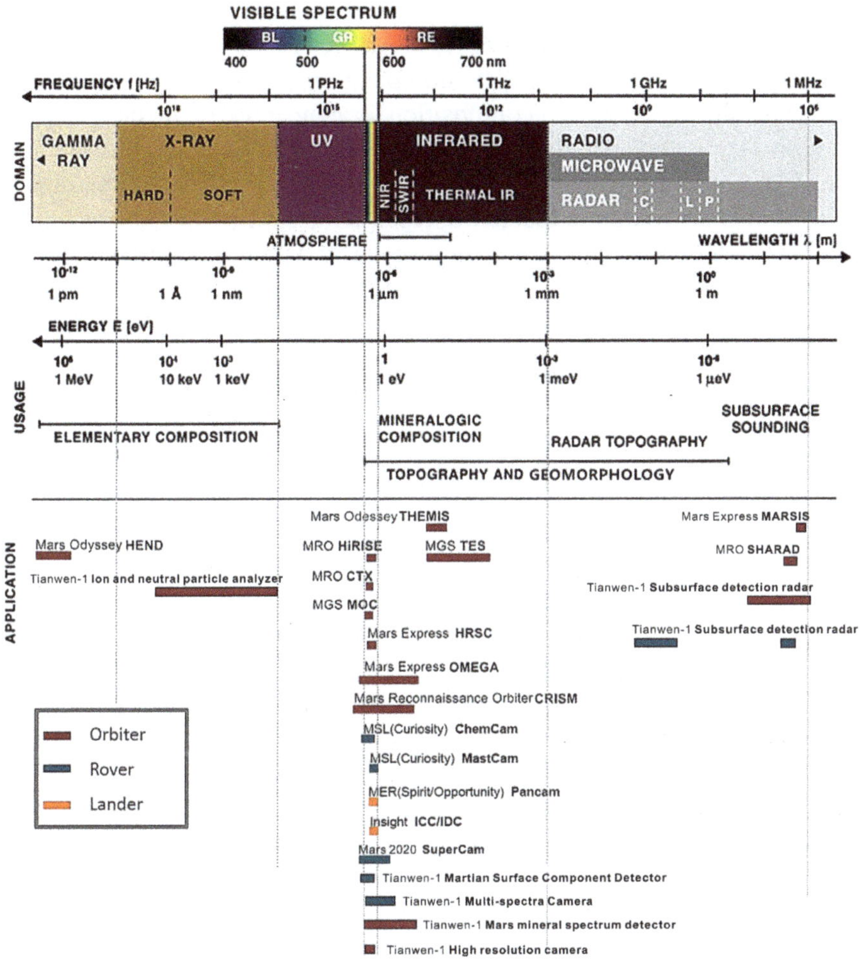

Fig. 1.1. Regions of the electromagnetic spectrum used in Martian remote sensing, and examples of applications on past and recent missions (Complied after Rossi and Gasselt, 2018).

The Thermal Emission Spectrometer (TES) was an instrument on board Mars Global Surveyor. TES collected two types of data, hyperspectral thermal infrared data from 6 to 50 micrometers (μm) and bolometric visible-NIR (0.3–2.9 μm) measurements. TES have been used to determine compositions and distributions of Martian low-albedo regions, and two surface spectral signatures were identified: a basaltic composition dominated by plagioclase feldspar and clinopyroxene, and an andesitic composition dominated by plagioclase feldspar and volcanic glass. The distribution of the two rock types is separated roughly along the planetary dichotomy. The basaltic composition is confined to older surfaces, and the more silicic composition is concentrated in the younger northern plains (Bandfield *et al.*, 2000), while this observation was later argued by Karunatillake *et al.* (2006) on the basis of GRS K/Th data that the lowlands do contain a distinct igneous unit, but that the Si concentration is not high enough for that unit to be andesitic in composition, as proposed by Bandfield *et al.* (Rogers *et al.*, 2015). TES also identified a large area of the Nili Fossae formation that contained the mineral olivine. Olivine was also discovered in many other small outcrops within 60 degrees north and south of the equator (Christensen *et al.*, 2001).

The Thermal Emission Imaging System (THEMIS) is a camera on board the 2001 Mars Odyssey orbiter. It images Mars in the visible and infrared parts of the electromagnetic spectrum in order to determine the thermal properties of the surface and to refine the distribution of minerals on the surface of Mars as determined by TES. Additionally, it helps scientists to understand how the mineralogy of Mars relates to its landforms, and it can be used to search for thermal hotspots in the Martian subsurface. The presence of minerals such as carbonates, silicates, hydroxides, sulfates, amorphous silica, oxides, and phosphates can be determined from THEMIS measurements. THEMIS has found a wide range of igneous rocks (e.g. olivine-rich rocks, granite-like rocks) and minerals (Christensen *et al.*, 2004b; Fergason *et al.*, 2006; Stockstill *et al.*, 2007). THEMIS also firstly confirmed the presence of water ice at the southern polar cap, melting snow carved gullies, evolved lavas, and helped to choose the landing site for Opportunity rover because THEMIS and TES showed signs of water in the rocks and surface materials in Meridiani Planum, a plain located 2° to the south of Martian equator (Andrews-Hanna *et al.*, 2007; Christensen *et al.*, 2004a; Williams *et al.*, 2017) (http://themis.asu.edu).

The Visible and Infrared Mineralogical Mapping Spectrometer (OMEGA) (Observatoire pour la Minéralogie, l'Eau, les Glaces et l'Activité) on Mars Express is used to determine mineral composition of the surface of Mars. OMEGA has covered almost the entire surface with a footprint 2 to 5 km in size, and some 5% at higher resolution (<400 m). On each pixel, OMEGA acquires the spectrum from 0.35 to 5.1 μm, enabling the identification of the composition (mineral, frost, ice). OMEGA has identified both pristine mafic mineral and their alteration products, in the form of anhydrous ferric oxides, clays, and other phases, widely spread over the bright regions, and hydrated minerals restricted to localized spots (Bibring *et al.*, 2006).

The Compact Reconnaissance Imaging Spectrometer for Mars (CRISM) is on the Mars Reconnaissance Orbiter (MRO). The instrument is a visible and near-infrared (VNIR) spectrometer that is used to produce detailed maps of the surface mineralogy of Mars. It operates from 0.37 to 3.92 μm, measures the spectrum in 544 channels (each 6.55 nm wide), and has a resolution of 18 m at an altitude of 300 km (Pelkey *et al.*, 2007). CRISM was used to identify minerals and chemicals indicative of the past or present existence of water on the surface of Mars. These materials include iron, oxides, phyllosilicates, and carbonates, which have characteristic patterns in their visible-infrared energy. In addition, CRISM is monitoring ice and dust particulates in the Martian atmosphere to learn more about its climate and seasons. CRISM has made great achievements in identifying aqueous minerals, detecting crustal compositions, and understanding modern climate of Mars (Bultel *et al.*, 2013; Carter *et al.*, 2013; Jain and Chauhan, 2015; Mustard *et al.*, 2008) (http://crism.jhuapl.edu/).

1.2.2 *Geomorphology and topography*

To investigate the surface features of Mars, various cameras and radar have been used to acquire geomorphological and topographic information. The cameras and radar cover wavelengths from visible to radio waves. The major ones are introduced as follows.

THEMIS, in addition to its function of composition detection, also has a visible imaging camera that acquires data in five spectral bands and takes images with a spatial resolution of 18 m. This resolution is intermediate between large-scale images from the Viking Orbiters (150–300 m per pixel) and the high-resolution images from the Mars Orbiter Camera (MOC) on board Mars Global Surveyor (1.5–3 m per pixel). Visible images from THEMIS are usually close to 20 km wide. The data have been used to determine the geological record of past liquid and volcanic environments on Mars (Christensen *et al.*, 2003; Fergason *et al.*, 2006; Osterloo *et al.*, 2007). Additionally, this dataset can be used in conjunction with the infrared data to identify potential landing sites for future Mars missions.

High Resolution Imaging Science Experiment (HiRISE) is a camera on the Mars Reconnaissance Orbiter (McEwen *et al.*, 2007). It consists of a 0.5 m aperture reflecting telescope, the largest so far of any deep space mission, which allows it to take pictures of Mars with resolutions of 0.3 m/pixel. It has returned abundant high-resolution imagery data and allows us to look at the targeted area in detail, including a closer look at fresh Martian craters, revealing alluvial fans, viscous flow features, and ponded regions of pitted materials containing breccia clast. This allows the study of the age of Martian features, looking for landing sites for future Mars landers (Cull *et al.*, 2010; Kirk *et al.*, 2008). By doing so, it is allowing better studies of Martian gullies, recurring slope lineae (RSL), channels and valleys, volcanic landforms, possible former lakes and oceans, sand dune fields, and other surface landforms on the Martian surface. Combined with CRISM data, both the textures and compositions of specific outcrops can be determined. Also, by comparing pictures taken in different times, the seasonal changes of some phenomena, such as sand dunes and slope lineae, can be observed (Auld and Dixon, 2016; Bridges *et al.*, 2010; Dundas *et al.*, 2015) (https://www.uahirise.org).

The Context Camera (CTX) provides grayscale images (500–800 nm) with a pixel resolution up to about 6 m (Malin *et al.*, 2007). CTX is designed to provide context maps for the targeted observations of HiRISE and CRISM, and is also used to mosaic large areas of Mars, monitor a number of locations for changes over time, and acquire stereo (3D) coverage of key regions and potential future landing sites (Shean *et al.*, 2011).

The High Resolution Stereo Camera (HRSC) is a camera experiment on Mars Express (Neukum and Jaumann, 2004). At an altitude of 250 km, HRSC can produce images with a resolution of 2.3 m/pixel of 2.35 km^2 of Mars terrain. The high-resolution imagery has covered almost the whole planet and provides an independent dataset to study the surface of Mars (Gwinner *et al.*, 2010; Jaumann *et al.*, 2007).

The Mars Orbiter Camera and Mars Observer Camera (MOC) were scientific instruments on board the Mars Observer and Mars Global Surveyor spacecraft, respectively. The scientific instrument consists of three elements: a black-and-white narrow-angle camera with a spatial resolution of 1.4 m/pixel (from an altitude of 378 km), and two pushbroom wide-angle cameras with resolution capabilities spanning 230 m/pixel to 7.5 km/pixel (Ormö, 2003).

The Mars Orbiter Laser Altimeter (MOLA) was on the Mars Global Survey (MGS) spacecraft. The MOLA instrument transmitted infrared laser pulses towards Mars at a rate of 10 times per second, and measured the time of flight to determine the range (distance) of the MGS spacecraft to the Martian surface. The range measurements resulted in precise topographic maps of Mars. Vertical resolution is 128 ppd/460 m. The precision maps are applicable to studies in geophysics, geology, and atmospheric circulation (http://pds-geosciences.wustl.edu/missions/mgs/mola.html) (de Pablo, 2003; Fassett and Head, 2008; Kaplan *et al.*, 2016).

In addition to these instruments mounted on now-completed or ongoing orbiters, the Tianwen-1 Mars exploration of China Space Agency (CNSA, 2020) will also feature an orbiter carrying two cameras, one intermediate resolution camera (~100 m/pixel at 400 km altitude) and another high resolution camera (2.5–10 m/pixel at 265 km altitude), to detect the geomorphology of Mars.

1.2.3 *Interior*

The magnetic field and gravity can help reveal the nature of planetary deep interior. The Mars magnetic field was only acquired by a magnetometer on Mars Global Surveyor. The data revealed stripes of opposing polarity that dominate the southern highlands (Plattner *et al.*, 2016; Wade *et al.*, 2017).

Gravity field measurements are used to acquire the bulk information of not only the static structure but also the dynamics of planetary interior. For Mars, kilometer-resolution gravity field has been obtained (Hirt *et al.*, 2012). The available data suggest that there is no obvious correlation between gravity anomalies and magnetic fields. Positive anomalies are related to large igneous provinces, but not to the topographic dichotomy (Neumann *et al.*, 2004; Turcotte *et al.*, 2001).

For detecting the relatively shallow subsurface of the planet Mars, two radar experiments named MARSIS (Mars Advanced Radar for Subsurface and Ionosphere Sounding) and SHARAD (Mars SHAllowRADar sounder) have been conducted by Mars Express and Mars Reconnaissance Orbiter, respectively.

MARSIS is a low frequency, pulse-limited radar sounder. It features ground-penetrating radar capabilities, which use synthetic aperture techniques and a secondary receiving antenna to isolate subsurface reflections. MARSIS identified buried basins on Mars. The MARSIS team measured a difference between the dielectric constant of the northern and southern high-latitude regions (Mouginot *et al.*, 2012). This is evidence that the material that fills the northern basin is a lower-density material, which could be interpreted as evidence of an ancient northern ocean. Orosei *et al.* (2018) discovered a subglacial lake on Mars, 1.5 km below the southern polar ice cap, and extending horizontally about 20 km (Orosei *et al.*, 2018).

SHARAD is a subsurface sounding radar that complements the MARSIS radar, providing lower penetration capabilities (some hundred meters) but much finer resolution (Seu *et al.*, 2007). The SHARAD radar penetrated the north polar layered ice deposits of Mars and revealed a relatively small (about 100 m) maximum deflection of the underlying rock, which suggests a strong lithosphere greater than 300 km thick. Radar results consistent with massive deposits of water ice in middle latitudes support a debris-covered glacier hypothesis (Holt *et al.*, 2008). In late 2016, NASA reported finding a large amount of underground ice in the Utopia Planitia region of Mars using SHARAD (Stuurman *et al.*, 2016). The volume of water detected has been estimated to be equivalent to the volume of water in Lake Superior ("Scalloped Terrain Led to Finding of Buried Ice on Mars", NASA.)

China's first Mars orbiter Tianwen-1 carries a radar with low frequency (10–50 MHz) to detect shallow subsurface structure of Mars, especially it's polar region where it is up to 1000 m deep.

1.3 *In Situ* Exploration

In situ exploration means robotic or human missions to specific sites on other planets. The selection of adequate landing sites is extremely important for both *in situ* exploration and, even more, for sample return. Landing sites are literally the locations for *ground truth* for extraterrestrial bodies. Any landing site has to satisfy scientific and engineering requirements, including safety, which can vary substantially depending on the mission scenario and the knowledge of the target planetary body: the selection of landing sites is a complex process and constrained by available data and information. In most recent cases for Mars, landing sites have been selected based on an increasingly large base of data used to make both engineering and scientific decisions. In some cases, landing sites are being selected with the help of instruments on associated orbital platforms and direct observations.

In that way, the landing sites for landers were selected and characterized as much as possible from orbit by the orbiter instrument suites before the landers were actually deployed. Different techniques and experiments can be mounted on robotic arms for easy access to the spot of investigation or hosted elsewhere on landers and rovers and fed by robotic appendices (arms, scoops, etc.).

Rover-based *in situ* analytics for the determination of compositions can include but are not limited to the following instruments (Rossi and Gasselt, 2018).

Alpha Particle X-ray Spectrometry (APXS) generates X-ray excitation by irradiating target materials with *a* particles produced by radioactive decay. Collected spectra are diagnostic of the chemical composition of the analyzed rock. This method has been applied in Mars Pathfinder, Spirit/Opportunity (MER) and Curiosity (MSL) Martian missions, and Rosetta comet mission.

X-Ray diffractometry, widely used in geology and mineralogy for terrestrial applications, has also been used on Mars (the CheMin on MSL). Powdered samples are irradiated with X-rays that are diffracted by the mineral structure, producing indicative spectra (Sarrazin *et al.*, 2005).

Mössbauer spectrometry uses μ-ray in order to provide mineralogical compositional information as well. It requires materials to contain iron. The spectrometers were included on both MER missions (Morris *et al.*, 2008).

Laser is used to investigate materials with Raman spectrometry: it can be used to identify both minerals and organic materials, thus, it is of interest for exobiology research (Wei *et al.*, 2014). This will be mounted on the Mars 2020 rover.

LIBS experiments measure spectra indicative of elemental composition by collecting spectra of surface materials vaporized by laser pulses. A LIBS experiment, ChemCam, is currently operated by Curiosity (MSL) on Mars. China's first Mars exploration (planned in 2020) will also use this technology (Williams *et al.*, 2013).

In situ explorations are not only used to detect composition, but also may be applied to do geophysical experiments that disclose subsurface and interior secrets. The InSight mission (2018) is conducting drilling for heat flow measurements and has placed a seismic equipment to acquire Martian quake signals. China's Mars rover plans to do ground penetrating radar experiments using both low (35–75 MHz) and high (800–1800 MHz) frequencies. A surface magnetic field detector is also mounted on the rover, and its range of magnetic field strength is ±65000 nT.

The 8 sites of landing exploration on Mars are shown in Fig. 1.2. Of these, 2 (MSL and InSight) were still active till the end of 2019.

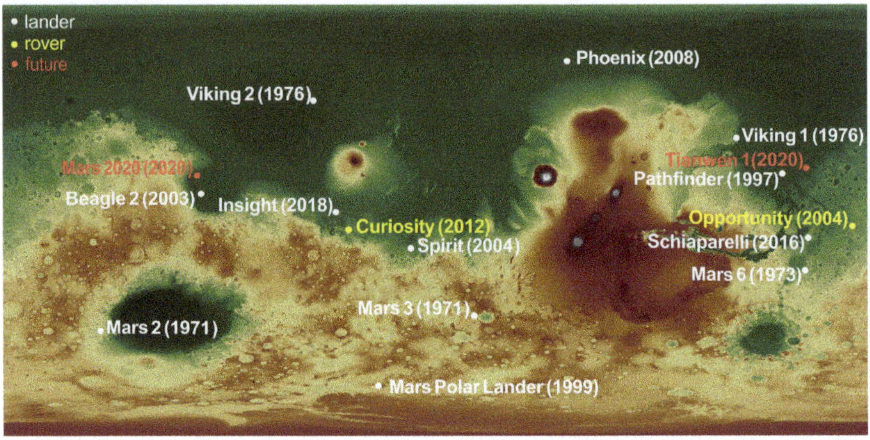

Fig. 1.2. Global topography of Mars, overlain with locations of Mars landers and rovers.

1.4 Sample Return

Returning samples from extraterrestrial bodies allows the application of advanced analytical techniques and sample processing and preparation far beyond what is feasible on any spacecraft. Returning extraterrestrial samples is effort-intensive and costly, though. Therefore, the choice of which samples to return is extremely important.

For Mars, although there is no sample returned yet, sample-return missions have been considered for many years, and one of those is presently under consideration. The Mars 2020 rover, operated by NASA, will seek signs of past life, set aside a returnable cache with the most compelling rock core and soil samples, and demonstrate technology needed for the future human and robotic exploration of Mars. The rover will collect and package samples of rock cores and surface soil for a later mission to bring back for definitive analysis on Earth.

For the Mars 2020 mission, Jezero crater (Fig. 1.3) within the Nili Fosse region of Mars was finally selected as the landing and sampling site, because it was thought to have once been flooded with water. The crater contains a fan-delta deposit rich in clays. The lake in the crater was present when valley networks were forming on Mars. Besides having a delta, the crater shows point bars and inverted channels (Goudge *et al.*, 2017).

Fig. 1.3. Jezero crater, the landing and sampling site of Mars 2020. Green refers to phyllosilicate-bearing materials, yellow indicates olivine materials, blue is low-calcium pyroxene-bearing materials, and purple-brown means no distinct spectral features (image: NASA/JPL/JHUAPL/MSSS/Brown University).

1.5 Analog Study

1.5.1 *The role of analogies*

There are different research strategies to interpret the various geological and geomorphological phenomena on the surface of Earth and extraterrestrial bodies. Geological research on Earth is usually carried out from the bottom up, that is, after detailed geological surveys of various locations are carried out, the geological facts obtained can be synthesized to analyze their causes, establish models and study their geological significance on a regional or even global scale. Geological studies of Earth-like planets are usually carried out from the top down, that is, after obtaining limited low-resolution surface images or global/regional compositional information, various models are established to explain the formation of such features. Then, through more detailed exploration (high-resolution remote sensing detection, landing detection and sample return, etc.), the models proposed above are tested and corrected. It should be noted that because field geological surveys on the surface of other Earth-like planets are not as convenient as they are on Earth, it is particularly important that detailed *in situ* surveys are limited. Thus, selecting representative areas is essential.

On Earth, the specific geological and geomorphological features mostly have specific formation mechanisms and formation environments. In the analysis of the genesis of some geomorphological and geological phenomena on the surface of an Earth-like object, the main basis is our understanding of similar geological phenomena on Earth. Therefore, a correct understanding of the genetic mechanism, causality, and/or "one cause, one effect" of specific geological phenomena on Earth is the basis for a reasonable explanation of the origin of planetary geological characteristics. From this point of view, analogy research is very important.

Unlike new geological phenomena found on other planets, geological phenomena on Earth are more likely to have both key features and causes. Thus, by recognizing key features similar to terrestrial analogs and their possible extraterrestrial analogs, geologists can infer potential causes of the latter. Earth has become the cornerstone or analog for understanding the geology of planets in the solar system. Terrestrial analogs can be environments, morphological features, structures, or processes that serve as a basis for comparison of similar features on other planetary bodies. Field analogs have been used for scientific research and training grounds for robotic and human space missions.

For experienced geologists, this form of reasoning from analogy is not a trivial exercise in form. Geologists use their extensive experience of Earth phenomena to build multiple working hypotheses and evaluate them by explaining their results in relation to a comprehensive understanding.

In 1886, Grove K. Gilbert pointed out that geologists are investigators, rather than theorists. The latter tests their theory by assessing the correspondence between theoretical results (usually inferences or theoretical predictions) and specific observations. Ideally, such observations are made during controlled experiments. In most cases, however, the subject of geology is not conducive to controlled experiments. Geologists/researchers cannot form glaciers, volcanoes, or mountains in the laboratory. In their scientific reasoning, geologists must be more like investigators than laboratory analysts. When testing a working geological hypothesis, geologists consider its consequences through its consistency, coherence, and consilience with related phenomena (Rossi and Gasselt, 2018).

Consistency implies a lack of contradiction, and the causal hypothesis of such a geological phenomenon is not contradicted by a clear historical development sequence or spatial relationship with other phenomena. Coherence requires a full and comprehensive interpretation in order to be consistent with the interpretation of other known phenomena. Finally, the initial hypothesis can be well assessed.

Although analogies by themselves do not fully account for the formation of flat landforms, they may lead to a series of studies that will allow researchers to find a reliable path to these explanations. The researchers hypothesize that a well-reasoned analogy might be correct, and then extrapolate the expected results that matched that assumption. In practice, the classification of phenomena, identification of potential analogs, verification of the consistency, coherence, and consilience of a working hypothesis all occur in the process of planetary mapping. The mapping process itself allows geologists to continuously assess the cause and age assumptions of the characteristics by examining the mapping relationship.

In fact, the first study to explain the origin of the ring structure on the surface of the Moon is based on the detailed comparative study of volcanic crater and impact craters on Earth (both can form such craters). Then it is concluded that the ring structures are mainly formed by the impacting, and is verified in subsequent lunar exploration missions. This analog study method was then widely accepted and used in the study of Earth-like bodies, and became an important method for the study of planetary geology.

Analogy research can be divided into two specific ways, one is to select the geological and geomorphological landscape formed naturally for detailed comparison, and the other is to carry out forward verification through experimental simulation. For Mars, the macroscopic geological and geomorphological landforms currently mainly adopt the analogy study of natural landscapes, and some of the microscopic geological processes adopt experimental verification.

The study of analogies has played an important role in the analysis of the geology and geomorphology of Mars and the planning of future Mars missions. Before Mariner 9 explored Mars, we didn't know whether Mars was a dead planet, like the Moon, or an active planet with life. Photographs from Mariner 9 show that the surface of Mars is neither as dead as the Moon, nor as viable as Earth for advanced life. Mariner 9 found that there were obvious internal and external dynamic geological traces on the surface of Mars, such as the large volcanic geomorphology formed by the internal dynamic forces and river channels and wind geomorphology formed by the external dynamic geological forces. Based on local coverage and low resolution and image data, NASA subsequently conducted a typical volcanic Earth and Mars surface comparison of the volcanic lava fields (such as Hawaii, the Snake river plain, Idaho, U.S. etc.), as well as multiple arid desert sand landforms on Earth (such as the Egyptian desert, Death Valley, etc.) and their aeolian counterparts on Mars.

Knowledge of the global geological landscape of Mars has become clearer with the data returned from the Viking orbiter and the landers, as well as later higher resolution data from Mars Reconnaissance Orbiter and the Mars Express missions. Multiple successful landing missions (Pathfinder, Opportunity + Spirit, Phoenix, Curiosity, InSight) have obtained detailed geological and composition data of their landing sites. These results together provide reliable evidence for verifying and correcting the proposed working hypotheses that were based on earlier orbiting exploration data. This further directs analog study.

Over the past decades, analogy has expanded to more types of geological landforms, such as volcanic, aeolian, glacial, and dry-salt lakes. The study of analogies can (1) help to analyze the causes of similar geological and geomorphological features, (2) provide a basis for landing site selection, and (3) provide evidence for future Mars missions, including landings, sampling returns and *in situ* resource utilization. For landing detection, analog research can also provide important reference for the rover's running, vehicle-mounted payload selection and working mode determination, sampling, and analysis device design, etc.

1.5.2 *Terrestrial analogies in Martian geology*

Several terrestrial analog sites have been established worldwide (Farr, 2004; Garry and Bleacher, 2011). They are chosen for their locations, which simulate a specific aspect of conditions that people could experience during a future mission to Mars: e.g. high altitudes can simulate low pressure of the Martian atmosphere (Chilean Altiplano, Garcia-Castellanos, 2007); high latitudes can provide solar radiation similar to the Martian surface (Flashline Mars Arctic Research Station, Bamsey *et al.*, 2009); hot and cold deserts such as the Atacama are an example of the former (Romig *et al.*, 2007) and the Antarctic Dry Valleys (Head and Marchant, 2014; Marchant and Head, 2007) and Svalbard in Norway (Reiss *et al.*, 2011; Hauber *et al.*, 2011) and Tuktoyaktuk Coastlands of northern Canada (Soare *et al.*, 2011) are examples of the latter, deserts (paleo-lakes) such as Death Valley and the Qaidam Basin (Mayer *et al.*, 2009;

Xiao *et al.*, 2017; Anglés and Li, 2017; Dang *et al.*, 2018; Wang *et al.*, 2018) are analogs of Martian paleo-lake deposits and later aeolian landforms, and volcanic regions like Hawaii (Hawaii Space Exploration Analogue and Simulation, Binsted, 2015), Iceland, Pinacate volcanic field in Mexico, and Axel Heiberg Island of Canadian high Arctic can provide similar geologic characteristics of Martian surface.

In addition, as one of the possibly habitable planets in the Solar System, Mars has been targeted for searching for life or life signatures. To solve this problem, assessing the habitable environments on Mars based on the direct linkage between life on Earth and specific environmental conditions became the theme of astrobiology and main mission goal for Mars exploration. Table 1.1 lists selected examples. Figure 1.4 shows locations of main examples of Earth analogs for Mars.

In this book, we choose the Qaidam Basin, NW China, as an example for a Martian paleolake basin analogy, for its unique and diverse Mars-like landforms. We will focus on a comparison study of dried lakes (playas), aeolian landforms, polygonal terrains, valleys, gullies, and habitability between the Qaidam Basin and their counterparts on Mars.

1.6 Martian Meteorites

Although there are no returned samples from Mars as yet, more than 125 Martian meteorites have been recovered, mainly from Antarctic and northwestern African deserts. These Martian meteorites are igneous in origin, mainly cumulates formed in magma chambers and basalt formed by volcanic eruption. According to the differences in their petrological features, Martian meteorites are formerly divided into: shergottites (named after Shergotty, fall in India, 1865), nakhlites (named after Nakhla, fall in Egypt, 1911), chassignites (after Chassigny, fall in France, 1815), a unique orthopyroxenite (ALH 84001, found in Antarctica, 1984), and an unusual ancient Mars breccia, NWA 7034 and paired samples NWA 7475, NWA 7533, etc. (Agee *et al.*, 2013).

The shergottites comprise several lithologies, including basalt, gabbro, and some lherzolite cumulates. The main minerals assemblages consist of pigeonite, augite, plagioclase, and accessory minerals, such as Fe-Ti oxide, sulfide, and phosphate. Some shergottites also contain olivine phenocrysts, or xenocrysts (Goodrich, 2003). The crystallization ages of these meteorites are very young, about 0.1–0.6 Ga (Papike *et al.*, 2009),

Table 1.1. Habitable environments on Mars and similar specific environmental condition on Earth (modified after Dohm *et al.*, 2011; Preston and Dartnell, 2014).

Environment	Earth locations	Features and Earth-environment condition	Mars locations	Mars-environment condition	References
Flood basalts and water related alteration	The Pilbara Region, West Australia. *Analog for flood basalts; water-related alteration minerals; preservation of early life.*	Undeformed Archaean 3.5–2.7 Ga, volcanic rocks. North Pole Dome hydrothermal activity. Abundant phyllosilicates within basaltic lavas. Stromatolites and putative microfossils >3 Ga old.	Nili Fossa, Elysium	Episodic, long-lived heat energy sources and associated interfacing in both subterranean cavernous and hydrothermal environments followed by extended periods of quiescence. Similar to those observed in Iceland.	Walter *et al.* (1980); Allwood *et al.* (2006); Brown *et al.* (2006); Van Kranendonk *et al.* (2008); Westall (2008)
Magmatic complexes and basal margins	Rodinia and Southern Pacific superplumes; The Pilbara Region, West Australia: DongwanziOphiolite Complex, China. *Analog for martian crustal rocks and Mars meteorites; chemolithoautotrophic biospheres.*	Magmatic-driven tectonic and hydrothermal activity; magmatically altered rock materials. The Pilbara Region, West Australia. DongwanziOphiolite Complex, China: Archaean ophiolite complex. Mafic and ultramafic rocks, water–rock–biology interactions. Bio-alteration on volcanic rocks.	Tharsis and Elysium	Episodic, long-lived heat energy sources and associated interfacing in both subterranean cavernous and hydrothermal environments followed by extended periods of quiescence	Kusky *et al.* (2001)
Hydrothermal environments	Yellowstone, Solfatara crater, caldera at summit of Mount Erebus in Antarctica. *Analog for silica-rich soils; hydrothermal activity; preservation of ancient life; thermophiles.*	Ranging pH, temperature, salinity, and anoxic aqueous conditions Solfatara: pH 1.7, temperature 40–160°C, anoxic-reducing aqueous conditions at subsurface and anoxic/oxic zones found in the rock fractures and at the surface environments.	Apollinaris Patera, Melas Chasma, and Toro crater	Ranging pH, temperature, salinity, and anoxic aqueous conditions; long-lived activity with major periods of quiescence	Parenteau & Cady (2010); Ruff *et al.* (2011)

		Yellowstone National Park, USA: Over 300 springs, e.g. Grand Prismatic and Chocolate Pots. Silica sinters, opaline-rich soils, iron deposits and microbial mats. Chemotrophic and phototrophic micro-organisms. Microfossils and stromatolites, permineralization.			
Caverns	The environs are numerous and diverse.	Ranging pH, temperature, salinity, and anoxic aqueous conditions, cavern temperature, and energy input.	In parts of Tharsis, as indicated by the chaotic terrains that head the circum-Chryse outflow channel system and the pit crater chains.	Protection against ultraviolet (UV) radiation; ranging pH, temperature, salinity, and aqueous conditions, cavern temperature, and energy input; magmatic-driven environmental fluctuations; preservation of water and icy bodies.	
Basin/aquifer systems	Many and diverse. Acid environment such as Rio Tinto. *Analog for past river channels; iron oxides and sulphates; preservation of life; acidophiles*	Diverse environmental conditions with water enrichment related to both surface runoff and infiltration and groundwater. Rio Tinto, SW Spain:. Natural acid rock drainage system. pH 3.2 and precipitates of iron sulphates and oxides. Chemolithotrophic community dominated by iron oxidizing acid tolerant bacteria. 2Ma iron oxide and sulphate-rich river terraces with microbial fossils.	Tharsis and Arabia Terra are distinct examples.	Catchment for sediments and water; long-term water-enrichment in the case of the Tharsis and Arabia Terra basins, the former of which would be dynamically related to the evolution of Tharsis and the latter which would remain relatively stable with differential erosion.	González-Toril *et al.* (1999, 2003); Fernández-Remolar *et al.* (2005); Amils *et al.* (2007); Preston *et al.* (2011)

(*Continued*)

Table 1.1. (*Continued*)

Environment	Earth locations	Features and Earth-environment condition	Mars locations	Mars-environment condition	References
Structurally controlled conduits and basins	Atacama Desert, Afar Depression are distinct locality types	Subsurface; conduits for the transferal of groundwater and other volatiles, as well as heat energy; protection from surface radiation; possible dynamic environments during events such as magmatic-related activity.	Magmatic- and impact-induced; within Tharsis and Elysium and on the margins of Hellas, Isidis, and Argyre.	Access to groundwater and other volatiles, as well as heat; protection from surface radiation; possible dynamic environments during events such as magmatic-related activity.	
Evaporite deposits such as salts	Atacama Desert; Qaidam Basin	Salt crusts and massive salt deposits formed after the evaporation of surface and subsurface water; pedogenic gypsum crusts.	Chloride-bearing deposits in the southern highlands, e.g. Terra Sirenum.	Access to liquid water under extremely dry conditions (deliquescence); protection from UV radiation; access to photosynthetic radiation.	
Possible marine and lacustrine deposits	Many and diverse; e.g. supraglacial and subglacial lakes, sea ice in the Arctic and Antarctica, as well as marine environments not covered by ice.	Water flow and ponding wave activity, sedimentation from incoming outlet channels, surface ice; varying aqueous (chemistry and temperature) and energy conditions; salinity and pH mainly depend on sedimentary and water inputs, climate, and overall hydrological balance; temperature depending on climate and water circulation; hydrothermal conditions are also possible.	Impact craters, troughs depressions, and other low-lying basins; examples include the northern plains, Eberswalde, and Jezero craters, and intravalley impoundments.	Uncertain and strongly dependent on the climatic conditions of early Mars and the duration of marine and/or lacustrine activity.	

Antarctic-likepaleosols	Dry Valleys of Antarctica	Paleosols with compositional variation and concentrations of secondary oxides/ coatings in <2 mm matrix material in tills; local concentrations of soluble salts; both salt-enriched horizons below Fe-enriched horizons in pedostratigraphic successions are niches.	Possible a really extensive and diverse localities; a type locality may be located on the eastern margin of the Hellas basin such as in the Reull Valles region.	Similar to those observed in the Antarctic.	
Vent structures	Dalhousie Springs Complex, Australia, parts of the Grand Canyon, and Azerbaijan mud volcanoes.	Varying aqueous (chemistry and temperature) and energy conditions.	Vernal crater, Arabia Terra; mounds with summit pits in the parts of the northern plains such as within the Vastitas Borealis Formation and Toro crater.	Aqueous chemistry and temperature conditions compatible with life; energy and heat sources.	
Ice bodies	Antarctic, Arctic, Greenland, and Iceland are distinct locality types *Analog for mineralogy and water chemistry at MeridianiPlanum and MawrthVallis; extremophiles*	Surface and subsurface ice; subsurface ice may be mantled by dry mantles. The Golden Deposit, Canadian High Arctic: Cold, acidic (pH 3) springs precipitating jarosite and natrojarosite. Permafrost environment. Microbial filaments, sulphate reducing bacteria.	North and south poles,south polar region, parts of the northern plains, possibly aureole deposits of the large shield volcanoes of Tharsis, and basin floor and rim materials of large impacts such as Hellas.	Subsurface ice with dry mantles.	Michel (1977); Battler *et al.* 2012)

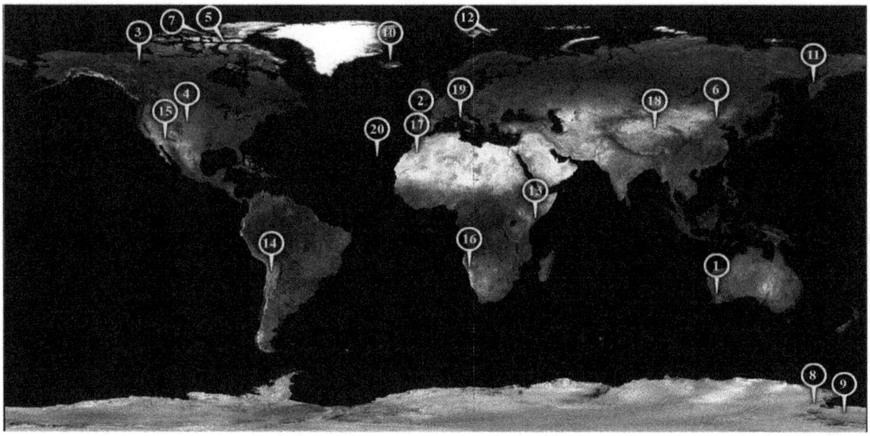

Fig. 1.4. Global map of the Earth with 33 analog sites identified (adapted from Preston and Dartnell, 2014).

Note: (1) Yilgarn Craton (Western Australia); (2) Rio Tinto (Spain); (3) The Golden Deposit (Canadian High Arctic); (4) Yellowstone National Park (USA); (5) Haughton Impact Structure (Devon Island); (6) Dongwanzi Ophiolite Complex (China); (7) Axel Heiberg Island (Canadian High Arctic); (8) The Antarctic Dry Valleys (Antarctica); (9) Antarctic North Victoria Land Mountains; (10) Sub-glacial Volcanism (Iceland); (11) Kamchatka (Russian Federation); (12) Bockfjord Volcanic Complex (Svalbard); (13) Kilimanjaro (Tanzania, Africa); (14) Atacama (South America); (15) The Mojave Desert (USA); (16) Namib Desert (Namibia, Africa); (17) Ibn Battuta Centre Sites (Morocco); (18) QaidamBasin (Tibetan Plateau, China); (19) Caves in Sardinia; (20) Caves in Tenerife and Lanzarote.

and only two special samples (NWA 7635 and NWA 8159, both augite basalts) have relatively older crystallization age at 2.4 Ga (Herd *et al.*, 2017). The geochemical characteristics of shergotties can be divided into the enriched and depleted types; they show enriched and depleted compatible trace element patterns, which are correlated with radiogenic isotopes and oxidation states. The depleted shergottites are the direct product of partial melting of the Martian mantle. Enriched shergottites may have originated by melting altered Martian mantle or by the contamination of depleted magma with enriched Martian crust (McSween *et al.*, 2001).

Nakhlites and chassignites represent olivine-bearing clinopyroxenite and dunite, respectively. Nakhlites are thought to represent a series of thick lava flows on the surface of Mars. The origins of nakhlites and chassignites are somewhat similar. Their crystallization ages were 1.3 Ga

(Nyquist *et al.*, 2001), their impact-eject ages were 11 Ma (Korochantseva *et al.*, 2011), and they had similar isotopic compositions. The significant difference is that the content of trace elements in pyroxene is different (Wadhwa and Crozaz, 1995), suggesting that meteorites like nakhlites experienced hydrothermal activity in the later stage of magmatism. Evidence for post-magmatic hydrothermal activity of chassignites is the difference in F and Cl content of phosphate minerals in melt inclusions or interstitial melts (McCubbin and Nekvasil, 2008).

Nakhlites are well known for significant fluid activity recorded in their minerals. Many alteration minerals have been found in such meteorites, including water-bearing Fe, Mg-layered silicates (montmorillonite, serpentine, opal, colloids of amorphous clay minerals, stilpnomelanite), carbonate minerals, sulphides and halides, etc. These minerals may have formed by water–rock interactions near the surface of Mars. One of the most popular models holds that all these secondary minerals formed from a relatively low temperature hydrothermal brine interaction (Changela and Bridges, 2011). Other models suggest that the dissolution and metasomatism of olivine can also form carbonates (Tomkinson *et al.*, 2013), and sulfide may have formed as an evaporative precipitation of acidic brine on the surface of Mars. The source of these fluids is controversial. H isotopes of phyllosilicate in different nakhlite samples indicate that the fluid is derived from Martian mantle (Gillet *et al.*, 2002). However, the metasomatic structure of olivine shown in serpentinization indicates that the fluid comes from the Martian crust or atmosphere (Lee *et al.*, 2015).

The time of alteration is also a paradox. Through the Ar-Ar dating of phyllosilicate minerals, some samples showed an alteration age of ~680 Ma (Swindle *et al.*, 2000), and some features displayed that the alteration occurred immediately following the magmatic process (~1.3 Ga). For example, the presence of Cl-enriched amphibole in residual melt of MIL 03346 meteorite (McCubbin *et al.*, 2009) and the assemblage of salt and residual melt in the Nakhla meteorite (Krzesiska *et al.*, 2016).

The ancient ALH84001 Martian meteorite (formed 4.1 billion years ago) is a coarse-grained cumulate orthopyroxenite (Romanek *et al.*, 1994). The rock underwent a strong impact metamorphism before it reached Earth. The meteorite has attracted the most attention because of its carbonate microspheres (Fig. 1.5), sparking a debate over whether ancient life existed on Mars. The origin of these carbonate spheres is still controversial, and one possible explanation is that they formed in the late hydrothermal active fluid deposition (Golden *et al.*, 2001).

Fig. 1.5. (a) ALH84001 on display at Smithsonian Museum of Natural History; (b) The electron microscope revealed chain structures in meteorite fragment ALH84001 (adapted from Wikipedia).

In addition, some breccia-type Martian meteorites have been found recently, such as NWA 7034 and its paired NWA 7533 (Agee *et al.*, 2013; Humayun *et al.*, 2015). These breccias contain clasts of basalt, andesite, and monzonite. They are also very old, formed approximately at 4.4 Ga.

Undoubtedly, the study of Martian meteorites has important implications for understanding the crust composition of Mars. However, because we do not know launch sites for these meteorites, we cannot know their geological context, and cannot adequately explain formation environment. In addition, there is a big difference between the meteorite samples (SNC) and the compositions obtained from remote sensing detection and landing detection (Fig. 1.6). How to explain this difference is an important scientific problem.

1.7 Simulation

Simulation is a means of planetary science and Mars science. Since the surface of Mars is difficult to reach, a large number of geological phenomena can only be remotely investigated and then interpreted by means of physical and/or computer simulations. For an example of physical simulation, with respect to the water–rock interaction processes and products on the surface of Mars, we can design a Mars environment simulation chamber, carry out known physical and chemical experiments in the artificial environment of Mars, and then test the validity of the recognition based

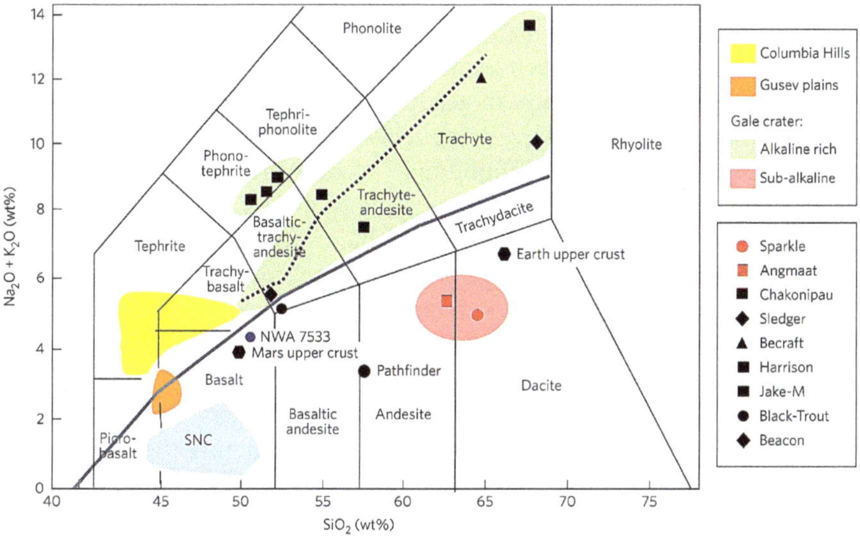

Fig. 1.6. TAS diagram showing the range of observed crustal compositions of Mars. The symbols in the legend on the lower right correspond to samples analyzed by the MSL rover, Curiosity, in Gale crater. Such trachy-andesitic, trachytic, and dacitic compositions may represent an early Martian crust that may be compared to continental-type compositions on Earth. The SNC green field includes all Martian meteorites except the Noachian breccia NW7533 (Sautter *et al.*, 2015).

on the general analogy of the Earth's environment. The formation mechanism of other geomorphic features such as aeolian landforms, valleys and channels, recurring slope lineae, and impact craters, can also be simulated. In addition, experiments on various organisms and microorganisms have been carried out in the Mars simulation chamber to understand the astrobiological significance.

Another kind of simulation is computer modeling, to simulate the impact of atmospheric circulation and evolution of Mars and early mantle convection on macro and large-scale geological and geomorphological formations. For example, previous studies proposed the possible formation mechanism of Tharsis Bugle, the largest volcanic province on Mars, through large-scale mantle convection models (Zhong and Zuber, 2001), and established several climate models to invert the atmospheric and climatic characteristics of early Mars (Palumbo and Head, 2018; Matsubara *et al.*, 2018).

1.8 Summary

Over forty missions have been launched to explore Mars, the most Earth-like planet in our solar system. In general, global orbiting probes mounted with cameras of different wave bands, laser ranger, gravity experiments, magnetometer, have obtained topography, compositional geomorphological, gravity, and magentic field characteristics at global scale of different resolutions and coverages. These data enable us to understand the surface geomorphological and interior geophysical properties. This also led us to choose specific regions and sites to conduct *in situ* exploration for knowing detailed geological and geochemical information, and provide ground truths for interpreting remotely sensed data.

A detailed study of the rock types, mineral assemblages, and geochemical characteristics of the Martian surface, including the assessment of its habitable environment, is the surest way to study Martian samples in ground laboratories. Therefore, the return of samples from the Martian surface is the next milestone for Mars exploration. The study of Martian meteorite samples is also important, although it is difficult to determine their original location and geological background. In addition, we need to use a variety of simulation experiments to verify the accuracy of geological interpretation, or to guide the layout and implementation of exploration plans.

In summary, the combination of the above-listed methods enables us to fully and better understand the geological and geomorphical features of Mars and their formation mechanism, as well as the evolution process over time. According to the contents of this book, the significance and specific research contents and methods of the analog study are introduced, and representative analog study areas or points on Earth and the geological environment they represent are listed.

Acknowledgments

We are grateful for funding from the Pre-research Project on Civil Aerospace Technologies of China Space Agency (D020101)), the Science and Technology Development Fund (FDCT) of Macau (Nos. 121/2017/A3), National Natural Science Foundation of China (41830214). Ms. Yutong Shi has helped to edit the figures, pictures, and references.

References

Agee, C. B., Wilson, N. V., McCubbin, F. M., Ziegler, K., Polyak, V. J., Sharp, Z. D., Asmerom, Y., Nunn, M. H., Shaheen, R., Thiemens, M. H. *et al.* (2013). Unique meteorite from early Amazonian Mars: Water-rich basaltic Breccia Northwest Africa 7034. *Science*, **339**, 780–785.

Andrews-Hanna, J. C., Phillips, R. J. and Zuber, M. T. (2007). Meridiani Planum and the global hydrology of Mars. *Nature*, **446**, 163–166.

Anglés, A. and Li, Y. (2017). The western Qaidam Basin as a potential Martian environmental analogue: An overview. *J. Geophys. Res. Planets*, **122**, 856–888. https://doi.org/10.1002/2017je005293.

Auld, K. S. and Dixon, J. C. (2016). A classification of martian gullies from HiRISE imagery. *Planet. Space Sci.*, **131**, 88–101.

Bamsey, M., Berinstain, A., Auclair, S., Battler, M., Binsted, K., Bywaters, K., Harris, J., Kobrick, R. and McKay, C. (2009). Four-month Moon and Mars crew water utilization study conducted at the Flashline Mars Arctic Research Station, Devon Island, Nunavut. *Adv. Space Res.*, **43**, 1256–1274.

Bandfield J. L., Hamilton V. E. and Christensen P. R. (2000). A global view of Martian surface compositions from MGS-TES. *Science*, **287**(5458), 1626–1630.

Bandfield, J. L. (2007). High-resolution subsurface water-ice distributions on Mars. *Nature*, **447**, 64–67.

Bibring, J. P., Langevin, Y., Mustard, J. F., Poulet, F. and Arvidson, R. (2006). Global mineralogical and aqueous mars history derived from OMEGA/Mars express data. *Science*, **312**(5772), 400–404.

Binsted, K. (2015). Hawai'i space exploration analog and simulation. http://hdl.handle.net/10125/37427.

Bocco, G. (1991). Gully erosion: Processes and models. *Prog. Phys. Geogr.*, **15**, 392–406.

Boynton, W. V., Feldman, W. C., Squyres, S. W., Prettyman, T. H., Bruckner, J., Evans, L. G., Reedy, R. C., Starr, R., Arnold, J. R., Drake, D. M., Englert, P. A., Metzger, A. E., Mitrofanov, I., Trombka, J. I., D'Uston, C., Wanke, H., Gasnault, O., Hamara, D. K., Janes, D. M., Marcialis, R. L., Maurice, S., Mikheeva, I., Taylor, G. J., Tokar, R. and Shinohara, C. (2002). Distribution of hydrogen in the near surface of Mars: Evidence for subsurface ice deposits. *Science*, **297**, 81–85.

Bridges, N., Banks, M., Beyer, R., Chuang, F., Dobrea, E. N., Herkenhoff, K., Keszthelyi, L., Fishbaugh, K., McEwen, A. and Michaels, T. (2010). Aeolian bedforms, yardangs, and indurated surfaces in the Tharsis Montes as seen by the HiRISE Camera: Evidence for dust aggregates. *Icarus*, **205**, 165–182.

Bultel, B., Quantin, C., Andreani, M. and Clenet, H. (2013). Identification of Phyllosilicates in Crustal Outcrops Between Hellas and Isidis Basins (Mars) Using Combinaison of Near 2.3–2.5 μm Absorptions on Crism Data, Lunar and Planetary Science Conference, p. 2078.

Carter, J., Poulet, F., Bibring, J. P., Mangold, N. and Murchie, S. (2013). Hydrous minerals on Mars as seen by the CRISM and OMEGA imaging spectrometers: Updated global view. *J. Geophys. Res. Planets*, **118**, 831–858.

Changela, H. G. and Bridges, J. C. (2011). Alteration assemblages in the nakhlites: Variation with depth on Mars. *Meteorit. Planet. Sci.*, **45**, 1847–1867.

Christensen, P. R., Bandfield, J. L., Hamilton, V. E., Ruff, S. W., Kieffer, H. H., Titus, T. N. *et al.* (2001). Mars Global Surveyor Thermal Emission Spectrometer experiment: Investigation, description and surface science results. *J. Geophys. Res.*, **106** (E10), 23823–23872.

Christensen, P. R., Bandfield, J. L., Bell III, J. F., Gorelick, N., Hamilton, V. E., Ivanov, A., Jakosky, B. M., Kieffer, H. H., Lane, M. D. and Malin, M. C. (2003). Morphology and composition of the surface of Mars: Mars Odyssey THEMIS results. *Science*, **300**, 2056–2061.

Christensen, P., Wyatt, M., Glotch, T., Rogers, A., Anwar, S., Arvidson, R., Bandfield, J., Blaney, D., Budney, C. and Calvin, W. (2004a). Mineralogy at Meridiani Planum from the Mini-TES experiment on the Opportunity Rover. *Science*, **306**, 1733–1739.

Christensen, P. R., Jakosky, B. M., Kieffer, H. H., Malin, M. C., McSween, J. H. Y., Nealson, K., Mehall, G. L., Silverman, S. H., Ferry, S., Caplinger, M. and Ravine, M. (2004b). The Thermal Emission Imaging System (THEMIS) for the Mars 2001 Odyssey Mission. *Space Sci. Rev.*, **110**, 85–130.

Cull, S., Arvidson, R. E., Mellon, M., Wiseman, S., Clark, R., Titus, T., Morris, R. V. and McGuire, P. (2010). Seasonal H_2O and CO_2 ice cycles at the Mars Phoenix landing site: 1. Prelanding CRISM and HiRISE observations. *J. Geophys. Res.*, **115**, E00D16, doi/full/10.1029/2009JE003340.

Dang, Y. N., Xiao, L., Xu, Y., Zhang, F., Huang, J., Wang, J., *et al.* (2018). The polygonal surface structures in the Dalangtan Playa, Qaidam Basin, NW China: Controlling factors for their formation and implications for analogous Martian landforms. *J. Geophys. Res. Planets*, **123**, https://doi.org/10.1029/2018JE005525.

de Pablo, M. (2003). MOLA topographic data analysis of the Atlantis paleolake basin, Sirenum Terrae, Mars, Sixth International Conference on Mars, p. 3037.

Dohm, J. *et al.* (2011). An inventory of potentially habitable environments on Mars: Geological and biological perspectives. *Geol. Soc. Am. Spec. Pap.*, **483**, 317–347.

Dundas, C. M., Diniega, S. and McEwen, A. S. (2015). Long-term monitoring of martian gully formation and evolution with MRO/HiRISE. *Icarus*, **251**, 244–263.

Farr, T. G. (2004). Terrestrial analogs to Mars: The NRC community decadal report. *Planet. Space Sci.*, **52**, 3–10.

Fassett, C. I. and Head, J. W. (2008). The timing of martian valley network activity: Constraints from buffered crater counting. *Icarus*, **195**, 61–89.

Fergason, R. L., Christensen, P. R. and Kieffer, H. H. (2006). High-resolution thermal inertia derived from the Thermal Emission Imaging System (THEMIS): Thermal model and applications. *J. Geophys. Res.*, **111**, E12004.

Garcia-Castellanos, D. (2007). The role of climate during high plateau formation, Insights from numerical experiments. *Earth Planet. Sci. Lett.*, **257**, 372–390.

Garry, B. and Bleacher, J. E. (eds.) (2011). *Analogs for Planetary Exploration. The Geological Society of America, Special Paper*, **483**, p. 567.

Gillet, P., Barrat, J. A., Deloule, E., Wadhwa, M., Jambon, A., Sautter, V., Devouard, B., Neuville, D., Benzerara, K. and Lesourd, M. (2002). Aqueous alteration in the Northwest Africa 817 (NWA 817) Martian meteorite. *Earth Planet. Sci. Lett.*, **203**, 431–444.

Golden, D. C., Ming, D. W., Schwandt, C. S., Lauer, H. V., Socki, R. A., Morris, R. V., Lofgren, G. E. and Mckay, G. A. (2001). A simple inorganic process for formation of carbonates, magnetite, and sulfides in Martian meteorite ALH84001. *Am. Mineral.*, **86**, 370–375.

Goodrich, C. A. (2003). Petrogenesis of olivine-phyricshergottitesSayh al Uhaymir 005 and Elephant Moraine A79001 lithology A. *Geochimica et Cosmochimica Acta*, **67**, 3735–3771.

Goudge, T. *et al.* (2017). Statigraphy and evolution of delta channel deposits, Jeaero crater, Mars. *Lunar and Planetary Science XLVIII* (2017), 1195.

Goudge, T. A., Mustard, J. F., Head, J. W., Fassett, C. I. and Wiseman, S. M. (2015). Assessing the Mineralogy of the Watershed and Fan Deposits of the Jezero Crater Paleolake System, Mars. *J. Geophys. Res.*, **120**, 775G, doi: 10.1002/2014JE004782.

Gwinner, K., Scholten, F., Preusker, F., Elgner, S., Roatsch, T., Spiegel, M., Schmidt, R., Oberst, J., Jaumann, R. and Heipke, C. (2010). Topography of Mars from global mapping by HRSC high-resolution digital terrain models and orthoimages: Characteristics and performance. *Earth Planet. Sci. Lett.*, **294**, 506–519.

Hauber, E., Reiss, D., Ulrich, M., Preusker, F., Trauthan, F., Zanetti, M., Hiesinger, H., Jaumann, R., Johansson, L., Johnsson, A., Olvmo, M., Carlsson, E., Johansson, H. A. B. and McDaniel, S. (2011). Periglacial landscapes on Svalbard: Terrestrial analogs for cold-climate landforms on Mars, in Garry, W. B. and Bleacher, J. E. (Eds.), *Analogs for Planetary Exploration. Geol. Soc. Am. Spec. Pap.*, **483**, p. 567.

Head, J. W. and Marchant, D. R. (2014). The climate history of early Mars: insights from the Antarctic McMurdo Dry Valleys hydrologic system. *Antarctic Sci.*, **26**(06), 774–800.

Herd, C. D. K., Walton, E. L., Agee, C. B., Muttik, N., Ziegler, K., Shearer, C. K., Bell, A. S., Santos, A. R., Burger, P. V., Simon, J. I. *et al.* (2017). The Northwest Africa 8159 martian meteorite: Expanding the Martian sample suite to the early Amazonian. *Geochimica et Cosmochimica Acta*, **218**, 1–26.

Hirt, C., Claessens, S. J., Kuhn, M. and Featherstone, W. E. (2012). Kilometer-resolution gravity field of Mars: MGM2011. *Planet. Space Sci.*, **67**(1), 147–154, doi: 10.1016/j.pss.2012.02.006.

Holt, J. W., Safaeinili, A., Plaut, J. J., Head, J. W., Phillips, R. J., Seu, R., Kempf, S. D., Choudhary, P., Young, D. A., Putzig, N. E., Biccari, D. and Gim, Y. (2008). Radar sounding evidence for buried glaciers in the Southern mid-latitudes of Mars. *Science*, **322** (5905), 1235–1238.

Humayun, M., Nemchin, A., Zanda, B., Hewins, R. H., Grange, M., Kennedy, A., Lorand, J. P., Göpel, C., Fieni, C., Pont, S. *et al.* (2013). Origin and age of the earliest Martian crust from meteorite NWA 7533. *Nature*, **503**, 513–516.

Jain, N. and Chauhan, P. (2015). Identification of Deposits of Aqueous Minerals in Northern Part of Hellas Planitia Region on Mars Using MRO-CRISM: Implications for Past Aqueous History of Mars, Lunar and Planetary Science Conference, p. 1838.

Jaumann, R., Neukum, G., Behnke, T., Duxbury, T., Eichentopf, K., Flohrer, J., Gasselt, S., Giese, B., Gwinner, K. and Hauber, E. (2007). The high-resolution stereo camera (HRSC) experiment on Mars Express: Instrument aspects and experiment conduct from interplanetary cruise through the nominal mission. *Planet. Space Sci.*, **55**, 928–952.

Jérémie, M., Pommerol, A., Beck, P. and Kofman, W. and Stephen, M. C. (2012). Dielectric map of the Martian northern hemisphere and the nature of plain filling materials. *Geophys. Res. Lett.*, **39**(2), 19 January 2012.

Kaplan, H. H., Milliken, R. E., Fernández-Remolar, D., Amils, R., Robertson, K. and Knoll, A. H. (2016). Orbital evidence for clay and acidic sulfate assemblages on Mars based on mineralogical analogs from Rio Tinto, Spain. *Icarus*, **275**, 45–64.

Kirk, R., Howington, K. E., Rosiek, M., Anderson, J., Archinal, B., Becker, K., Cook, D., Galuszka, D., Geissler, P. and Hare, T. (2008). Ultrahigh resolution topographic mapping of Mars with MRO HiRISE stereo images: Meter-scale slopes of candidate Phoenix landing sites. *J. Geophys. Res. Planets (1991–2012)*, **113**.

Korochantseva, K., Schewenzer, S. P. and Buikin, A. I. (2011). 40Ar — 39Ar and cosmic — ray exposure ages of nakhlites — Nakhla, Lafayette, Governador Valadares — and Chassigny, doi: 10.1111/j.1945-5100.2011.01240.x.

Krzesińska, A. M., Michalski, J. R., Smith, C. L. and Schofield, P. F. (2016). Secondary minerals in Nakhla: A 3D reconstruction of alteration sequences. In *Meteorit. Planet. Sci.*, **51** (S1) p. 6226.

Lee, M. R., Tomkinson, T., Hallis, L. J. and Mark, D. F. (2015). Formation of iddingsite veins in the martian crust by centripetal replacement of olivine: Evidence from the nakhlite meteorite Lafayette. *Geochimica et Cosmochimica Acta*, **154**, 49–65.

Malin, M. C., Bell, J. F., Cantor, B. A., Caplinger, M. A., Calvin, W. M., Clancy, R. T., Edgett, K. S., Edwards, L., Haberle, R. M., James, P. B., Lee, S. W., Ravine, M. A., Thomas, P. C. and Wolff, M. J. (2007). Context camera investigation on board the Mars reconnaissance orbiter. *J. Geophys. Res.*, **112**, E05S04.

Marchant, D. R. and Head, J. W. (2007). Antarctic dry valleys: Microclimate zonation, variable geomorphic processes, and implications for assessing climate change on Mars. *Icarus*, **192**(1): 187–222.

Matsubara, Y., Alan, D. H., and Rossman P. I. (2018). Constraints on the Noachain paleoclimate of the martian highlands from landscope evolution modeling. *J. Geophys. Res. Planets*, https://doi.org/10.1029/2018JE005572.

Mayer, D., Arvidson, R., Wang, A., Sobron, P. and Zheng, M. (2009). Mapping minerals at a potential Mars analog site on the Tibetan Plateau, Lunar and Planetary Science Conference, p. 1877.

McCubbin, F. M. and Nekvasil, H. (2008). Maskelynite-hosted apatite in the Chassigny meteorite: Insights into late-stage magmatic volatile evolution in Martian magmas. *Am. Mineral.*, **93**, 676–684.

McCubbin, F. M., Tosca, N. J., Smirnov, A., Nekvasil, H., Steele, A., Fries, M. and Lindsley, D. H. (2009). Hydrothermal jarosite and hematite in a pyroxene-hosted melt inclusion in martian meteorite Miller Range (MIL) 03346: Implications for magmatic-hydrothermal fluids on Mars. *Geochimica et Cosmochimica Acta*, **73**, 4907–4917.

McEwen, A. S., Eliason, E. M., Bergstrom, J. W., Bridges, N. T., Hansen, C. J., Delamere, W. A., Grant, J. A., Gulick, V. C., Herkenhoff, K. E., Keszthelyi, L., Kirk, R. L., Mellon, M. T., Squyres, S. W., Thomas, N. and Weitz, C. M. (2007). Mars Reconnaissance Orbiter's high resolution imaging science experiment (HiRISE). *J. Geophys. Res.*, **112**, E05S02.

McSween, H. Y., Grove, T. L., Lentz, R. C. F., Dann, J. C., Holzheid, A. H., Riciputi, L. R. and Ryan, J. G. (2001). Geochemical evidence for magmatic water within mars from pyroxenes in the shergotty meteorite. *Nature*, **409**, 487–490.

Morris, R. V., Klingelhöfer, G., Schröder, C., Fleischer, I., Ming, D.W., Yen, A. S., Gellert, R., Arvidson, R. E., Rodionov, D. S., Crumpler, L. S., Clark, B. C., Cohen, B. A., McCoy, T. J., Mittlefehldt, D. W., Schmidt, M. E., de Souza, P. A. and Squyres, S. W. (2008). Iron mineralogy and aqueous alteration from husband hill through home plate at Gusev Crater, Mars: Results from the Mössbauer instrument on the spirit Mars exploration Rover. *J. Geophys. Res.*, **113**, E12S42.

Mustard, J. F., Murchie, S., Pelkey, S., Ehlmann, B., Milliken, R., Grant, J., Bibring, J.-P., Poulet, F., Bishop, J. and Dobrea, E. N. (2008). Hydrated silicate minerals on Mars observed by the Mars Reconnaissance Orbiter CRISM instrument. *Nature*, **454**, 305–309.

Neukum, G. and Jaumann, R. (2004). In: Mars Express: the scientific payload. Ed. by Andrew Wilson, Scientific coordination: Agustin Chicarro. ESA SP-1240, Noordwijk, Netherlands: ESA Publications Division, ISBN 92-9092-556-6, 2004, pp. 17–35.

Neumann, G., Zuber, M., Wieczorek, M., McGovern, P., Lemoine, F. and Smith, D. (2004). Crustal structure of Mars from gravity and topography. *J. Geophys. Res. Planets*, **109**, E08002.

Nyquist, L. E., Bogard, D. D., Shih, C. Y., Greshake, A., Stöffler, D. and Eugster, O. (2001). Ages and geologic histories of Martian meteorites. *Chronology and Evolution of Mars*, **96**, 105–164.

Ormö, J. (2003). Mars Orbiter Camera observation of linear and curvilinear features in the Hellas basin: Indications for multiple processes of formation. *J. Geophys. Res.*, **108**.

Orosei, R., Lauro, S. E., Pettinelli, E., Cicchetti, A., Coradini, M., Cosciotti, B., Di Paolo, F., Flamini, E., Mattei, E., Pajola, M., Soldovieri, F., Cartacci, M., Cassenti, F., Frigeri, A., Giuppi, S., Martufi, R., Masdea, A., Mitri, G., Nenna, C., Noschese, R., Restano, M. and Seu, R. (2018). Radar evidence of subglacial liquid water on Mars. *Science*, doi:10.1126/science.aar7268.

Osterloo, M., Anderson, F., Hamilton, V. and Glotch, T. (2007). Analysis of a Spectrally Distinct Surface Feature in the Terra Sirenum Region of Mars from THEMIS and TES, Lunar and Planetary Science Conference, p. 1814.

Palumbo, A. M. and Head, J. W. (2018). Early mars climate history: Characterizing a "Warm and Wet" Martian climate with a 3-D global climate model and testing geological predictions. *Geophys. Res. Lett.*, https://doi.org/10.1029/2018GL079767.

Papike, J. J., Karner, J. M., Shearer, C. K. and Burger, P. V. (2009). Silicate mineralogy of Martian meteorites. *Geochimica et Cosmochimica Acta*, **73**, 7443–7485.

Pelkey, S. M., Mustard, J. F., Murchie, S., Clancy, R. T., Wolff, M., Smith, M., Milliken, R., Bibring, J. P., Gendrin, A., Poulet, F., Langevin, Y. and Gondet, B. (2007). CRISM multispectral summary products: Parameterizing mineral diversity on Mars from reflectance. *J. Geophys. Res.*, **112**, E08S14.

Peplowski, P. N., Jack, T. W., Andrew, W. B., Morgan, B., John, O. G. and David, J. L. (2018). Characterizing near-surface elemental layering on Mars using gamma-ray spectroscopy: A proof-of-principle experiment, Nuclear Instruments and Methods in Physics Research Section B: Beam Interactions with Materials and Atoms, **415**, 89–99, https://doi.org/10.1016/j.nimb.2017.11.016.

Plattner, A., Simons, F. and Golabek, G. (2016). The crustal magnetic field of Terra Sirenum and Cimmeria, Mars, A Spectral Perspective, AGU Fall Meeting Abstracts.

Preston, L. J. and Dartnell, L. R. (2014). Planetary habitability: Lessons learned from terrestrial analogues. *Intern. J. Astrobio.*, **13**(1), 81–98, doi: 10.1017/S1473550413000396.

Reiss, D., Hauber, E., Hiesinger, H., Jaumann, R., Trauthan, F., Preusker, F., Zanetti, M., Ulrich, M., Johansson, L., Johansson, L., Olvmo, M., Carlsson, E., Johansson, H. A. B. and McDaniel, S. (2011). Terrestrial gullies and debris-flow tracks on Svalbard as planetary analogs for Mars, in Garry, W. B., and Bleacher, J. E. (Eds.), *Analogs for Planetary Exploration. Geol. Soc. Am. Spec. Pap.*, **483**, p. 567.

Rogers, A. D. and Hamilton, V. E. (2015). Compositional provinces of Mars from statistical analyses of TES, GRS, OMEGA and CRISM data. *J. Geophys. Res. Planets*, **120**(1), 62–91.

Romanek, C. S., Grady, M. M., Wright, I. P., Mittlefehldt, D. W., Socki, R. A., Pillinger, C. T. and Gibson, E. K. (1994). Record of fluid-rock interactions on Mars from the meteorite ALH84001. *Nature*, **372**, 655–657.

Romig, B., Kosmo, J., Ross, A., Bernard, C., Aitchison, L., Eppler, D. and Splawn, K. (2007). Desert research and technology studies 2006 report, SAE Technical Paper. No. 2007-01-3131.

Rossi, A.P. and van Gasselt S. (eds.) (2018). *Planetary Geology*, Springer Praxis Books, doi: 10.1007/978-3-319-65179-8_3.

Sarrazin, P., Blake, D., Feldman, S., Chipera, S., Vaniman, D. and Bish, D. (2005). Field deployment of a portable X-ray diffraction/X-ray flourescence instrument on Mars analog terrain. *Powder Diffr.*, **20**, 128–133.

Sautter, V., Toplis, M. J. and Wray, J. J. (2015). In situ evidence for continental crust on early Mars. *Nat. Geosci.*, **8**, 605–609.

Seu, R., Phillips, R. J., Biccari, D., Orosei, R., Masdea, A., Picardi, G., Safaeinili, A., Campbell, B. A., Plaut, J. J., Marinangeli, L., Smrekar, S. E. and Nunes, D. C. (2007). SHARAD sounding radar on the Mars Reconnaissance Orbiter. *J. Geophys. Res.*, **112**.

Shean, D., Fahle, J., Malin, M., Edwards, L. and Posiolova, L. (2011). MRO CTX stereo image processing and preliminary DEM quality assessment, Lunar and Planetary Science Conference Abstracts 42, 2646.

Soare, R. J., Sejourne, A., Pearce, G., Costard, F. and Osinski, G. R., The Tuktoyaktuk coastlands of northern Canada: A possible "wet" periglacial analog of Utopia Planitia, Mars, in Garry, W. B. and Bleacher, J. E. (Eds.) (2011). *Analogs for Planetary Exploration. Geol. Soc. Am. Spec. Pap.*, **483**, p. 567.

Stockstill, K. R., Moersch, J. E., McSween, H. Y., Piatek, J. and Christensen, P. R. (2007). TES and THEMIS study of proposed paleolake basins within the Aeolis quadrangle of Mars. *J. Geophys. Res.*, **112**.

Stuurman, C. M., Osinski, G. R., Holt, J. W., Levy, J. S., Brothers, T. C., Kerrigan, M. and Campbell, B. A. (2016). SHARAD detection and characterization of subsurface water ice deposits in Utopia Planitia, Mars. *Geophys. Res. Lett.*, **43**, 9484–9491.

Tomkinson, T., Lee, M. R., Mark, D. F. and Smith, C. L. (2013). Sequestration of Martian CO_2 by mineral carbonation. *Nat. Commun.*, **4**, 1–6.

Turcotte, D., Shcherbakov, R., Malamud, B. and Kucinskas, A. (2001). Analyses of Mars Topography, Gravity and Aroid: Implications for Tharsis and Hellas, Lunar and Planetary Science Conference, p. 1006.

Wade, J., Dyck, B., Palin, R. M., Moore, J. D. P. and Smye, A. J. (2017). The divergent fates of primitive hydrospheric water on Earth and Mars. *Nature*, **552**, 391–394.

Wadhwa, M. and Crozaz, G. (1995). Trace and minor elements in minerals of nakhlites and Chassigny: Clues to their petrogenesis. *Geochimica et Cosmochimica Acta*, **59**, 3629–3645.

Wang, J., Xiao, L., Reiss, D., Hiesinger, H., Huang, J., Xu, Y. *et al.* (2018). Geological features and evolution of Yardangs in the Qaidam Basin, Tibetan Plateau (NW China): A terrestrial analogue for Mars. *J. Geophys. Res. Planets*, **123**, https://doi.org/10.1029/2018JE005719.

Wei, J., Wang, A., Lambert, J., Wettergreen, D., Cabrol, N. and Warren-Rhodes, K. (2014). Automated Core Sample Analysis by the Mars Microbeam Raman Spectrometer (MMRS) On-Board the Zoe Rover in Atacama: A Terrestrial Test for Mars Exploration, Lunar and Planetary Science Conference, p. 2428.

Williams, R. M., Grotzinger, J. P., Dietrich, W. E., Gupta, S., Sumner, D. Y., Wiens, R. C., Mangold, N., Malin, M. C., Edgett, K. S., Maurice, S., Forni, O., Gasnault, O., Ollila, A., Newsom, H. E., Dromart, G., Palucis, M. C., Yingst, R. A., Anderson, R. B., Herkenhoff, K. E., Le Mouelic, S., Goetz, W., Madsen, M. B., Koefoed, A., Jensen, J. K., Bridges, J. C., Schwenzer, S. P., Lewis, K. W., Stack, K. M., Rubin, D., Kah, L. C., Bell, J. F., 3rd, Farmer, J. D., Sullivan, R., Van Beek, T., Blaney, D. L., Pariser, O., Deen, R. G. and Team, M. S. L. S. (2013). Martian fluvial conglomerates at Gale crater. *Science*, **340**, 1068–1072.

Williams, R. M. E., Chuang, F. C. and Berman, D. C. (2017). Multiple surface wetting events in the greater Meridiani Planum region, Mars: Evidence from valley networks within ancient cratered highlands. *Geophys. Res. Lett.* **44**(4), 1669–1678.

Xiao, L., Wang, J., Dang, Y. N., Cheng, Z. Y., Huang, T., Zhao, J. N., Xu, Y., Huang, J., Xiao, Z. Y. and Komatsu, G. (2016). A new terrestrial analogue site for Mars research: The Qaidam Basin, Tibetan Plateau (NW China). *Earth–Sci. Rev.*, doi:10.1016/j.earscirev. 2016.11.003.

Zhong, S. J. and Maria, T. Z. (2001). Degree-1 mantle convection and the crustal dichotomy on Mars. *Earth Planet. Sci. Lett.*, **189**, 75–84.

https://doi.org/10.1142/9789811224027_0002

Chapter 2

The Surface of Mars: An Overview

Long Xiao

State Key Laboratory of Geological Processes and Mineral Resources, Planetary Science Institute, School of Earth Sciences, China University of Geosciences, Wuhan 430074, PR China

2.1 Introduction

Mars is the fourth planet of our solar system. Its size is about half and mass is about ten percent of our Earth. Although it is smaller than Earth, Mars's surface presents diverse landforms based on Earth-based, orbiter- and *in situ* observations. These originated by internal or external forces, and formed endogenic and exogenic landforms, respectively, the same as Earth.

At global scale, the surface of Mars displays an obvious dichotomy in topography (Fig. 2.1). The southern hemisphere is much higher and has more impact craters than the northern hemisphere. These are also termed the southern highlands and northern lowlands, and the boundary lies roughly along the equator. Other large remarkable edifices include giant volcanoes on the Tharsis bulge at the western part and large impact basins, such as Hellas Basin, Argyre Basin, and Isidis Basin on the southern highland (Carr, 2006).

The south pole and north pole are also distinct in altitude, and both have swirled ice (water ice and dry ice) or ice and dust formed caps (Byrne, 2009). The sizes of these caps usually change with seasons, being larger in winter and smaller in summer.

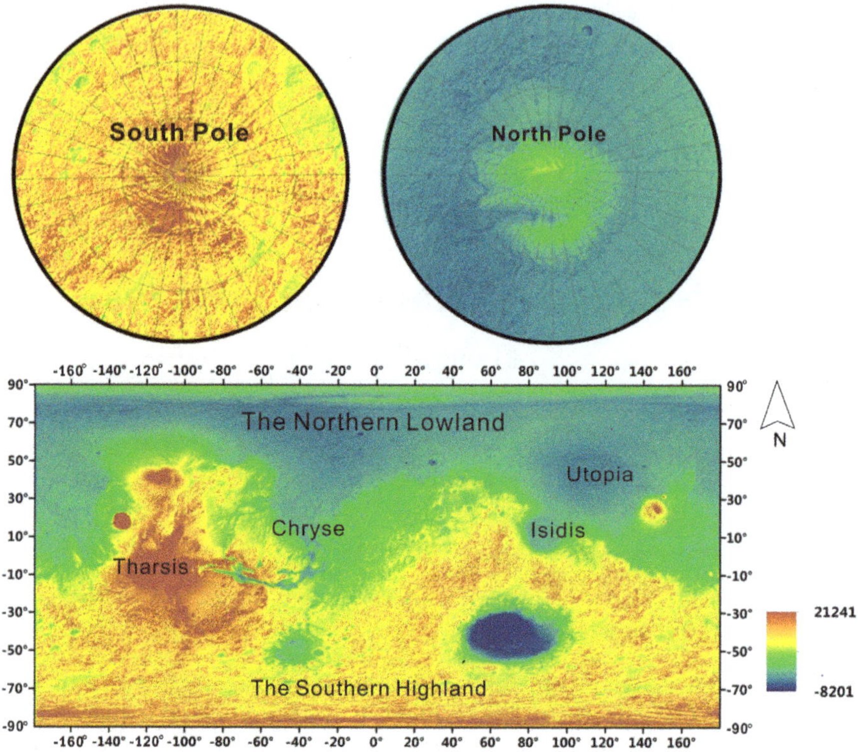

Fig. 2.1. The topography map of Mars, created by using MOLA (Mars Orbiter Laser Altimeter, Mars Global Surveyor spacecraft, NASA) data. Bright red and yellow colors indicate high, while green and blue mean low.

Specific endogenic landforms consist mainly of tectonic and volcanic topographies. Exogenic landforms are more diverse. The driving forces include water, ice, atmosphere, impact, and gravity. In general, almost all landforms found on Earth are present on Mars (Xiao *et al.*, 2017).

2.2 Endogenic Geomorphology

2.2.1 *Tectonic features*

Similar to Earth, there are both extensional and compressional deformation features on Mars driven by internal forces. Extension resulted in rifts and grabens, which are more common than compressional features, such as ridges and thrust faults (Anderson *et al.*, 2008, Basilevsky *et al.*, 2006, Head *et al.*, 2002).

2.2.1.1 *Extensional structures*

The largest rift on Mars is Valles Marineris. With its length over 4000 km, width of 200 km and depth up to 7 km, the Valles Marineris is also the largest canyon in the solar system. From west to east, it consists of Noctis Labyrinthus, Tithonium and Ius chasmata, Melas, Candor and Ophir chasmat, Coprates Chasma, Gandes, Capri, and Eos Chasmata (Figs. 2.2 and 2.3) (Carr, 2012). Finally, it empties into an outflow channel region containing chaotic terrain that terminates in the basin of Chryse Planitia. The most agreed upon theory today is that Valles Marineris was formed by rift faults like the East African Rift, later made bigger by erosion and collapsing of the rift walls (De Blasio 2011, Jackson *et al.*, 2011, Le Deit *et al.*, 2010, Rodriguez *et al.*, 2015a).

There are quite a few large extensional faults, and some of them then formed as elongated closed depression/trough or out-flow channel connected linear structures (Anderson and Dohm 2011; Carey *et al.*, 2016; Egea-Gonzalez *et al.*, 2017; Rice *et al.*, 2011). The former is represented by the Tithonium Chasma, Hebes Chasma (rifting with the same mechanism as the Valles Marineris), and Cerberus Fossae and Memnonia Fossae (rifting plus dike emplacement, and cryospheric cracking) (Fig. 2.4). Most of them are connected with outflow channels, and these strongly indicate that large amounts of groundwater poured out through these fractures, accompanied by dike emplacement, volcanic events, and possibly impacts (Carr, 2006; Clifford 1993; Davis and Tanaka, 1995; Head *et al.*, 2003).

Other smaller rifts/grabens mostly occurred in the Tharsis rise region, forming the Tharsis Radial feature system as shown in Fig. 2.5 (Wilson and Head, 2002). The Valles Marineris also starts from this huge bulge.

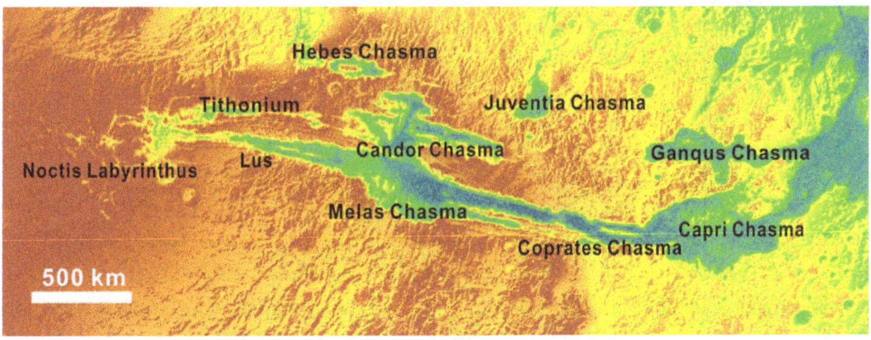

Fig. 2.2. The Valles Marineris. It is a gigantic system of canyons on Mars and has been interpreted as resulted from extensional rifting/crack with lateral offsets.

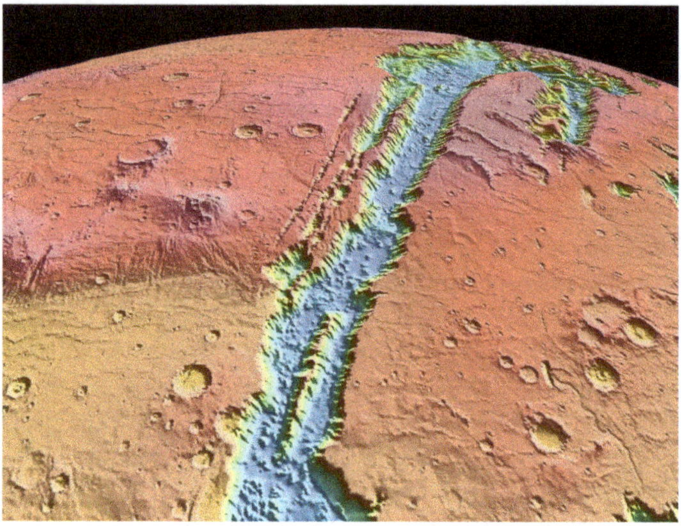

Fig. 2.3. Valles Marineris topographic view constructed from MOLA altimetry data. Image shows Coprates Chasma, with Melas Chasma at the top, Candor Chasma at top right, and part of Capri Chasma at the bottom (adapted from wikipedia).

The densest rift/graben cluster is around the Alba Patera (Fig. 2.6). There are numerous SN-NW linear fractures and circular grabens cross-cutting the patera (Ivanov and Head, 2006). The Laba Patera is a low shield volcano. Summit calderas are visible. On the flank, lava flows (black arrows) were cut by NNE trending grabens (white arrows). The later appear as pit chains, thought to have formed as fractures developed underneath and then surface materials collapsed; alternatively, these are vents or passages of lavas, but their orientation and surface lava flow features do not support this hypothesis (Cattermole, 1987; Crown and Greeley, 1993; Korteniemi *et al.*, 2010).

2.2.1.2 *Compressional structures*

Typical compressional features include ridges and lobate scarps. Similar to the surface of the Moon, the most obvious compression deformation on the surface of Mars is found in the form of wrinkle ridges (Anderson and Herkenhoff, 2012; Bleacher *et al.*, 2017; Gullikson *et al.*, 2016). They are asymmetrical linear structures, tens to hundreds of kilometers long and thousands of meters wide. Most of these ridges consist of a broad, gentle

(a)
(b)
(c)

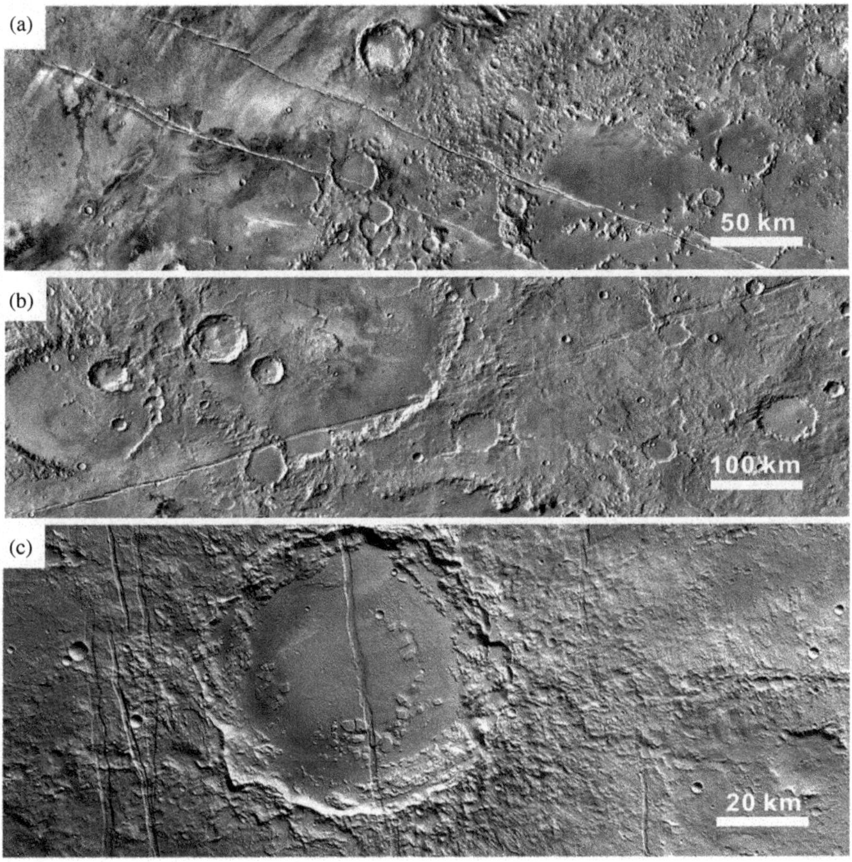

Fig. 2.4. Selected typical rifts and faults. (a) Cerberus Fossae at 10°N, 157°E; (b) The Sirenum Fossae, (c) Memnonia Fossae.

arched body (generally 100~200 m high) and long, narrow wrinkles, the latter of which may be at the top of the arch, or at the edges, which may be continuous or discontinuous (Bleacher *et al.*, 2013). Such ridges are generally distributed on the volcanic plains. In the western hemisphere, ridge rings surround the Tharsis rise. In the eastern hemisphere, the ridges are distributed around several volcanic plains: Lunae Planum (Fig. 2.7(a)), Isidis Planitia, Hesperia Planum, and Malea Planum regions. In addition, ridges can also be seen at the bottom of some large impact craters, such as Gusev, and the intercrater plain. The formation of the ridges in this hemisphere appears to be disordered and unrelated to Tharsis.

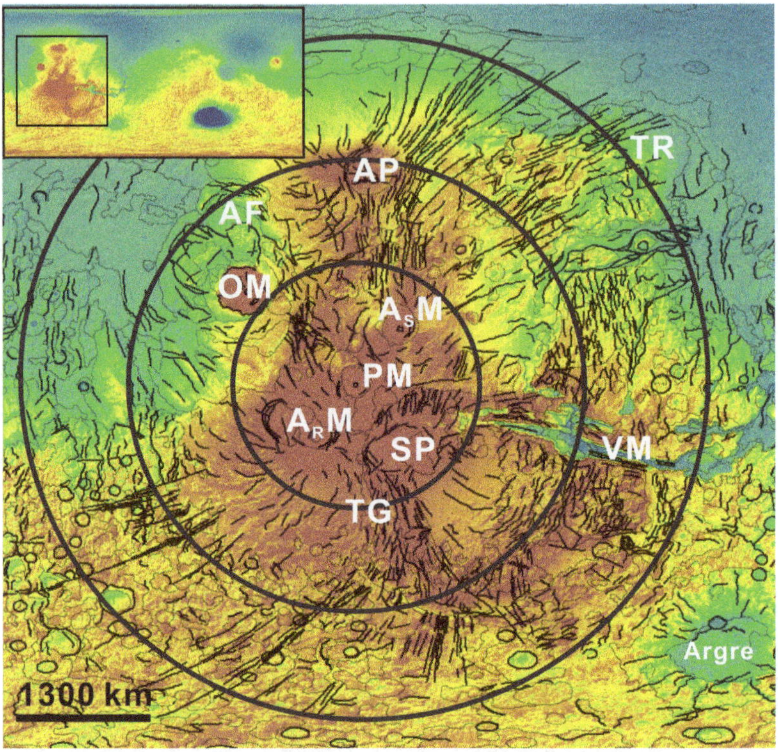

Fig. 2.5. Tharsis radial feature system.

Lobate scarps are found in the ancient highlands of the eastern hemisphere of Mars (Chuang *et al.*, 2014, Fastook *et al.*, 2014, Plaut *et al.*, 2009). They are interpreted to be thrust faults. Lobate scarps occur near and oriented roughly parallel to the Martian crustal dichotomy. The typical lobate scarp occurs in the northern Terra Cimmeria-Amenthes region (Fig. 2.7(b)).

2.2.2 *Volcanic features*

There are a variety of volcanic features on Mars, including shield volcanoes formed by extensive effusive eruptions, plains filled with lava flows, and edifices formed by explosive eruptions. Many factors govern the morphology of volcanic landforms, such as planetary variables (gravity, lithosphere pressure, atmospheric properties, surface/subsurface liquids, planetary radius, planetary composition, temperature), magma properties

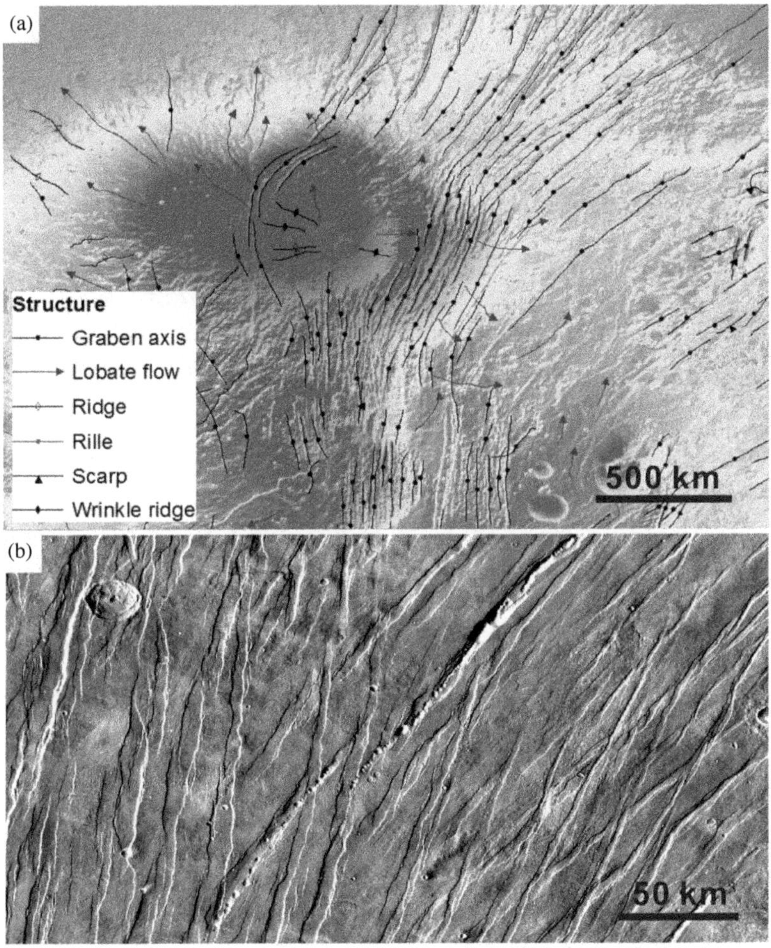

Fig. 2.6. Fractures in the Alba Patera region. (a) LOLA image (the scale bar is 250 km). (b) Is THEMIS image enlarged from upper dashed box (the scale bar is 100 km).

controlling rheology (viscosity, temperature, density, composition, volatiles, amount of solids, yield strength, shear strength), and eruption properties (eruption rate, eruption volume, eruption duration, vent characteristics, topography, ejection velocity) (Khalaf and Hammed, 2016, Williams *et al.*, 2009, Wilson and Head, 1994).

On Mars, the most remarkable large central volcanoes are located in the Tharsis, Elysium, circum-Hellas, and Nili Fossa regions. Most of the surface of Mars is covered by lava flows (Greeley and Spudis, 1981) (Table 2.1).

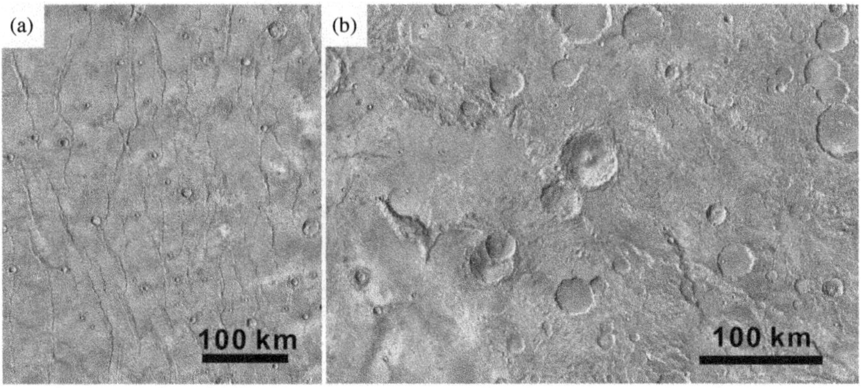

Fig. 2.7. (a) MOLA, Lunae Planitia, a ridged plain. These ridges are deformed thick basalt flows and similar to those seen on the Moon, Mercury, and Venus. (b) lobate scarp occurs in the northern Terra Cimmeria-Amenthes region.

Table 2.1. Classification and extent of Martian volcanic features (Greeley and Spudis, 1981).

Type	Example	Characteristics	Extent (10^6km^2)	Percentage of Mars' surface
Central volcanoes				
Alba	Alba Patera	Unique, extremely low-relief shield-like volcano; sheet and tube-fed flows	1.13	0.78
Highland paterae	Tyrrhena Patera	Low-relief, degraded, radially textured volcanoes	0.23	0.16
Shields	Olympus Mons	Broad, moderate-relief central vent volcanoes; tube-fed flows	1.23	0.85
Domes	Tharsis Tholus	Steep-sided central volcanoes	0.08	0.05
Total area of central volcanoes			2.67	1.84
Volcanic plains				
Simple flows	Hesperia Planum	Regional plains, wrinkle ridges only; no-flow-lobes	42.08	29.22
Complex flows	Tharsis Plains	Complex flow units; flow-lobes abundant; rare wrinkle ridges	9.16	6.36
Undifferentiated	Northern plains	Plains of uncertain origin (flow-lobes rare), although probably volcanic	27.37	19.00
Questionable	Aureole materials	Modified units associated with volcanic units	5.92	4.11
Total area of volcanic plains			84.53	58.69
Total volcanic surface area on Mars			85.20	60.53

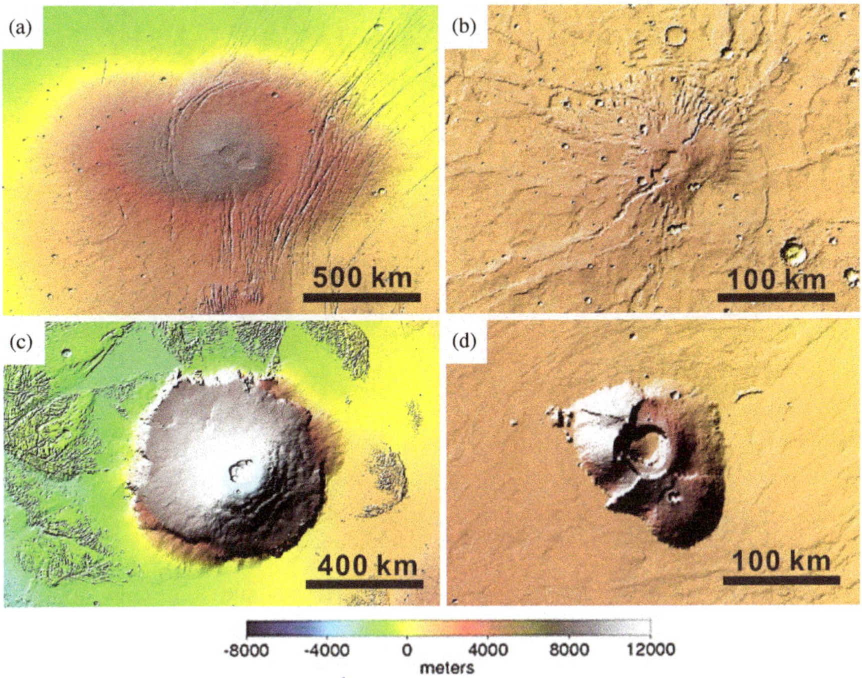

Fig. 2.8. Typical large central volcanoes on Mars. (a) Alba Patera, (b) Tyrrhena Patera, (c) Olympus Mons, (d) Tharsis Tholus.

Central volcanoes on Mars are largely due to long-lasting volcanic accumulation at fixed sites, because Mars lacks Earth-like plate tectonics. Olympus Mons and Alba Patera are two of the largest central volcanoes (Fig. 2.8). They are also among the largest volcanoes in our solar system.

There are also numerous small shield volcanoes or shield volcano swarms, such as those on the flanks of large shield volcanoes (e.g. Pavonis Mons, Mon Olympus, Elysium and Syria Planum).

In addition to these outstanding large central volcanoes, most of the Martian surface is covered by flooded lavas, namely *volcanic plains*. These plains are generally similar to the mare on the Moon, flat in topography and showing wrinkled ridges (Fig. 2.7(a)) and flow lobes. These ridged plains are most visible in the southern highlands, representing flood eruptions that buried their sources. In contrast, complex flows are composed of multiple flow features (Williams *et al.*, 2010) (Fig. 2.9).

Undifferentiated flows constitute much of the northern lowland plains (Greeley, 2011). Due to a wide variety of post-volcanic modification, such

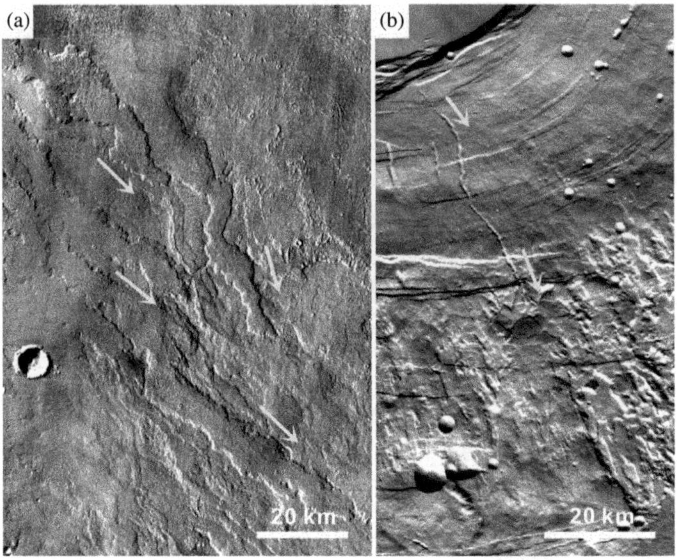

Fig. 2.9. Flow lobes in Arisa Mons area showing multiple flow-lobes and lava channels.

as eolian, periglacial, fluvial, and lacustrine processes, typical flow features are lacking, but flow lobes are seen in some areas. Remote sensing data are indicative of basaltic compositions.

Volcanism styles are not all effusive. Explosive eruption could have played a major role in early Martian history (Xiao *et al.*, 2012) (Huang *et al.*, 2015). The presence of water and ice on and near the surface could have resulted in extensive magma–water interactions. Such interactions resulted in the formation of small cones (Fig. 2.10), mud volcanoes (Fig. 2.11), and other volatile-rich exploration features (Xiao and Wang, 2009).

On Earth, quite a few regions have active mud volcanoes, such as Baku of Azerbaijan (Fig. 2.11(a)), Taiwan and Xinjiang (Fig. 2.11(b)) of China, and Yellowstone of the United States. In general, a mud volcano is a pile of mud (and gas) that rises above the ground. Its appearance is a conical mound. The tip of the mound often sags, and from this mud and gas are intermittently ejected. Mud volcanoes are usually related to normal volcanism and tectonism, and they could appear on land and on the ocean floor.

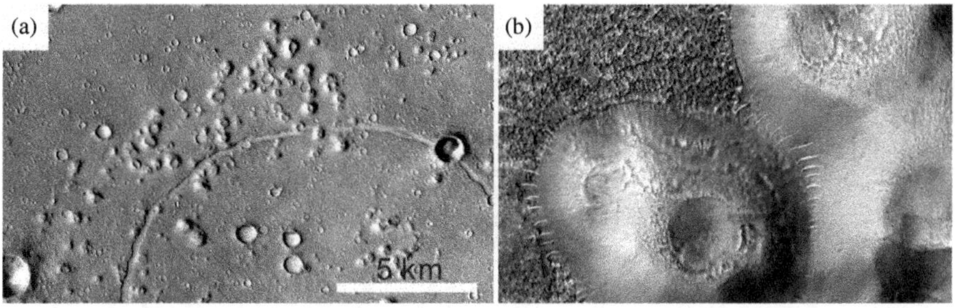

Fig. 2.10. (a) Cones in the Isidis plain, and (b) mud volcanoes (credit: NASA/JPL-Caltech/UA).

Fig. 2.11. Mud volcanoes in Baku, Azerbaijan.
Source: http://www.sohu.com/a/317946404_681282.

2.3 Exogenic Geomorphology

2.3.1 *Water-related landforms*

At present, water carved landforms on Mars are divided into outflow channels, valley networks, and gullies (Carr, 2006, 2012). Another type, the fretted valley, is found in mid-to-high latitudes and is associated with the presence and movement of underground ice.

2.3.1.1 *Outflow channels*

The size of outflow channels varies greatly. For example, the largest, Kasei Vallis, is 400 km wide at the mouth and 2.5 km deep in some places (Fig. 2.12), while some small outflow channels are less than 1 km wide

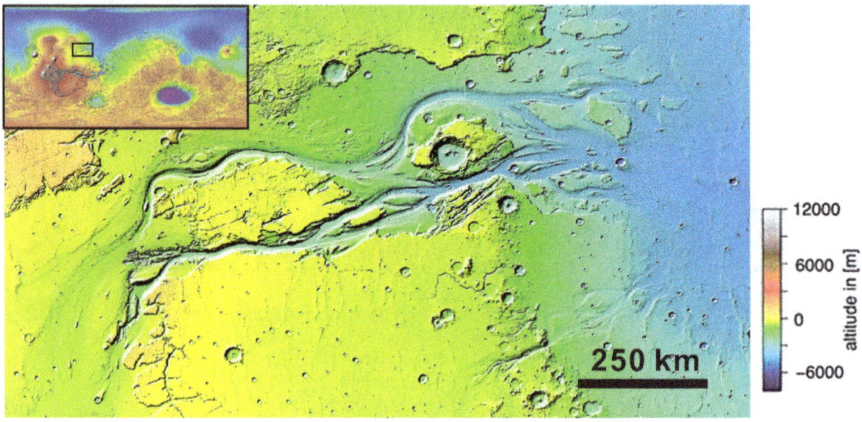

Fig. 2.12. Outflow channels in Kasei region.

(Rodriguez *et al.*, 2015b). In general, most outflow channels have obvious geomorphological features formed by water erosion, such as small curvature, large length–width ratio and remarkable branching and recombination. Teardrop-like islands are often seen in the middle of channels. These are similar to the landforms formed by floods on the Earth's surface.

Another typical outflow channel is Mangala Vallis (Ghatan *et al.*, 2005; Wilson and Head, 2004), which is hundreds of kilometers long and over 100 km wide locally (Fig. 2.13). It originates from a fault graben in the south.

In addition to the Kasei Vallis Plain and the Mangala Vallis, many other places have such outflow channels, albeit on a smaller scale, such as the Tharsis, Amazonian, and Elysium Planitia. The outflow channels here clearly originated in the graben. Two models can be used to explain their causes. One is the rapid outflow of liquid water from deep underground caused by tensional fracture, resulting in flood (Andrews-Hanna *et al.*, 2008; Carr and Clow, 1981). Another is the intrusion of magma along a dyke into the cryosphere, which causes the melting of frozen water and outflow along the fissure (Head *et al.*, 2003).

2.3.1.2 *Valley networks*

The valley networks are usually found in the southern highlands, and their spatial distribution is irregular. They are sparsely distributed in the

Fig. 2.13. Mangala Vallis (18°S, 151°W) outflow channel, which originates from a graben up to 7 km wide in the south (lower part of the figure), which gradually widens to the north and extends up to 1,000 km, flowing into the northern plain. Image of MOLA superimposed on THEMIS Day IR.

northwestern part of Arabia Terra and on the southwest and southeast sides of the Hellas Basin (Carr, 2006). Most valley networks are relatively short (less than 200 km) and end up in locally low-lying areas. Given that most valley networks are found on the ancient Noachian terrains, there is reason to believe that early Mars surface was warm, moist, and flooded with water (Luo *et al.*, 2017). Extensive running water indicates that the erosion rate is fast and the deposition rate is fast, too. In fact, large deposits were also found in the Noachian formation. At the same time, a large amount of running water is bound to pool in low-lying areas, forming lakes and even oceans (Carr and Head, 2019; Chan *et al.*, 2018; Di Achille and Hynek, 2010; Head, 1999). As examples, there are many outward and inward valley networks in the Huygens impact crater (Fig. 2.14).

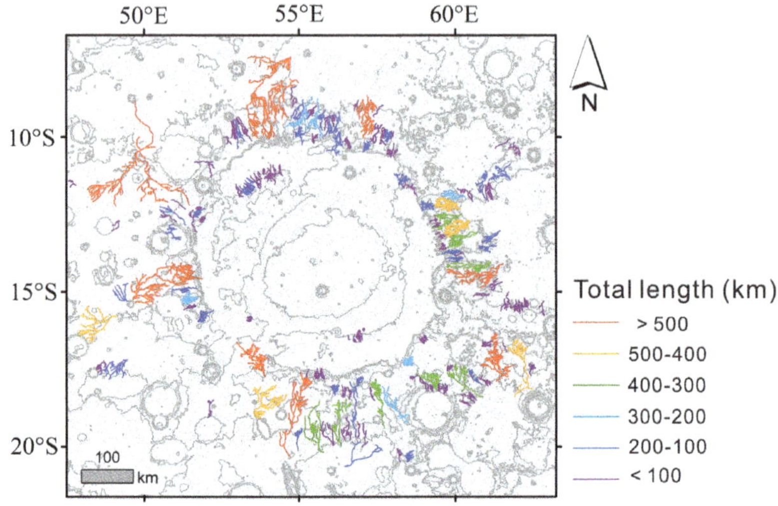

Fig. 2.14. Valley networks around Huygens impact crater, with 1 km contour as background. Color of the mapped valley suggests valley length.

Fig. 2.15. Warrego Vallis (42°S, 267°E). Image of MOLA superimposed on THEMIS Day IR.

Warrego Vallis is another typical example, probably the densest valley network on the surface of Mars (Fig. 2.15), very similar to the river network on the surface of our Earth, strongly suggesting that it should also have been formed by surface drainage system (Ansan and Mangold, 2006).

Although these valley network features indicate there were surface running water activities on Mars, an unanswered question is how did these streams form: Rain? Groundwater leakage? Underground ice melt and leak? Melting snow or ice? So far, there has been no confirmation that Mars ever had long periods of rain. It is clear that Noachian formations have been subjected to intense erosion and carving by running water. Younger formations experienced little erosion (in addition to the local formation of the outflow channels), showing that there was a dramatic environmental change in the late Noachian Era, although the cause is unknown (Alemanno *et al.*, 2018).

Gullies are young erosion patterns that develop on steep slopes of the Martian surface. They usually bifurcate at the top, combine into one or more gullies downward, and eventually disappear into triangular accumulations (Heldmann and Mellon, 2004). The width of such gullies is generally several to dozens of meters, and the length is several hundred meters. So, they are much smaller than valley networks. These gullies are also newly formed because impact craters are barely visible on their surfaces, and they cut through all other structural features along the way, including dunes. The significance of studying these gullies is that their formation may be related to recent erosion by liquid water or another kind of liquid (Conway *et al.*, 2015, 2019, de Haas *et al.*, 2015; Dundas *et al.*, 2019).

The gullies are distributed locally, mainly on steep slopes of latitude 30° in the southern hemisphere, such as impact crater walls, central uplift of impact craters, channel flanks and depressions, as well as on steep slopes of dunes (Auld and Dixon, 2016) (Fig. 2.16), and all of them are distributed on the southward slope. In contrast, there is little in the northern hemisphere, possibly because there are few steep slopes.

Due to the mass wasting, there are a lot of gully-like features named RSL (recurring slope lineae) on the crater walls (Fig. 2.17). Observed from HiRISE image, RSL are narrow linear features extending directly down the topographic gradient which are commonly up to 40% darker than the surroundings (McEwen *et al.*, 2011), with width of 0.5 to 5 meters and length up to hundreds of meters. The occurrence of RSL differs with seasons, showing increasing of length down steep slopes during warmer seasons and fading during cold seasons. Some people think that RSL may be associated with the activity of water, but their true origin remains unclear (Dundas *et al.*, 2017).

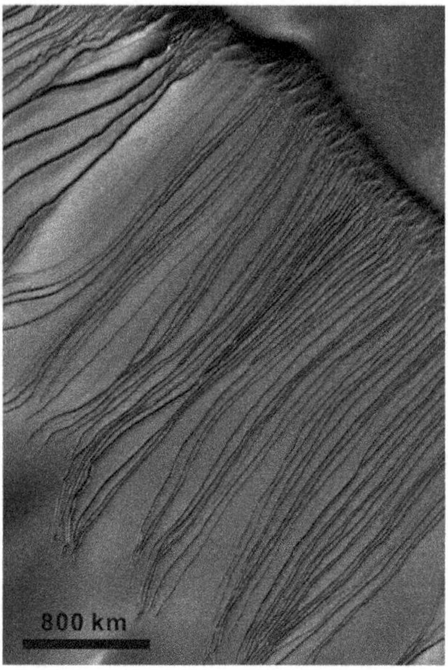

Fig. 2.16. Gullies developed on a large dune in Russell impact crater (MGS MOC No. Moc2-343. image width of about 3 km, sunlight from upper left, center position of 54.5°S, 347.3°W). The dune slopes down to the left. The formation of these gullies may be related to fluid or gas-fluid mixtures flowing down slopes.

Fig. 2.17. Recurring slope lineae on the slope of impact crater wall. The light-toned ones indicate older RSLs and dark ones are more recently formed. (Credit: HiRISE ESP_012410_1835_RED).

2.3.1.3 *Paleocrater lake basins*

From the various landforms formed by surface water activity on the surface of Mars, it is conceivable that water would eventually converge in low-lying areas to form lakes and/or oceans. These low-lying areas that collect surface water can be impact craters or other broad depressions. The former may form lakes, while the latter may form seas or even oceans (McEwen *et al.*, 2007).

Some impact craters had been filled with water, with evidence of inlet flow channels, delta layered deposits (Fig. 1.3), and outlet flow channels in some cases (Fig. 2.18).

All of the crater lake basins are located in the southern highlands and were formed during the Noachian Era, the same time as the valley network systems. But studies have also found some younger paleolakes, formed in the Amazonia Era (Cabrol, 1997; Di Achille and Ori, 2008; Melwani Daswani and Kite 2017; Soare *et al.*, 2008). It is worth further examination to disclose their relationships with young valley networks.

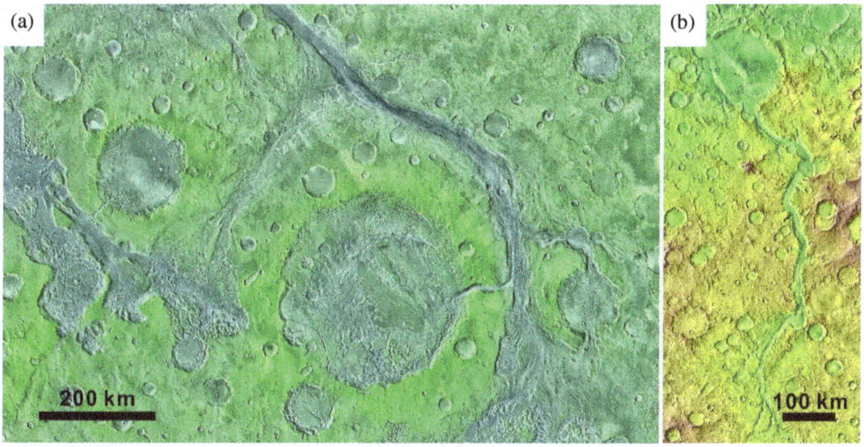

Fig. 2.18. (a) A closed depression in the Terra Sirunum (35°S, 185°E) region, may be the site of an ancient lake. Some sediments can be seen at the bottom of the depression, which may be from inter-basin upland and mountains (Irwin *et al.*, 2002). (b) A possible paleolake at region 4°S, 111°E. It can be seen that the Tinto Vallis River cuts through the edge of the Palos impact crater with a diameter of about 60 km and flows into the impact crater and flows out from the other side of the impact crater. The channel originates from the south of the linear distribution of pits, supporting its source of underground water. Smaller channels also connect the two craters on the right. The images are MOLA superimposed on THEMIS Day IR.

2.3.2 *Wind and aeolian landforms*

2.3.2.1 *Dust storm and Wind tracks*

Mars has very strong wind activity, although its atmosphere is very thin. Fine materials formed by impact gardening and weathering could be moved by wind (McCauley, 1973). It is evident that dust storms occur about half the time each year (Greeley *et al.*, 1981) (Fig. 2.19). Wind erosion and aeolian deposition are also evident on the surface of Mars. These features can be directly observed by orbiters and landers (Rogers *et al.*, 2018; Squyres *et al.*, 2004; Zimbelman, 2010). The wind mainly carries and redeposits loose layered sediments. Long-lasting wind operation on the surface of Mars will cause landform alteration, although the wind force is small.

Dust storms on Mars usually begin in the spring and continue throughout the summer. Sometimes dust storms occur in small areas, but there are also dust storms that cover the entire planet (Fig. 2.18). Dust storms are most likely to occur and be observed at the edge of the polar cap and at latitudes from 10° to 30° south (Greeley *et al.*, 1981). Other regions are the Hellas Basin, between the Noachis and Hellas, the Argyre Basin, and the Solis-Syria-Sinai planitia in the southern hemisphere, and the

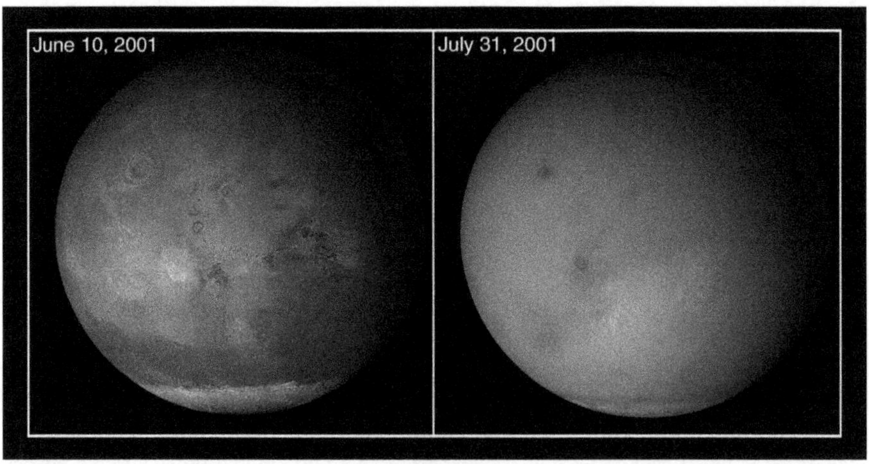

(a) (b)

Fig. 2.19. A quiet Mars in June of 2011 (a) compared with a global dust storm Mars in July 2001 (b).

Source: NASA, J. Bell, M. Wolff and Hubble Heritage Team (STSci/AURA).STSci-PRC01-31.

Chryse-Acidalia, Isidis-Syrtis, and Cerberus fossae regions in the northern hemisphere. The formation of dust storms is closely related to the circulation of Martian atmosphere, and detailed studies are investigating its triggers and processes (Bhardwaj *et al.*, 2017; Fedorova *et al.*, 2018).

Dust devils are an efficient dust-raising mechanism and leave tracks on the surface of Mars. As the dust devil moves across the surface, it scours the fine dust particles, revealing the darker rock surface below (Fig. 2.20).

2.3.2.2 *Aeolian landform Features*

Wind movement will cause erosion, transportation, and deposition. Although the present Martian atmosphere pressure is far lower than before, long-term observations show that the sand storms are still active during nearly half of a Martian year, forming various aeolian landforms, e.g. wind streaks, sand dunes, and sand ripples. Compared with wind deposition, there are not many landforms formed by wind erosion. One of the reasons is that wind force is not strong enough at present. However, some yardang fields have been observed in the Medusae Fossae Formation (MFF) (Jacobsen and Burr, 2013; Lefort *et al.*, 2012).

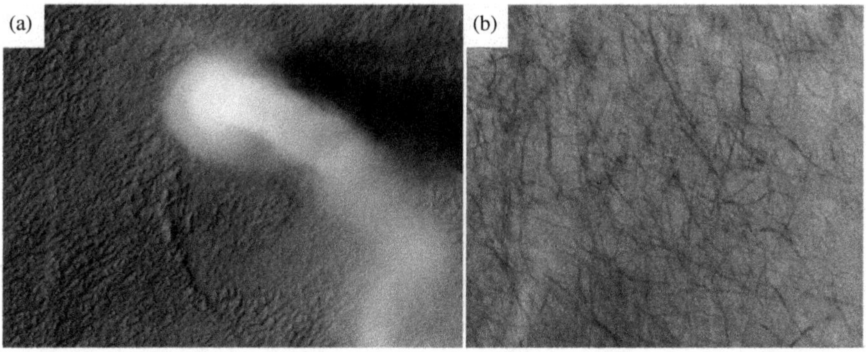

Fig. 2.20. A Martian dust devil roughly 20 km high was captured winding its way along the Amazonis Planitia region of northern Mars on March 14, 2012, by the High Resolution Imaging Science Experiment (HiRISE) camera on NASA's Mars Reconnaissance Orbiter. Despite its height, the plume is little more than three-quarters of a football field wide ((a), Image credit: NASA/JPL-Caltech/UA), dust devil tracks on the plains of Aonia Terra, Image Width17 km ((b), THEMIS images credit: NASA/JPL-Caltech/Arizona State University).

As the wind speed slows down, carried particles will settle to the surface and form aeolian deposits. The most common aeolian deposition landforms are dunes and ripples. They are distributed almost everywhere and are very similar to those on Earth in shape, but are usually lager in size (Carr, 2007).

2.3.2.3 *Yardangs and dunes*

Martian yardangs mostly occur in the equatorial regions, especially in the Medusae Fossae Formation region, other areas include Ares Valles, Sacra Sulci, Candor Chasma, Iapygia, and some paleolakes within impact craters (e.g. Gale crater, Nicholson crater). Typical yardang forms display as mesa yardangs, whaleback yardangs, saw toothed yardangs, and long-ridge yardangs. Yardangs on mars usually develop in friable materials, such as volcanic ashes and lacustrine sediments (Day and Kocurek, 2016; Desai and Murty, 2016; Greeley and Iversen, 1987; Kite *et al.*, 2013; Tanaka *et al.*, 2014; Ward, 1979).

The surface of Mars is rich in wonderful sand dunes that form in low-lying places, such as crater floors, large basin bottoms, and channels and valleys. Typical dune forms include crescent-shaped dunes, linear dunes, and star-shaped dunes (Fig. 2.21). Dune formations on the Martian surface are most common in the north polar region, and there are swarms of dunes between latitude 40° and 50° south, mainly on crater floors (Jouannic *et al.*, 2015, Kossacki and Markiewicz, 2010).

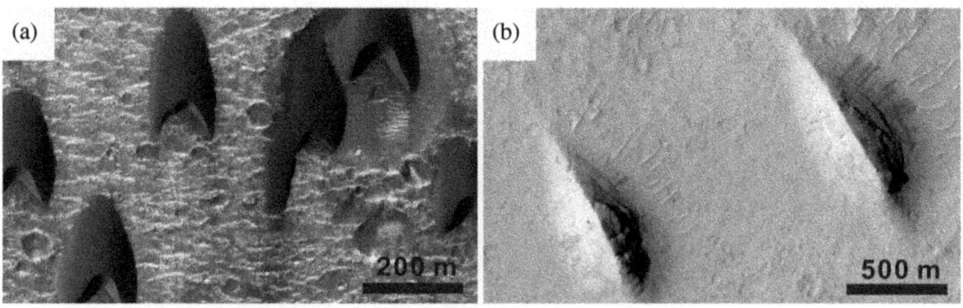

Fig. 2.21. Yardangs (b) and dunes (a) on Mars. Black dunes at the Mawrth Vallis, which are made up of basalt particles. Credit: HiRISE. Yardangs at the Zephyria Planum, which are surrounded by sand ripples. Credit: HiRISE.

2.3.3 *Ice and polar deposits*

2.3.3.1 *Ice distribution: Surface and subsurface*

Early Viking missions identified water ice at the poles of Mars, which was later suspected to be dry ice (CO_2). Recent Mars exploration has further confirmed the presence of large amounts of water ice at the poles (Bibring *et al.*, 2004), and radar data also indicate the widespread presence of underground ice (Picardi *et al.*, 2005). The volume of water ice at the poles varies with seasons and polar wander. Ice originally located at the poles moves to lower latitudes and is mainly distributed on poleward slopes due to true polar wander. There is indirect evidence of subsurface ice, too, and the vast systems of outflow channels and gullies on the Martian surface appear to have been formed less by rainfall than by groundwater outflow. The outflow of water is most likely caused by the melting of underground ice sheets by heat, such as magma activity, tectonism, and impact cratering (Cassanelli *et al.*, 2015). There are other ice-bearing features on the surface in high latitude regions that resemble frozen or permafrost regions of the Earth. Even in middle latitude regions, HiRISE observation has found exposed fresh water ice layers (Dundas *et al.*, 2014, 2018) (Fig. 2.22).

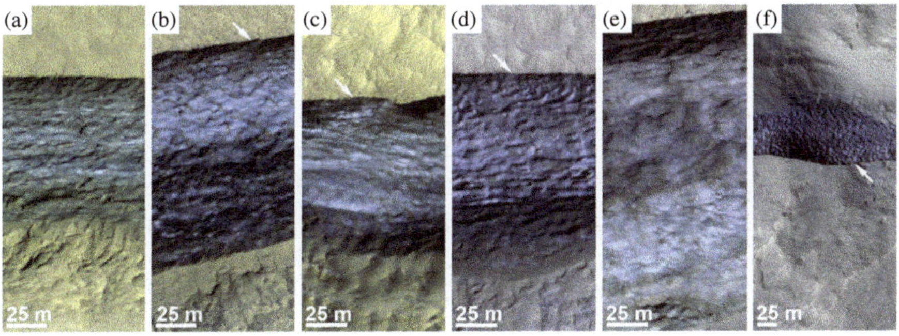

Fig. 2.22. Enhanced-color transverse sections of icy scarps in late spring/early summer ((a)–(f)). Arrows indicate locations where relatively blue material is particularly close to the surface. Downhill is to the bottom in (a) to (e) (Scarps 1 to 5, respectively) and to the top in (f) (Milankovič Crater). Banding or layering is visible in several scarps, and (e) shows a distinct change in color as well as multiple fractures cutting the ice (Dundas *et al.*, 2018).

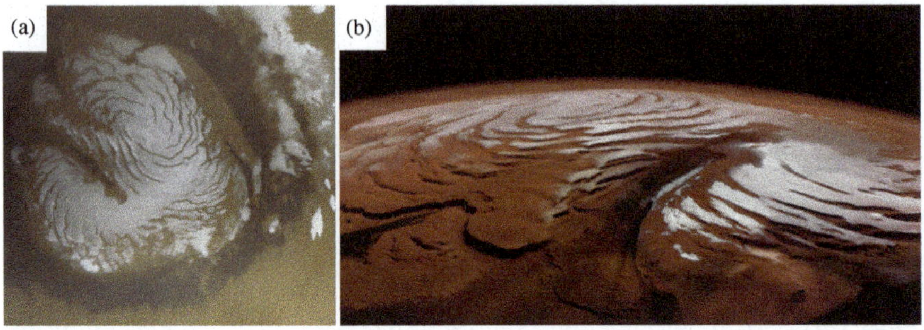

Fig. 2.23. (a) Bird view, (b) Side-view. Ice cap at the north pole of Mars. The size of Mars's north polar ice cap varies with the seasons, decreasing in summer and increasing in winter. (Image credit: NASA)

2.3.3.2 *Polar deposits: Ice cap and layering deposits*

Like Earth, Mars's north and south poles are covered by ice sheets. The north polar ice cap on Mars is larger than that at south pole, and the white top of both ice sheets is a mixture of dry ice and water ice (Fig. 2.23) (Byrne, 2009). Sediments in the north pole region differ: the upper part is light-colored fine-grained sediments, and the lower part is dark coarse-grained sediments, both of which develop very typical sedimentary bedding structures.

2.3.3.3 *Ice-sublimation formed landforms*

The south polar ice cap of Mars is small, but permafrost covers a very large area at high latitudes, because the surface is permanently frozen to depths of several kilometers. Within the permafrost region, both water ice and carbon dioxide ice are major components. When the season changes, carbon dioxide in the permafrost will solidify and sublimate in cycles that form bizarre patterns like hard cheese and spider webs (Fig. 2.24).

2.3.4 *Impact cratering formed landforms*

Impact cratering has played a major role for shaping the surface of planets. Large impacts have produced basins with diameters up to thousands of kilometers, while small impacts resulted in craters several meters to kilometers in size. There are several impact formed large basins on Mars (Frey, 2008, Schultz and Glicken, 1979). The representative ones are

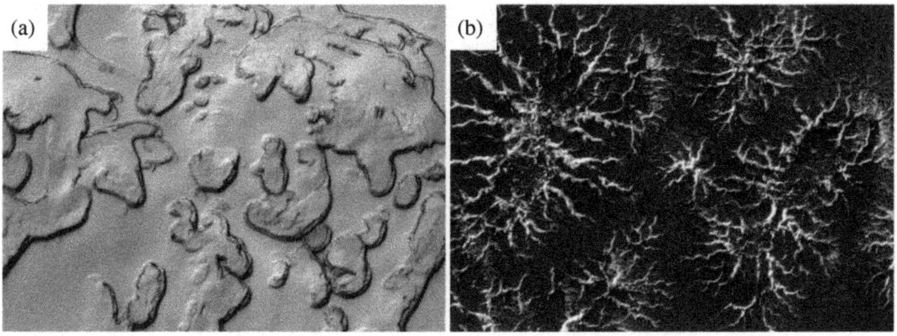

Fig. 2.24. The strange depressions left by the sublimation of carbon dioxide dry ice at Mars's South Pole (a) have never been seen on Earth. Spider web patterns formed as carbon dioxide dry ice sublimates during spring in the Martian north pole region (b).

Hellas Basin and Argyre Basin in the southern highlands, Isidis Basin located at the dichotomy boundary, and Utopia Basin in the northern lowlands. All of them form remarkable circular negative topography and are visible in global scale topography map.

Impact craters are common features of the surface of all solid planets, most of which are circular, oval, or formed into radial and annular structures. The craters on Mars can be divided into simple craters, complex craters, multi-ring basins, and buried craters according to their morphology, size, and characteristics of ejecta (Strom *et al.*, 1992).

Simple impact craters are usually bowl-shaped, with relatively small diameters, generally less than 5 km, and depth-to-diameter ratios of about 0.2. Complex craters have diameters of more than 5 to 8 km, with a range of variations between Earth's (2–5 km) and lunar (15–25 km) craters. Multi-ring basins are larger than 100 km in diameter, and the central bulge is often replaced by a central peak-ring, or for larger diameters, by a set of concentric rings similar to the Orientale Basin on the Moon. Figure 2.25 shows some typical impact craters on Mars, and some of them are unique, not found on other planets.

2.3.5 *Mass wasting*

Mass wasting, also known as slope movement or mass movement, is the geomorphic process by which soil, sand, regolith, and rock move downslope typically as a solid, continuous or discontinuous mass, largely

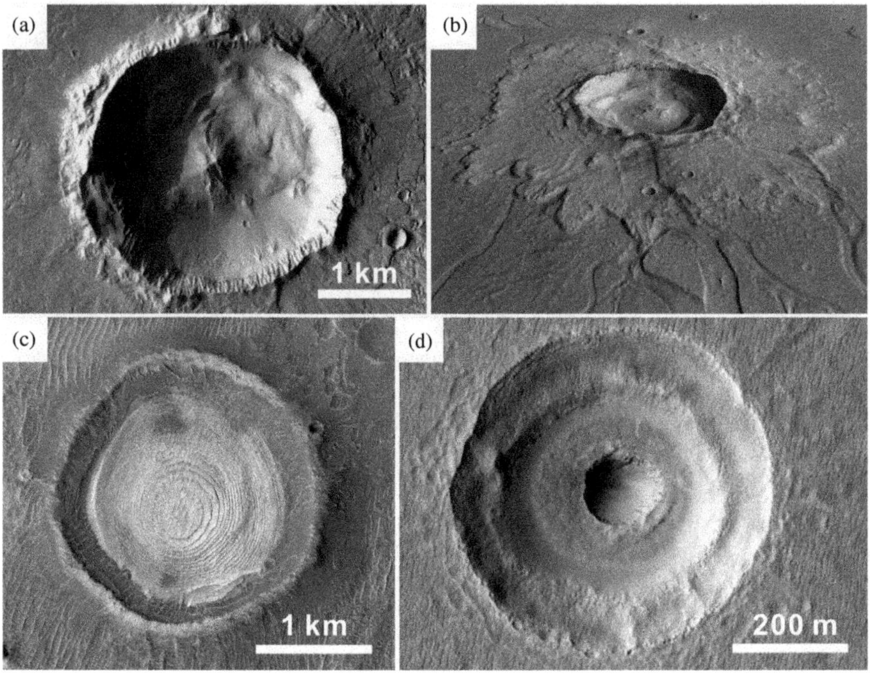

Fig. 2.25. Some typical impact craters on Mars. (a) A young crater with central peak and ejecta rays (*Sources*: NASA); (b) A rampart crater, it has lobate ejecta blankets and is usually formed with water. In this case, impact targeted at outflow channels and should have excavated underground auriferous materials forming the fluid ejecta; (c) Schiaparelli multi-ring crater (0.9_S, 346.2_W, diameter 2.3 km). The crater presumably was completely filled with layered sediments, and then the material was partly eroded to reveal stacked layers of sediments. MGS MOC2-403 (NASA/JPL/MSSS); (d) Central pit crater. What caused the central pit in the crater floor is not clear yet.

under the force of gravity, but frequently with characteristics of a flow as in debris flows and mudflows (Wincander and Monroe, 2005). Transport of rock or debris by a transporting medium (e.g. wind, water, ice) is not included (e.g. aeolian deposits; distributary system deposits), but flows with high fluid content where the pore fluid mediates (but does not completely dominate) grain-to-grain interactions are included in this category (Heldmann *et al.*, 2007). Types of mass wasting include creep, slides, flows, topples, and falls, each with its own characteristic features, and taking place over timescales from seconds to hundreds of years. Mass wasting has been observed on all terrestrial bodies, such as Earth,

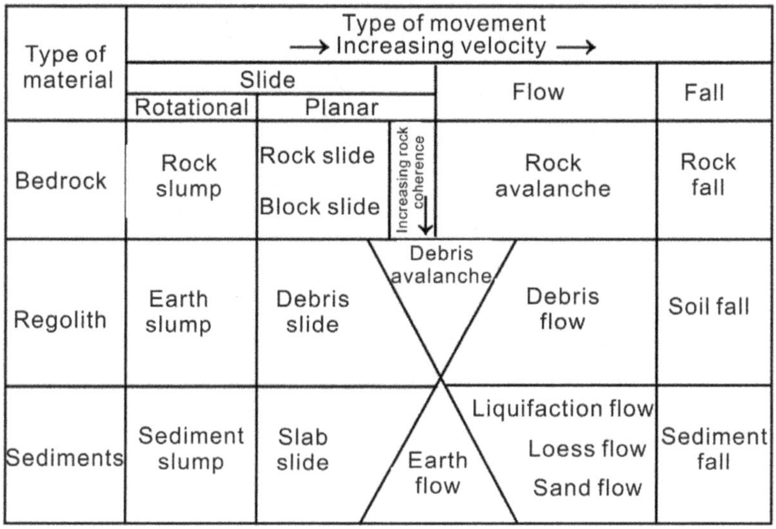

Fig. 2.26. Classification of mass movements (After Malin, 1992. Modified from Coates, 1977).

Mercury, Mars, Venus, and Jupiter's moon Io, etc. Typical mass wasting formed landforms on Mars include rock fall/landslide, debris flow and recurring slope lineae (partly).

Mass-wasting features can be classified according to the material(s) involved, the movement type, and the movement velocity (Varnes, 1978) (Fig. 2.26)

On Mars, although we put the gullies and RSL as related to water activities, both of them are subject to debris flow and belong to mass wasting. Typical mass wasting features on Mars include gullies on the steep crater wall, and landslides on the wall over valleys and channels, such as Chaotic landslide deposits in Ius Chasma, Ganges Chaos and Ganges Chasma, and Valles Marineris (Fig. 2.27).

2.4 Geological History

Like the other planets in our solar system, Mars was formed about 4.5 billion years ago due to accretion and differentiation shortly after its formation to form a core, mantle, and crust. Then subsequent impact cratering, volcanism, water/glacial and wind forces extensively modified the

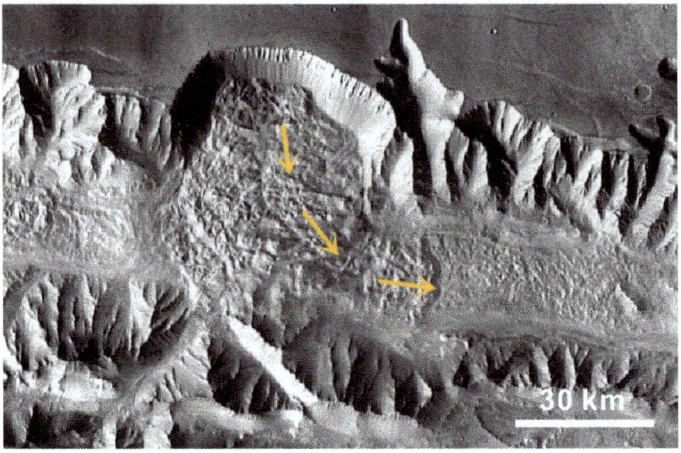

Fig. 2.27. Mass wasting on Mars. Valley Mariners and crater wall.

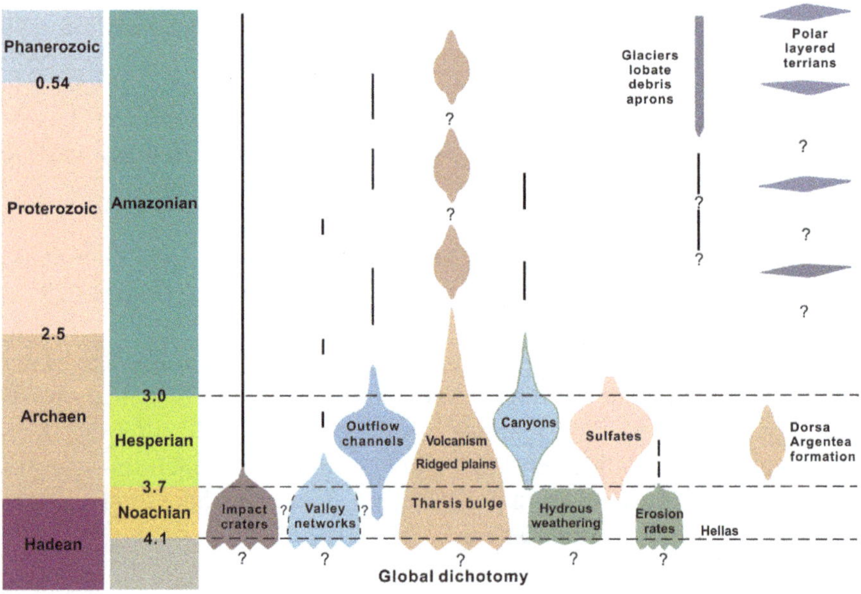

Fig. 2.28. The geological history of Mars (modified after Carr and Head, 2010).

planet. Mars's history has been classified into four eras: Pre-Noachian, Noachian, Hesperian, and Amazonian (Fig. 2.28).

The Pre-Nochian Era was the oldest history period of Mars, from its formation to ~4.1 Ga ago. Due to long-lasting modifications on the surface of Mars, its surface geology apparently dates back only to the beginning of the Early Noachian, at ~4.1 Ga, leaving ~400 Myr of Mars's earliest evolution effectively unconstrained. This is somewhat similar to our Earth's Pre-Archean, as there is little geological record remaining. Recent study suggests a pre-Noachian Mars that was less dynamic than Noachian Mars in terms of impacts, geodynamics, and hydrology. However, the pre-Noachian may have had the greatest astrobiological potential of any time period on Mars (Bottke and Andrews-Hanna, 2017).

The Noachian Era, 4.1 to 3.7 billion years ago, was subdivided into the early, middle, and late Noachian. Volcanic activity was high, meteorite impacts were frequent, the atmosphere was thicker (at least in the early years), it was probably warmer, and there was more water, possibly lakes and even oceans, which eroded and formed valley networks, and the water carried sediments. The Tharsis plateau was thus formed. This period is named after the ancient Noachis Terra in the southern hemisphere.

The Hesperian Era, 3.7 to 3.1 billion years ago, is divided into early and late Hesperian eras. This period was a transition from the present, when a large amount of water began to seep into the ground and freeze. Due to the decrease of water, erosion and transportation were reduced, although sometimes underground water burst out and caused local collapse and flooding. Geological processes were reduced, mainly taking the form of large lava plains. The period was named for the middle-aged Hesperia Planum in the southern hemisphere.

The Amazonian Era, from 3.1 billion years ago to the present, is divided into the early, middle, and late Amazonian. This period was similar to the present, with fewer, but more varied, dry and cold geological processes and meteorite impacts, and occasional water spills from the rocks into the atmosphere or the surface, creating gullies. Olympus Mons and lava plains formed at this time, named after a young, lava-filled Amazonis Planitia in the northern hemisphere.

2.5 Summary

There are plenty and diverse geomorphological features on the surface of Mars. They are genetically grouped into endogenic and exogenic. Driven by internal forces and heat, long lasting volcanism created several volcanic provinces, representing by the Tharsis, Elysium, and circum-Hellas regions. Olympus Mons is the largest single volcano and it is close to three other large volcanoes, making the Tharsis the biggest volcanic province in the solar system. Elysium province is composed of a main volcano and two small ones on its flanks. Circum-Hellas Basin volcanoes are distinct from those in the Tharsis and Elysium provinces, in that they feature low flank slopes. Plain volcanism is wide-spread in southern highland and northern lowland, making Mars a volcanic world along with those large central volcanoes. These volcanic features were formed due to different eruption types, mainly the early stage explosive and late stage effusive eruptions that created different volcanic structures and are globally distributed except for the no-polar regions. Crustal deformation was dominated by extensional rifts (e.g. Valles Marineris), while compressional features (e.g. wrinkle ridges) mostly occurred on lava plains and are topographically less remarkable than rifts. The Tharsis bulge is unique, which was made of extensive volcanism and tectonic deformation. Mantle plume induced uplifting, volcanism, radial rifting, and circle compressional deformations together built the huge bulge.

Exogenic forces such as water, air/wind impacting all played roles on the formation and degradation of landforms on Mars. Liquid water has carved the surface of Mars and formed widespread valleys, channels, channel networks, and gullies that are mostly distributed in southern highland. In contrast, water is believed to be stored in subsurface as ice or even liquid water. The most remarkable water erosion landforms are outflow channels in northern Tharsis, crossing the dichotomy boundary. Wind has also played an important role in shaping the current surface of Mars since it was changed to a dry and cold world, and formed plenty and diverse yardangs, dunes, and wind streaks. Wind has transferred large amounts of fine materials, and filled in low areas (e.g. crater floors) and deposited at polar regions. Thick fine-grained sands and dry ice mixture deposits are major components in polar regions. Sublimation of dry ice has also created unique patterned terrains. Furthermore, impact cratering was always one of the major forces reshaping and resurfacing the surface of Mars. All of the large basins, including Hellas, Isidis, Argyre, and numerous craters were formed by impact events. Due to the diversity of surface properties of Mars and the projectile types and impact dynamic regimes, a variety of impact craters have been created. Many craters had been filled with water and formed basins, and then dried out later, leaving lacustrine deposits visible in many paleolake basins, among which some are rich in salts and featured polygonal terrains similar to terrestrial playas. These regions have been listed as top priority candidate sites for searching life on Mars.

As mentioned, Mars experienced long-lasting geological activities and is generally divided into four eras: Pre-Noachian, Noachian, Hesperian, and Amazonian. Based on geological environment records, the early stage including Pre-Noachian and Noachian was geologically most active, wet, and more habitable than present. The Hesperian Era became drier and volcanism was most remarkable. The Amazonian Era was cold and dry, most endogenic geological activities were caused and replaced by exogenic wind, glacial, and meteorite impact cratering activities. Layered polar deposits formed in these periods were good records of the climate change history of Mars.

Acknowledgments

We are grateful for funding from the Pre-research Project on Civil Aerospace Technologies of China Space Agency (D020101), National Natural Science Foundation of China (41830214), and the Science and Technology Development Fund (FDCT) of Macau (Nos. 121/2017/A3). Ms. Yutong Shi has helped to edit the figures, pictures and references.

References

Alemanno, G., Orofino, V. and Mancarella, F. (2018). Global map of Martian Fluvial Systems: Age and total eroded volume estimations. *Earth Space Sci.*, **5**, 560–577.

Anderson, R. and Dohm, J. (2011). Unraveling the Complex History of Faulting for the Terra Sirenum Region; Mars, Lunar and Planetary Science Conference, p. 2221.

Anderson, R. and Herkenhoff, K. (2012). A Survey of Sinuous Ridges and Inferred Fluvial Discharge Rates in Northwest Hellas, Mars, AGU Fall Meeting Abstracts, p. 1832.

Anderson, R. C., Dohm, J. M., Haldemann, A. F. C., Pounders, E., Golombek, M. and Castano, A. (2008). Centers of tectonic activity in the eastern hemisphere of Mars. *Icarus*, **195**, 537–546.

Andrews-Hanna, J. C., Zuber, M. T. and Phillips, R. J. (2008). Early Mars hydrology: Valley networks and evaporites, Lunar and Planetary Science Conference, p. 1993.

Ansan, V. and Mangold, N. (2006). New observations of Warrego Valles, Mars: Evidence for precipitation and surface runoff. *Planet. Space Sci.*, **54**, 219–242.

Auld, K. S. and Dixon, J. C. (2016). A classification of martian gullies from HiRISE imagery. *Planet. Space Sci.*, **131**, 88–101.

Basilevsky, A., Werner, S., Neukum, G., Head, J., van Gasselt, S., Gwinner, K. and Ivanov, B. (2006). Geologically recent tectonic, volcanic and fluvial activity on the eastern flank of the Olympus Mons volcano, Mars. *Geophys. Res. Lett.*, **33**.

Bhardwaj, A., Sam, L., Martin-Torres, F. J., Zorzano, M. P. and Fonseca, R. M. (2017). Martian slope streaks as plausible indicators of transient water activity. *Sci. Rep.*, **7**, 7074.

Bibring, J.-P., Langevin, Y., Poulet, F., Gendrin, A., Gondet, B., Berthé, M., Soufflot, A., Drossart, P., Combes, M. and Bellucci, G. (2004). Perennial water ice identified in the south polar cap of Mars. *Nature*, **428**, 627–630.

Bleacher, J., Orr, T., Garry, W., Hamilton, C., Zimbelman, J. and de Wet, A. (2013). Sinuous Ridges and Plateaus as Evidence for Lava Flow Inflation in the Tharsis Plains of Mars: Insights from Analogous Features on the Coastal Plain of Kilauea Volcano, HI, Lunar and Planetary Institute Science Conference Abstracts, p. 2090.

Bleacher, J. E., Orr, T. R., de Wet, A. P., Zimbelman, J. R., Hamilton, C. W., Brent, G. W., Crumpler, L. S. and Williams, D. A. (2017). Plateaus and sinuous ridges as the fingerprints of lava flow inflation in the Eastern Tharsis Plains of Mars. *J. Volcanol Geoth Res.*, **342**, 29–46.

Bottke, W. F. and Andrews-Hanna, J. C. (2017). A post-accretionary lull in large impacts on early Mars. *Nat. Geosci.*, **10**, 344–348.

Byrne, S. (2009). The polar deposits of Mars. *Annu. Rev. Earth Planet Sci.*, **37**, 535–560.

Cabrol, N. (1997). Early Amazonian lake in Gale Crater (Mars), Lunar and Planetary Science Conference, p. 191.

Carey, B. E. and Dawers, N. (2016). Dike-Related Fault Morphology from Structural Mapping of Sirenum Fossae, Lunar and Planetary Science Conference, p. 1858.

Carr, M. (2006). *The Surface of Mars*. Cambridge University Press, Cambridge, UK.

Carr, M. and Head, J. (2019). Mars: Formation and fate of a frozen Hesperian Ocean. *Icarus*, **319**, 433–443.

Carr, M. H. (2007). *The Surface of Mars*. Cambridge University Press, Cambridge, UK.

Carr, M. H. (2012). The fluvial history of Mars, Philosophical transactions, Series A. *Math. Phys. Eng. Sci.*, **370**, 2193–2215.

Carr, M. H. and Clow, G. D. (1981). Martian channels and valleys: Their characteristics, distribution, and age. *Icarus*, **48**, 91–117.

Cassanelli, J. P., Head, J. W. and Fastook, J. L. (2015). Sources of water for the outflow channels on Mars: Implications of the Late Noachian "icy highlands" model for melting and groundwater recharge on the Tharsis rise. *Planet. Space Sci.*, **108**, 54–65.

Cattermole, P. (1987). Sequence, rheological properties, and effusion rates of volcanic flows at Alba Patera. Mars, *J. Geophys. Res. Solid Earth*, **92**, E553–E560.

Chan, N.-H., Perron, J. T., Mitrovica, J. X. and Gomez, N. A. (2018). New evidence of an ancient Martian Ocean from the global distribution of valley networks. *J. Geophys. Res. Planets*, **123**, 2138–2150.

Chuang, F., Crown, D., Berman, D. and Joseph, E. (2014). Lobate Debris Aprons and Ice-Rich Flow Features in Eastern Hellas, Mars: Mapping Using THEMIS IR and CTX Data, Lunar and Planetary Science Conference, p. 2081.

Clifford, S. M. (1993). A model for the hydrologic and climatic behavior of water on Mars. *J. Geophys. Res.*, **98**, 10973–11016.

Conway, S. J., Balme, M. R., Kreslavsky, M. A., Murray, J. B. and Towner, M. C. (2015). The comparison of topographic long profiles of gullies on Earth to gullies on Mars: A signal of water on Mars. *Icarus*, **253**, 189–204.

Conway, S. J., de Haas, T. and Harrison, T. N. (2019). Martian gullies: A comprehensive review of observations, mechanisms and insights from Earth analogues. *Geol. Soc. London Spec. Pub.*, **467**, 7–66.

Crown, D. A. and Greeley, R. (1993). Volcanic geology of Hadriaca Patera and the eastern Hellas region of Mars. *J. Geophys. Res.*, **98**, 3431.

Davis, P. and Tanaka, K. (1995). Curvilinear Ridges in Isidis Planitia, Mars—The Result of Mud Volcanism, Lunar and Planetary Science Conference.

De Blasio, F. V. (2011). Landslides in Valles Marineris (Mars): A possible role of basal lubrication by sub-surface ice. *Planet. Space Sci.*, **59**, 1384–1392.

de Haas, T., Ventra, D., Hauber, E., Conway, S. J. and Kleinhans, M. G. (2015). Sedimentological analyses of martian gullies: The subsurface as the key to the surface. *Icarus*, **258**, 92–108.

Di Achille, G. and Hynek, B. M., (2010). Ancient ocean on Mars supported by global distribution of deltas and valleys. *Nat. Geosci.*, **3**, 459–463.

Di Achille, G. and Ori, G. (2008). Complex Intermontaine Glacial Systems in Arabia Terra, Mars: Evidence for an Amazonian Proglacial Lake with Associated Glacilacustrine Deposits, Lunar and Planetary Science Conference, p. 2096.

Dundas, C. M., Bramson, A. M., Ojha, L., Wray, J. J., Mellon, M. T., Byrne, S., McEwen, A. S., Putzig, N. E., Viola, D., Sutton, S., Clark, E. and Holt, J. W. (2018). Exposed subsurface ice sheets in the Martian mid-latitudes. *Science*, **359**, 199–201.

Dundas, C. M., Byrne, S., McEwen, A. S., Mellon, M. T., Kennedy, M. R., Daubar, I. J. and Saper, L. (2014). HiRISE observations of new impact craters exposing Martian ground ice. *J. Geophys. Res. Planets*, **119**, 109–127.

Dundas, C. M., McEwen, A. S., Chojnacki, M., Milazzo, M. P., Byrne, S., McElwaine, J. N. and Urso, A. (2017). Granular flows at recurring slope lineae on Mars indicate a limited role for liquid water. *Nat. Geosci.*, **10**, 903–907.

Dundas, C. M., McEwen, A. S., Diniega, S., Hansen, C. J., Byrne, S. and McElwaine, J. N. (2019). The formation of gullies on Mars today. *Geol. Soc. London Spec. Pub.*, **467**, 67–94.

Egea-Gonzalez, I., Jiménez-Díaz, A., Parro, L. M., López, V., Williams, J.-P. and Ruiz, J. (2017). Thrust fault modeling and Late-Noachian lithospheric structure of the circum-Hellas region, Mars. *Icarus*, **288**, 53–68.

Fastook, J. L., Head, J. W. and Marchant, D. R. (2014). Formation of lobate debris aprons on Mars: Assessment of regional ice sheet collapse and debris-cover armoring. *Icarus*, **228**, 54–63.

Fedorova, A., Bertaux, J.-L., Betsis, D., Montmessin, F., Korablev, O., Maltagliati, L. and Clarke, J. (2018). Water vapor in the middle atmosphere of Mars during the 2007 global dust storm. *Icarus*, **300**, 440–457.

Frey, H. (2008). Ages of very large impact basins on Mars: Implications for the late heavy bombardment in the inner solar system. *Geophys. Res. Lett.*, **35**, L13203.

Ghatan, G. J., Head, J. W. and Wilson, L. (2005). Mangala Valles, Mars: Assessment of early stages of flooding and downstream flood evolution, *Earth, Moon, and Planets*, **96**, 1–57.

Greeley, R. and Spudis, P. D. (1981). Volcanism on Mars. *Rev. Geophys.*, **19**, 13.

Greeley, R., White, B. R., Pollack, J. B., Iversen, J. D. and Leach, R. N. (1981). Dust storms on Mars: Considerations and simulations. *Geol. Soc. London Spec. Pub.*, **186**, 101–122.

Gullikson, A., Anderson, R. and Williams, R. (2016). Mapping Sinuous Ridges in Northwest Hellas, Mars, Lunar and Planetary Science Conference, p. 2376.

Head, J. W., Kreslavsky, M. A. and Pratt, S. (2002). Northern lowlands of Mars: Evidence for widespread volcanic flooding and tectonic deformation in the Hesperian Period. *J. Geophys. Res. Planets*, (1991–2012), **107**, 3.

Head, J. W., Mustard, J. F., Kreslavsky, M. A., Milliken, R. E. and Marchant, D. R. (2003). Recent ice ages on Mars. *Nature*, **426**, 797–802.

Heldmann, J. L., Carlsson, E., Johansson, H., Mellon, M. T. and Toon, O. B. (2007). Observations of martian gullies and constraints on potential formation mechanisms. *Icarus*, **188**, 324–344.

Heldmann, J. L. and Mellon, M. T. (2004). Observations of martian gullies and constraints on potential formation mechanisms. *Icarus*, **168**, 285–304.

Ivanov, M. A. and Head, J. W. (2006). Alba Patera, Mars: Topography, structure, and evolution of a unique late Hesperian–early Amazonian shield volcano. *J. Geophys. Res.*, **111**, E09003.

Jackson, M. P. A., Adams, J. B., Dooley, T. P., Gillespie, A. R. and Montgomery, D. R. (2011). Modeling the collapse of Hebes Chasma, Valles Marineris, Mars. *Geol. Soc. Am. Bull.*, **123**, 1596–1627.

Jacobsen, R. E. and Burr, D. M. (2013). Local-scale stratigraphy of inverted fluvial features in Aeolis Dorsa, Western Medusae Fossae formation, Mars, 44th Lunar and Planetary Science Conference, LPI Contribution, p. 2165.

James W. H. III, H. H., Mikhail A. I., Mikhail A. K., Stephen, P., Bradley, J. T. (1999). Possible ancient oceans on Mars: Evidence from Mars Orbiter Laser Altimeter Data. *Science*, **286**(5447), pp. 2134–2137.

Jouannic, G., Gargani, J., Conway, S. J., Costard, F., Balme, M. R., Patel, M. R., Massé, M., Marmo, C., Jomelli, V., Ori, G. G., (2015). Laboratory simulation of debris flows over sand dunes: Insights into gully-formation (Mars). *Geomorphology*, **231**, 101–115.

Khalaf, E. E. D. A. H. and Hammed, M. S. (2016). Morphology and development of pahoehoe flow-lobe tumuli and associated features from a monogenetic basaltic volcanic field, Bahariya Depression, Western Desert, Egypt. *J. African Earth Sci.*, **113**, 165–180.

Korteniemi, J., Raitala, J., Aittola, M., Ivanov, M. A., Kostama, V.-P., Öhman, T. and Hiesinger, H. (2010). Dike indicators in the Hadriaca Patera–Promethei Terra region, Mars. *Earth Planet. Sci. Lett.*, **294**, 466–478.

Kossacki, K. J. and Markiewicz, W. J. (2010). Interfacial liquid water on Mars and its potential role in formation of hill and dune gullies. *Icarus*, **210**, 83–91.

Le Deit, L., Bourgeois, O., Mège, D., Hauber, E., Le Mouélic, S., Massé, M., Jaumann, R. and Bibring, J.-P. (2010). Morphology, stratigraphy, and mineralogical composition of a layered formation covering the plateaus around Valles Marineris, Mars: Implications for its geological history. *Icarus*, **208**, 684–703.

Lefort, A., Burr, D. M., Beyer, R. A. and Howard, A. D. (2012). Inverted fluvial features in the Aeolis-Zephyria Plana, western Medusae Fossae Formation, Mars: Evidence for post-formation modification. *J. Geophys. Res.*, **117**, E03007.

Luo, W., Cang, X., Howard, A. D. (2017). New Martian valley network volume estimate consistent with ancient ocean and warm and wet climate. *Nat. Commun.*, **8**, 15766.

McCauley, J. F. (1973). Mariner 9 evidence for wind erosion in the equatorial and mid-latitude regions of Mars. *J. Geophys. Res.*, **78**, 4123–4137.

McEwen, A. S., Hansen, C. J., Delamere, W. A., Eliason, E. M., Herkenhoff, K. E., Keszthelyi, L., Gulick, V. C., Kirk, R. L., Mellon, M. T., Grant, J. A., Thomas, N., Weitz, C. M., Squyres, S. W., Bridges, N. T., Murchie, S. L., Seelos, F., Seelos, K., Okubo, C. H., Milazzo, M. P., Tornabene, L. L., Jaeger, W. L., Byrne, S., Russell, P. S., Griffes, J. L., Martinez-Alonso, S., Davatzes, A., Chuang, F. C., Thomson, B. J., Fishbaugh, K. E., Dundas, C. M., Kolb, K. J., Banks, M. E. and Wray, J. J. (2007). A closer look at water-related geologic activity on Mars. *Science*, **317**, 1706–1709.

McEwen, A. S., Ojha, L., Dundas, C. M., Mattson, S. S., Byrne, S., Wray, J. J., Cull, S. C., Murchie, S. L., Thomas, N. and Gulick, V. C. (2011). Seasonal flows on warm Martian slopes. *Science*, **333**, 740–743.

Melwani, D. M. and Kite, E. S. (2017). Paleohydrology on Mars constrained by mass balance and mineralogy of pre-Amazonian sodium chloride lakes. *J. Geophys. Res. Planets*, **122**, 1802–1823.

Picardi, G., Plaut, J. J., Biccari, D., Bombaci, O., Calabrese, D., Cartacci, M., Cicchetti, A., Clifford, S. M., Edenhofer, P., Farrell, W. M., Federico, C., Frigeri, A., Gurnett, D. A., Hagfors, T., Heggy, E., Herique, A., Huff, R. L., Ivanov, A. B., Johnson, W. T., Jordan, R. L., Kirchner, D. L., Kofman, W., Leuschen, C. J., Nielsen, E., Orosei, R., Pettinelli, E., Phillips, R. J., Plettemeier, D., Safaeinili, A., Seu, R., Stofan, E. R., Vannaroni, G., Watters, T. R. and Zampolini, E. (2005). Radar soundings of the subsurface of Mars. *Science*, **310**, 1925–1928.

Plaut, J. J., Safaeinili, A., Holt, J. W., Phillips, R. J., Head, J. W., Seu, R., Putzig, N. E. and Frigeri, A. (2009). Radar evidence for ice in lobate debris aprons in the mid-northern latitudes of Mars. *Geophys. Res. Lett.*, **36**(2), https://doi.org/10.1029/2008GL036379.

Rice, M. S., Gupta, S., Bell, J. F. and Warner, N. H. (2011). Influence of fault-controlled topography on fluvio-deltaic sedimentary systems in Eberswalde crater, Mars. *Geophys. Res. Lett.*, **38**(16), https://doi.org/10.1029/2011GL048149.

Rodriguez, J. A. P., Kargel, J. S., Baker, V. R., Gulick, V. C., Berman, D. C., Fairen, A. G., Linares, R., Zarroca, M., Yan, J., Miyamoto, H. and Glines, N. (2015a). Martian outflow channels: How did their source aquifers form, and why did they drain so rapidly? *Sci. Rep.*, **5**, 13404.

Rodriguez, J. A. P., Kargel, J. S., Baker, V. R., Gulick, V. C., Berman, D. C., Fairén, A. G., Linares, R., Zarroca, M., Yan, J. and Miyamoto, H. (2015b). Martian outflow channels: How did their source aquifers form, and why did they drain so rapidly? *Sci. Rep.*, **5**.

Rogers, A. D., Warner, N. H., Golombek, M. P., Head, J. W., 3rd, and Cowart, J. C. (2018). A really extensive surface bedrock exposures on Mars: Many are Clastic Rocks, not Lavas. *Geophys. Res. Lett.*, **45**, 1767–1777.

Schultz, P. H. and Glicken, H. (1979). Impact crater and basin control of igneous processes on Mars. *J. Geophys. Res. Solid Earth, (1978–2012)*, **84**, 8033–8047.

Soare, R. J., Osinski, G. R. and Roehm, C. L. (2008). Thermokarst lakes and ponds on Mars in the very recent (late Amazonian) past. *Earth Planet. Sci. Lett.*, **272**, 382–393.

Squyres, S. W., Grotzinger, J. P., Arvidson, R. E., Bell, J. F., 3rd, Calvin, W., Christensen, P. R., Clark, B. C., Crisp, J. A., Farrand, W. H., Herkenhoff, K. E., Johnson, J. R., Klingelhofer, G., Knoll, A. H., McLennan, S. M.,

McSween, H. Y., Jr., Morris, R. V., Rice, J. W., Jr., Rieder, R. and Soderblom, L. A. (2004). In situ evidence for an ancient aqueous environment at Meridiani Planum, Mars. *Science*, **306**, 1709–1714.

Strom, R. G., Croft, S. K. and Barlow, N. G. (1992). The Martian impact cratering record. *Mars*, 383–423.

Williams, D. A., Greeley, R., Fergason, R. L., Kuzmin, R., McCord, T. B., Combe, J.-P., Head, J. W., Xiao, L., Manfredi, L., Poulet, F., Pinet, P., Baratoux, D., Plaut, J. J., Raitala, J. and Neukum, G. (2009). The Circum-Hellas Volcanic Province, Mars: Overview. *Planet. Space Sci.*, **57**, 895–916.

Williams, D. A., Greeley, R., Manfredi, L., Raitala, J. and Neukum, G. (2010). The Circum-Hellas Volcanic Province, Mars: Assessment of wrinkle-ridged plains. *Earth Planet. Sci. Lett.*, **294**, 492–505.

Wilson, L. and Head, J. W. (1994). Mars: Review and analysis of volcanic eruption theory and relationships to observed landforms. *Rev. of Geophys.*, **32**, 221–263.

Wilson, L. and Head, J. W. (2002). Tharsis-radial graben systems as the surface manifestation of plume-related dike intrusion complexes: Models and implications. *J. Geophys. Res.*, **107**, 1-1–1-24.

Wilson, L. and Head, J. W. (2004). Evidence for a massive phreatomagmatic eruption in the initial stages of formation of the Mangala Valles outflow channel, Mars. *Geophys. Res. Lett.*, **31**, L15701.

Wincander, M. (2005). *The Changing Earth: Exploring Geology and Evolution*, Thomson Brooks/Cole.

Xiao, L., Huang, J., Christensen, P. R., Greeley, R., Williams, D. A., Zhao, J. and He, Q. (2012). Ancient volcanism and its implication for thermal evolution of Mars. *Earth Planet. Sci. Lett.*, **323-324**, 9–18.

Xiao, L. and Wang, C. (2009). Geologic features of Wudalianchi volcanic field, northeastern China: Implications for Martian volcanology. *Planet. Space Sci.*, **57**, 685–698.

Xiao, L., Wang, J., Dang, Y., Cheng, Z., Huang, T., Zhao, J., Xu, Y., Huang, J., Xiao, Z. and Komatsu, G. (2017). A new terrestrial analogue site for Mars research: The Qaidam Basin, Tibetan Plateau (NW China), *Earth-Science Rev.*, **164**, 84–101.

Zimbelman, J. R. (2010). Transverse Aeolian Ridges on Mars: First results from HiRISE images. *Geomorphology*, **121**, 22–29.

https://doi.org/10.1142/9789811224027_0003

Chapter 3

The Qaidam Basin, NE Tibetan Plateau

Long Xiao[*,†,‡], Jiang Wang[*], Yanan Dang[†,‡], Jiannan Zhao[*] and Mingjie Zhang[†,‡]

[*]*State Key Laboratory of Geological Processes and Mineral Resources, Planetary Science Institute, School of Earth Sciences, China University of Geosciences, Wuhan 430074, China*

[†]*State Key Laboratory of Lunar and Planetary Sciences, Macau University of Science and Technology, Macau, PR China*

[‡]*CNSA Macau Center for Space Exploration and Science, Macau, PR China*

3.1 Introduction

The Qaidam Basin, located between 90°16' E to 99°16' E and 35°00' N to 39°20' N, is the largest intermontane sedimentary basin along the north-eastern Tibetan Plateau. It covers about 120,000 km^2 (~700 km long and up to ~300 km wide) in area, including the largest yardang fields (~38,800 km^2) (Kapp *et al.*, 2011a) in China. The average elevation is ~2800 m above sea level, being one of the highest deserts in the world (Rohrmann *et al.*, 2013). Topographically, it is bounded by the Altyn-Tagh Mountain Ranges to the northwest, the East Kunlun-Qimantag Mountain Ranges to the south-southwest, the Qilian Mountain Ranges to the north-northeast, and E'lashan to the east (Fig. 3.1).

The Qaidam is a tectonically controlled sedimentary basin (Yin *et al.*, 2002). Since the middle to late Miocene, it has been compressed and actively shortened in the NE–SW direction in response to the ongoing

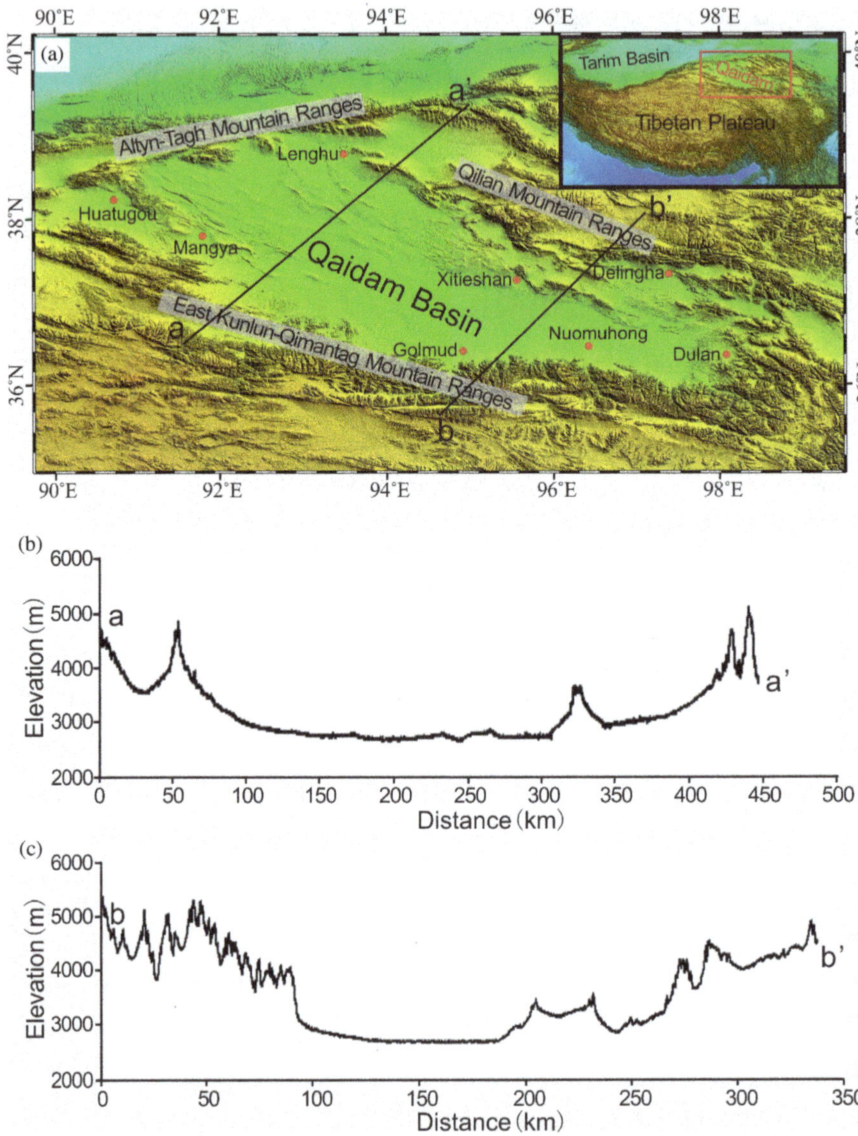

Fig. 3.1. (a) Topographic map (generated from 90 m SRTM data) of Qaidam Basin and surrounding mountain ranges, red dots indicate the main cities. (b) Topographic profile along the line a-a' shown in figure (a). (c) Topographic profile along the line b-b' indicated in figure (a).

convergence between India and Eurasia plates (Tapponnier *et al.*, 2001; Wu *et al.*, 2014; Zhang *et al.*, 2004) and the NW-SE and NNW-SSE trending folds are, therefore, widely developed in the sediments of western basin (Zhang *et al.*, 2013). These folds separated the basin into several sub-depressions and lakes (Fang *et al.*, 2007; Han *et al.*, 2014; Wang *et al.*, 2006) (Fig. 3.2).

The basement of the Qaidam Basin consists of Precambrian–Silurian metamorphic rocks and they are mainly exposed along the rims of the basin (e.g. Huang *et al.*, 1996). The overlain strata within the basin are Devonian–Cenozoic sediments that were originated from weathering and denudation of the surrounding mountains since the Paleocene (Yang *et al.*, 2001; Yin *et al.*, 2007, 2008). The total thickness of Cenozoic sediments within the basin can be up to about 12,000 m (Fang *et al.*, 2007; Xia *et al.*, 2001). In the northwestern part, the surface layer of the sub-depressions is Pleistocene strata, consisting of lacustrine sediments and salt deposits, while most of the outcropped folded structures are Tertiary strata that are mainly composed of mudstones, siltstones, and sandstones, with interbedded gypsum and sulfate layers. Holocene deposits, including aeolian (e.g. dunes), seasonal fluvial, and modern saline lake sediments, are distributed on the north alluvial-diluvial clinoplains of the East Kunlun-Qimantag Mountains (Li *et al.*, 2016) or interspersed in the low-lying areas within the basin (Fig. 3.1).

3.2 Geological Setting

3.2.1 *Stratigraphy*

Systematic litho-stratigraphy, from Precambrian to Cenozoic periods, has been developed in the periphery and interior of the Qaidam Basin, in which Precambrian metamorphic rocks are exposed in the surrounding mountain ranges of the basin, and sedimentary strata of Paleozoic, Mesozoic, and Cenozoic are developed in the basin. The analog landforms comparable with Mars are mainly developed on the Cenozoic strata. Here we simply divided the strata into pre-Cenozoic and Cenozoic strata and their main characteristics are summarized as follows:

3.2.1.1 *Pre-Cenozoic Stratigraphy*

The basement of the Qaidam Basin is composed of Precambrian–Silurian metamorphic rocks, which were overlain by Devonian–Cenozoic

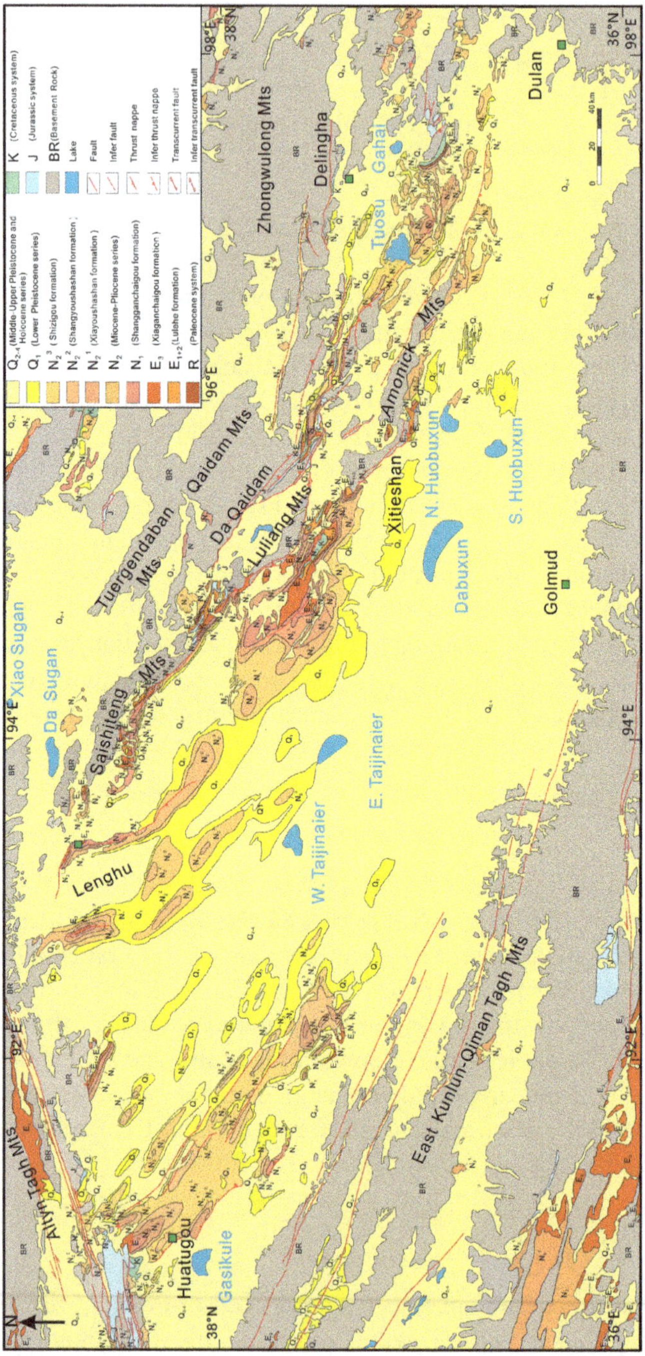

Fig. 3.2. Sketched Geological map of the Qaidam Basin (Jin and Zhang, 1997; Yu *et al.*, 2018).

Fig. 3.3. Well-developed layering Quaternary stratigraphy, part of the Nanyishan anticline in the Qaidam Basin (38.14°N, 91.86°E).

sedimentary strata (Huang *et al.*, 1996). The metamorphic rocks are exposed along the rims of the basin, as major rock components of surrounding mountain ranges.

The Triassic strata are only distributed in the Kumkuli–Qimantag–Burhanbuda–Dulan belt and are mainly terrestrial volcanic rocks. The lower Jurassic is mainly composed of variegated conglomerate, gray medium coarse sandstone, gray-black carbonaceous shale with coal seam, gray-green sandstone, mudstone, and oil shale. The upper part of the middle Jurassic is composed of gray-yellow mudstone with fine conglomerate, sandstone, gray-black carbonaceous mudstone, coal seam, oil shale, etc. The lower part is purplish red, dark brown-red and gray-purple, gray-green sandstone, sandy mudstone, siltstone, and sometimes gray-white conglomerate at the bottom.

Cretaceous is only found in the front zone of Altyn Mountain and part of the northern margin of Qaidam Basin. The lower cretaceous in the front of Altyn and the northern margin of the Qaidam Basin is mainly composed of red-brown conglomerate, pebbly sandstone, and medium and fine sandstone.

3.2.1.2 *Cenozoic stratigraphy*

The Qaidam Basin preserves a complete record of Cenozoic sedimentation, which is over 6,000 m in general, with a maximum thickness of over 15 km. From the bottom up, they were the Lulehe Formation, the lower Gangchaigou Formation, the upper Gangchaigou Formation, the lower Youshashan Formation, the upper Youshashan Formation, the Shizigou Formation, and the Qigequan Formation (Table 3.1).

Lulehe Formation: It is composed of conglomerate, gravel sandstone, argillaceous siltstone, and mudstone. There are outcrops in Yuka, Lulehe, Dahonggou, Wuchaigou, Saixi, Hongshuigou, Quanyagou, and Wucai mountain areas.

Xiaganchaigou Formation: It is mainly exposed in the western margin, northeastern margin, and eastern margin of the basin. In the west, Ganchaigou is dominated by the gray conglomerate, sandstone, and mudstone, with sandy mudstone or silty belt. The eastern facies is dominated by black-gray calcareous marlstone, while the northern edge of the Eastern Kunlun Mountain is dominated by gray-green, gray-yellow, brown and red-brown mudstone, with mixed colored

Table 3.1. Cenozoic stratigraphy of the Qaidam Basin (after Yin, 2018; Song and Wang, 1993).

Geologic time	Ages	Unit names	Thickness (m)	Facies
Holocene	0.01 Ma-present	DabuxunYanqiao Fm.		
Pleistocene	1.8–0.01 Ma	Qigequan Fm.		
Pliocene	5.3–1.8 Ma	Shizigou Fm.	>1500	Lacustrine
Late Miocene	10–5.3 Ma	Shangyoushashan Fm.	>822	Lacustrine
Early-middle Miocene	23–10 Ma	Xiayoushashan Fm.	1243	Lacustrine/ Fluvial
Late Oligocene	28–23 Ma	Shangganchaigou Fm.	848	Fluvial/ Lacustrine
Early Oligocene–Middle Eocene	48–28 Ma	Xiaganchaigou Fm.	1011	Fluvial/ Lacustrine
Paleocene-Early Eocene	60–48 Ma	Lulehe Fm.	1043	Fluvial/ Lacustrine

marlstone. In the east, there is a set of yellowish-green giant thick layers of sandstone mixed with colored sandy mudstone. The standard section is Xichagou section at Lenghu town, with 183–1232 m surface exposure, which is in conformable contact with the underneath Lulehe Formation.

Shanggaichaigou Formation: The formation appears discontinuously in the southwest margin and northeast margin of the basin. The northeast is interbedded with yellow-green sandstone and red-brown mudstone, and the southwest is interbedded with green-gray calcareous shale. The standard profile of this formation is Xichagou profile at Lenghu town, with a surface exposure thickness of 172–968 m, which is in contact with the underneath Ganchaigou Formation.

Youshashan Formation: The lower Youshashan Formation in the west is interbedded with red-brown argillaceous siltstone, sandy mudstone and brown, green and yellow gray sandstone, with a small amount of conglomerate, tufted marl, dolomite, and sandy gypsum. In the east, there are yellow-green and gray-green giant thick layers of sandstone with brown ash, red-brown sandy mudstone, siltstone, conglomerate, and marl. In contrast, the upper member is interbedded with gray conglomerate, sandstone and brown siltstone and sandy mudstone in the west, and is interbedded with red-brown, gray-green, and brown sandy mudstone and siltstone in the east. The standard profile of this formation is the Youshashan profile, close to Lenghu town.

Shizigou Formation: The Shizigou Formation is more widely distributed than the Youshashan Formation. The basin margin is dominated by conglomerates and in the center of the basin is sandy mudstone, fine sandstone, and siltstone interbedded with marl and carbonaceous sandy belt. The argillaceous rocks in the center of the lake basin are mostly composed of gypsum crystals with thin layers of gypsum and oolitic sandstone. The thickness is generally thickened from the margin to the center. In the basin center, this formation is in continuous contact with the underlying Youshashan Formation, and in the basin margin it overlays unconformities and overlying strata. The standard section of this formation is at the eastern mound of Lenghu town, which is in conformity with the underlying Youshashan Formation and in angular unconformity with the overlying Qigequan Formation.

Qigequan Formation: The Qigequan Formation is mainly distributed in Sebei, Tuofengshan Yanhu areas. The strata in Sebei, Tainan, and Senie lake areas are mainly composed of gray, light gray silty mudstone, dark gray mudstone, and gray argillaceous siltstone and siltstone interbedded. At the top of the formation, it is usually dominated by grayish and light gray silty mudstone, with more grayish-black carbonaceous mudstone, and grayish-white marl. At the bottom, the formation is usually dominated by brownish-grayish-green mudstone and sandy mudstone, with a small amount of grayish-black carbonaceous mudstone.

Dabuxun Formation: It is the youngest sediment in the Qaidam Basin and only located at some small low-lying areas. Halite is the dominant component of this formation, and silty clays are only seen in low-lying with fresh-water supplying near the Xitaijinaier Lake.

3.2.2 *Structure*

Both faults and folds are developed within the basin. The fault system of the Qaidam Basin is very well developed, which is shown as an interactive cutting diagram making the NW direction and the NNE direction faults. In which the dominated faults are in NW direction. These faults, from north to south, mainly include the Oulongbuluke-Yuniushan fault zone, the Saishiteng-Xitieshan Mountain-front fault zone, the Hubei-Lingjian-Dahuo fault zone, the Pingdong fault zone, the Yingbei fault zone, and the Alar fault zone (Fig. 3.4).

The regional fault system around the Qaidam Basin plays an important role in controlling the development and evolution of the basin. Regional fracture systems control plane geometry of the Qaidam Basin, in which the Qilianshan-northern marginal fault system of the Qaidam Basin marked the northern margin of Qaidam Basin, the east Kunlun, the southern marginal fault system of the basin, defines the distribution direction of the southern basin margin. The Altyn strike-slip fault system and Elashanstrike-slip fault system limit the east and west of the Qaidam Basin boundary, respectively.

As primary structures, these regional fault systems also control the formation and distribution of secondary structures in the basin. For example, the secondary fault zone at the western Qaidam Basin is mainly distributed in the NW direction, which is the same as that of the primary

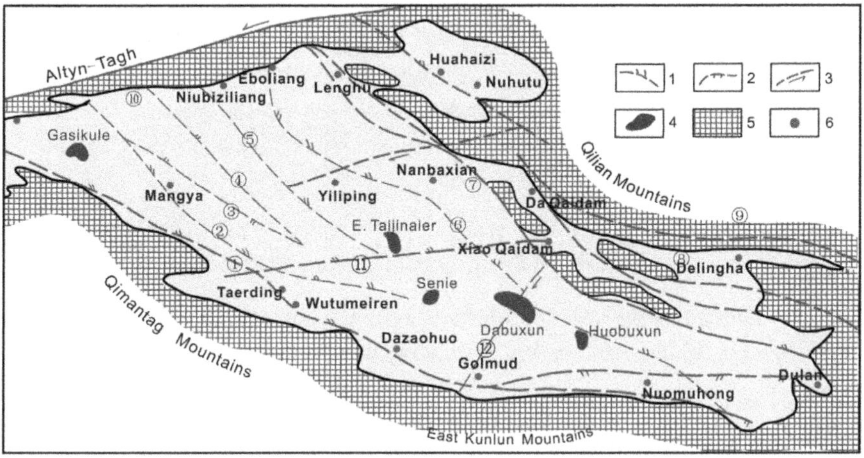

Fig. 3.4. Substrate structure of the Qaidam Basin. (1) Compression-shear fault; (2) Compression fault; (3) Ductile strike-slip fault; (4) Lakes; (5) Ancient surrounding mountains; (6) Cities; ① North Kunlun Fault; ② No.XI Fault; ③ Youbei Fault; ④ Fengnan Fault; ⑤ Jianbei Fault; ⑥ Hubei-Lingjian-Dahuo Fault; ⑦ North Qaidam Fault; ⑧ South Qilian Fault; ⑨ North Zongwulong Mountain Fault; ⑩ Anan Fault; ⑪ the Central Fault Zone; ⑫ Golmud-Xitieshan Fault.

Kunbei fault zone. The regional fault system also controlled the migration of the sedimentary center of the basin from the Tertiary period in the west to the Quaternary period in the east.

In addition to faults, the fold is also developed. There are quite a few anticlines within the basin, showing as uplifts and small mountains in landforms (Fig. 3.2). These folds were resulted from the compression between the Indian plate and the Tibetan Plateau, the same as most of the regional faults, although they may not form at the same time. Typical anticlines include Nanyishan, Jiandingshan, and Dafengshan, etc. We have also found remarkable faults developed upon some of the anticlines. On the Nanyishan and an anticline to the north of Dongtaijinaier Lake, NE direction faults are visible, which cut the layered strata and caused an offset of 200–300 m (Fig. 3.5).

We have also found some faults that cut yardangs and formed fault triangular facets (Fig. 3.6). In fact, yardangs are one of the youngest landforms in the basin, thus these faults could have been formed very recently and may still be active.

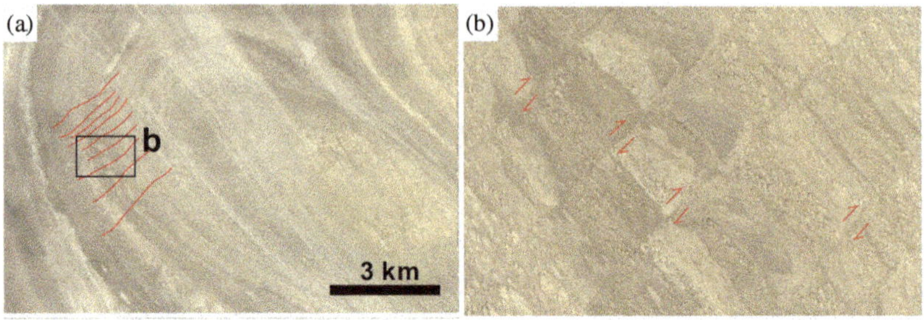

Fig. 3.5. Located on the anticline structure above the northeast of East Taijinaire lake (38.14°N, 91.86°E), the southwestern wing develops a set of left-handed fault structures.

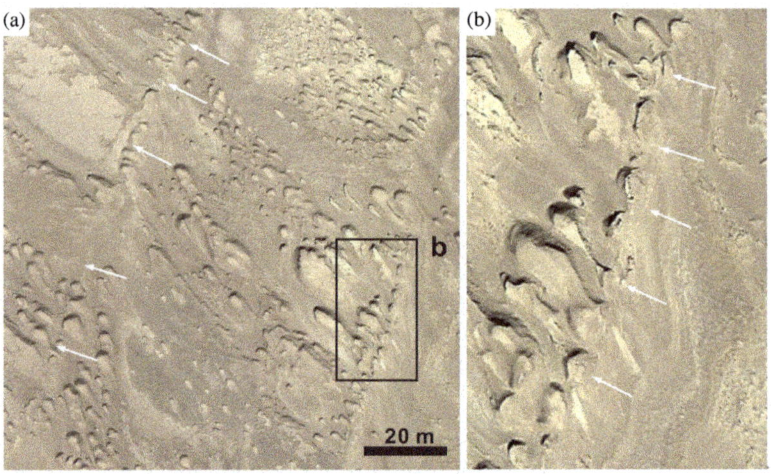

Fig. 3.6. A group of faults cut-through the yardangs on an anticline, forming linear triangular facets (37.69°N, 93.99°E).

Recent study also supports our observation. According to (Yu *et al.*, 2018), all Early to Middle Pleistocene strata in the basin have been involved in deformation, caused remarkable variation in the rates of deposition along the basin margin, and very thick layers of conglomerates presented in Quaternary sediments. These all indicate tectonic activities were very strong in the Quaternary era and should have played an important role in shaping the geography and landforms at various scales.

3.2.3 *Geological history*

Prior to the Hercynian orogeny (late Paleozoic, about 250 million years ago), the Qaidam region was part of the ancient Tethys sea that traversed southern Eurasia. The Hercynian movement caused the uplift of the Kunlun Mountains, which gradually separated the Qaidam area from the transgression, and the area became a denudation area in the early Mesozoic era. Since the Hercynian movement, each orogenic movement has intensified the activities of the deep and large faults in the basin margin. The basin was formed from the Indochina movement in the early stage of Yanshanian tectonic cycle. In the early Cenozoic era, the Indian plate collided with the Eurasian plate and continued to drift northward, causing the Qinghai-Tibet Plateau to rise gradually. The overall tectonic trend of the Qaidam area was that the mountains around the basin were uplifted, and the main part of the basin was an uneven depression, from which the Qaidam Basin was formed. Altyn Mountain, Qilian Mountain, Kunlun Mountain and three groups of deep and large faults, Qilian Mountain fault, and Kunlun Mountain fault constitute the tectonic pattern, which obviously controls the formation of the Qaidam Basin. It evolved into a diamond-shaped basin, and some deep and large faults developed in it, which played a crucial role in controlling the formation and evolution of the basin.

The proto-shape of the Qaidam Basin has been changing since the Mesozoic era when it was formed initially. The Jurassic and Cretaceous paleotopographical features were high in the south and low in the north, and entered Cenozoic, Paleocene, and Eocene, the above highs experienced erosion and formed piedmont sediments along the hill front of Qilian and Altyn Mountain ranges. During the Oligocene, the ancient Qaidam Basin entered a depression stage and deposited more than 2,600 m of piedmont, fluvial, and lacustrine sediments. The paleotopography gradually changed from the original high in the south and low in the north, to high in the east and low in the west, high in the north and low in the south, and to late high in the northwest and low in the southeast. During the whole Cenozoic period, the basin expanded first to the south and then to the east and the subsidence centers migrated accordingly.

From the beginning of the Oligocene to the end of the late Pleistocene of the Quaternary, the Qaidam Basin experienced the four stages of occurrence, development, stable settlement, and shrinkage and decline. The evolution of these stages is different in different parts of the basin. In the western part of the basin, the early lake basin in the Oligocene entered the stage of occurrence and development. At this time, the eastern part of the basin was subsiding and accepting sedimentation until the early Pleistocene. With the rise of the Himalayas and the rise of the Tibetan Plateau, the Qaidam Basin shrank sharply and declined rapidly in the late Pleistocene.

In the process of lake basin evolution, at the end of the late Pliocene and Pleistocene, the Qaidam Basin has experienced two strong tectonic activities. For the first time, the western basin formed angular unconformity between the widely distributed Pleistocene sediments and its underlying Pliocene series, resulted in uplifting of the basin floor, and formed many sub-lakes, strong depressions in the east. The second intense activity started from the Himalayan movement in the middle Pleistocene, resulting in a wide distribution of "dry salt lakes" in the basin at the end of the Pleistocene, forming 25 brine lakes in the basin.

The present Qaidam Basin is affected by the surrounding mountains, and the climate is extremely dry and often suffers from sandstorms. A large number of lakes have been transformed into salt lakes or dry salt flats, and the original lacustrine deposits have formed various types of wind erosion landforms under the erosion of wind and water, such as yardangs, desert, sand dunes, and dry salt flats. Life is hard to see in most of the basin, and just like on another planet. Because these features are very similar to the surface of Mars, they have become a natural laboratory for studying the environment on Mars, providing a rare testing ground for understanding the changes in the geological and climatic environment on Mars, exploring life on Mars, and exploring Mars.

3.3 Geographical and Geomorphological Features

3.3.1 *Surrounding mountain ranges*

The Qaidam Basin is bounded by the Kunlun Mountain range in the south, Altyn Mountain in the west, the Qilian Mountain range in the northeast, and the Elashan Mountain in the east (Fig. 3.1).

3.2.1.1 *The Kunlun Mountain range*

The Kunlun Mountain range is about 2500 km long, with an average altitude of 5,500–6,000 m and a width of 130–200 km. It is narrow in the west and wide in the east, with a total area of more than 500,000 square kilometers. The highest peak is Gonger peak in the west Kunlun region (about 7,649 m above sea level). Its latitudinal range divides southern and northern China.

The northern slope of the Kunlun Mountains is close to the center of the driest Asian continent, namely the Tarim Desert in the west of the warm temperate zone and the Qaidam Desert in the east. Mountain precipitation is less than 100 mm, 60 mm in the west, 20 mm in the east, and only 15 ~ 20 mm in Ruoqiang area at the southern Tarim Basin. Originated from the northern Slope of Kunlun Mountains, these rivers have a long history and converge in the Tarim and the Qaidam Basins.

The Kunlun Mountains are high in the west and low in the east in general, and can be divided into three sections, the west, middle, and east, according to the topography. There are three peaks above 7,000 in the west Kunlun Mountains, and seven peaks above 6000 m, with an average altitude of 5,500–6,000 m above sea level. There are 8 peaks in middle Kunlun Mountains with an altitude of over 6000 m, with an average altitude of 5,000–5,500 m and the snow line on the north slope of 5,100–5,800 m. The east Kunlun Mountains have 4 peaks above 6,000 m and 8 peaks above 5,000 m, with an average altitude of 4,500–5,000 m. Snow is distributed over 5,800 m.

The eastern section of the Kunlun Mountains forms the southern boundary of the Qaidam Basin. From west to east, Qimantag Mountain, Bokaretag Mountain, and Burhanbuda Mountain extend for about 900 km. In east Kunlun, the snow line in the pass of Kunlun Mountain is 5,200 meters in the north slope and 5,400 Meters in the south slope. The annual average temperature near the snow line is −9 ~ −8°C. The annual average temperature in the Xidatan area of the mountain valley (4,200 m) is lower than −3°C, and the annual precipitation is about 350 mm. The annual precipitation at the top of the mountain has increased slightly, and the lower limit of the modern permafrost on the northern slope of the Tibetan Plateau is about 4,200 m.

The Kunlun Mountains are separated from Tarim Basin and Qaidam Basin by deep faults (see Fig. 3.7). The basement is made of Presinian system metamorphic rocks. During the Paleozoic, the Paleo-Tethys sea area was strongly sinking and accompanied by volcanic activities. At the end of the Paleozoic, the fold of the Variscian movement rose and formed the central axis of Kunlun and the mid-ridge of the mountain range. Over 4,000 m of mountains on both sides of the main ridge were formed by Mesozoic Yanshanian movement.

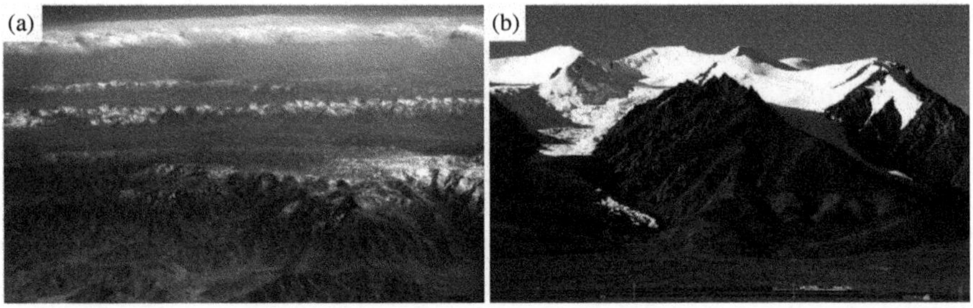

Fig. 3.7. (a) Bird-view of Kunlun-Hoh Xil (credit: Visual China Group); (b) Kunlunshan (August 10, 2019).

3.3.1.2 *The Altyn Mountain range*

The Altyn Mountain range lies between the Qaidam Basin and the Tarim Basin (see Fig. 3.8), and is connected with Qilian Mountain at the west and with Kunlun Mountain from the east. It extends in NEE, with a length of about 700 km and a width of about 60–100 km. The eastern and western ends of the mountain range are high and the central part is lower (between 4,000 m and 4,200 m). There are modern glaciers at an altitude of 5,000 m or more.

The highest peaks in the western section of the Altyn Mountain range are Suramutag Peak (6,295 m) and YusupuAlek Peak (6,062 m). The highest mountain in the eastern section is also known as Altyn Mountain, 5,828 m above sea level. The climate of the Altyn region is arid with poor vegetation and no perennial rivers.

3.3.1.3 *Qilian Mountain range*

Qilian Mountain range is one of the main mountain ranges in China. It consists of several parallel mountains and wide valleys running from northwest to southeast (see Fig. 3.9). It is 800 km long from east to west, 200–400 km wide from north to south, 4,000–6,000 m above sea level, and has 3,306 glaciers.

The western end meets the Altyn Mountains at the mouth of the Dangjinshan Mountain. The eastern end to the Yellow River valley, and connected with Qinling and Liupan Mountains. It's almost 1,000 km long. The widest site is between Jiuquan city and Qaidam Basin, up to 300 km.

Fig. 3.8. Altyn Mountain range (credit: China National Geography).

Fig. 3.9. Qilian Mountain ranges (credit: China National Geography).

Qilian Mountain range extends from northwest to southeast, including Xueshan (snow) Mountain, Tolai Mountain, Tolainan Mountain, Yemanan Mountain, Shulenan Mountain, Danghenan Mountain, Tuergendaban Mountain, Qaidam Mountain, and Zongwulong Mountain. Its peak is 4,000–5,500 m above sea level, and its highest peak, Tuanjie peak, is 5,808 m high. The mountains above the altitude of 4,000 m are covered with snow all the year round, and the valleys are between 3,000 m and 3,500 m high.

From the west to the east, the small mountain ranges forming the northern margin of Qaidam Basin include Saishiteng Mountain, Lvliang

Mountain, Xitie Mountain, Amnick Mountain, and Maoniu Mountain, all of which show a north-west-south-east trend on the whole. These mountains are relatively low in altitude and they are separated and form many small intra-mountain basins. Their features are briefly summarized as follows.

Saishiteng Mountain is located between Sugan Lake Basin and Mahai Basin, 15 km east of Lenghu town in the west, and Aolao Valley in the east. The upper part of the mountain slope is steep, the lower part is gentle, the ridge continuity is poor, both ends have formed isolated peaks. The highest peak is located in the east, 4,576 m above sea level.

Lvliangshan Mountain is located in the northern Qaidam Basin, south of the Daqaidan Lake, from the Yuka Valley in the west to the Tatareng Valley in the east. It is 60 km long and 10 km wide. Most of the mountain peaks are 3500–4000 m above sea level. The highest peak is located in the western section, 4111 m high.

Xitieshan Mountain, also located in the northern Qaidam Basin, south of the Xiaoqaidan Lake. It is a typical block mountain, length and width both are ~15 km. The mountain slopes are steep and the peak is 4037 m above sea level. The mountains are mainly composed of ancient metamorphic rocks and have a variety of metal deposits.

Amonick Mountain, located in the north of the Baihobson lake basin. It starts from Quanji River valley from the west, to the western edge of the Toso lake basin in the east. It is about 100 km long and 10–20 km wide. The elevation of the peak is ~3,000–3,500 m, with gentle slopes, rounded peaks, and not very continuous ridges. The highest peak is 4,166 m above sea level.

Maoniu Mountain is located in the northeast corner of Qaidam Basin, southwest of Delingha — Xili Valley Basin. It is ~70 km long and generally 5–10 km wide. Most of the peaks are between 3,500 m and 4,000 m above sea level, and the ridges are not very continuous. The lower part of the western section is also called Nanshan Mountain, while the higher part of the eastern section is the main part of Maoniu Mountain. The highest peak is located in the eastern section, with an altitude of 4,472 m.

3.3.2 *Inside the basin*

The average elevation of the Qaidam Basin is about 2,800 m, higher in the west than in the east, lower in the middle. The surrounding area is piedmont alluvial-diluvial fan deposition, with an average slope of 6‰ ~ 9‰ and an elevation of 2800 ~ 3850 m. The elevation close to the mountain is generally more than 3850 m, and there are scattered sand dunes. Rivers develop on alluvial-diluvial fans. The sediments of these fans mostly consist of sand and gravel, which contain pore water and are runoff areas of groundwater. Wind erosion remnants and depressions are distributed in the western plain area, and in the northern part of the plain area, there are more isolated mountain ranges, such as Qilian Mountain and Altyn Mountain derived mountain ranges. These mountains also cut off several basins, which developed rivers and lakes. The lakes in the basin are arranged in a beaded pattern along the middle part of the basin, and the isolated mountain and derived mountains are distributed in the northern plain area of the basin, as well as the intermountain basins, valleys, and lakes related to these mountains, which are closely related to the geological structure.

The surface sediments of the basin, from the margin to the center, show a regular succession of diluvial gravel fan (Gobi), alluvial-diluvial silty clayey plain, lacustrine alluvial-silty clayey plain, and lacustrine silt saline plain. Salt lakes and marshes are widespread in low-lying areas.

The inner margin of the Gobi belt in the northwest of the basin is relatively high. The southeast subsidence of the basin is intense, and the alluvial and lacustrine plain is broad. The lower reaches of Qaidam River, Suringole River, and Golmud River and the surrounding lakes have a large area of marsh.

Due to the uplift of a series of low mountain fault blocks in the northeastern part of the basin, a sub-class small intermountain basin was formed between the basin and the Qilian Mountains, from the west to the east, there were Huhaizi, Daqaidan, Xiaoqaidan, Delingha, and Ulan Basins, and the rivers in these basins were fed into the lakes in the low-lying center.

3.3.3 *Water system and lakes*

The distribution of the water system in the Qaidam Basin is significantly affected by the topography, mainly the inflow system. The inflow rivers are affected by the topography of the basin, and are centered on the low-lying lands of the local lakes, forming concentric radial water systems. The rivers originating from the surrounding hills, after crossing the alluvial fan and the Gobi, accumulate in the lower reaches of the basin into many lakes. The lakes are mainly distributed in the basin along the outer edge of the Gobi (Fig. 3.10). Most of the rivers are intermittent rivers. Among the total of 100 rivers, there are only more than 10 perennial rivers, mainly distributed in the eastern part of the basin, while the western water network is extremely sparse. There are more than 20 salt lakes and the water quality of the lakes has been salty (Fig. 3.10).

For a long time in the inland closed natural geographical environment, the Qaidam basin has created very favorable conditions for the formation of the salt lake. There are 48 lakes with an area of more than 1 km^2 in the basin. Among them, there are 26 salt lakes with an area of ~1330 km^2. Qarhan salt lake is one of the world's highest altitude salt lakes with an elevation of 2675–3171 m. Most of the water is shallow, generally tens of centimeters, conducive to evaporation, so the salinity is extremely high, with the average salt content of 332.4 g/l, 10 times the salt content of seawater, up to 526.5 g/l. "Dry salt lakes" with no or little surface water are widespread, covering an area of ~10,000 km^2.

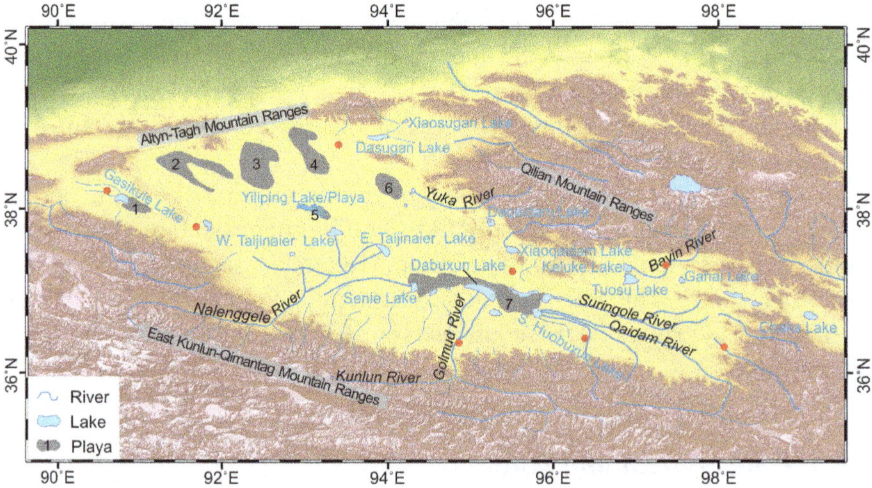

Fig. 3.10. Distribution of major inland rivers, lakes, and playas in the Qaidam Basin, red dots indicate the main cites shown in Fig. 3.1. Playas are marked as: (1) Gasikule; (2) Dalangtan; (3) Chahansilatu; (4) Kunteyi; (5) Yiliping; (6) Mahai; (7) Qarhan. Yiliping is a periodical lake: it is a lake sometimes, but becomes dry in other times. Base image is generated from 90 m SRTM data.

The larger lakes (Fig. 3.11) in the basin include Dabuxun Lake, Taijinaier Lake, Huobuxun Lake, Tuosu Lake, Keluke Lake, Daqaidam Lake (between Dagendaban Mountain and Lvliang Mountain), Sugan Lake (covering a total area of ~120 m^2), Gasikule Lake, and Xiaoqaidam Lake.

The eastern subsidence area forms a series of large lakes connecting east and west, such as Hobuxun Lake, Dabuxun Lake, and Taijinaier Lake. The western lakes are quite untidy. In the north, a series of small intermountain basins are distributed with lakes of different sizes, such as Sugan Lake, Qaidam Lake, Keluke Lake, and Tuosu Lake. Due to the arid climate and vigorous evaporation, the lake surface shrank and the salt continuously became concentrated, many lakes gradually changed from salty lake to salt lake and finally developed into dry salt lake or playa.

All the lakes are sourced from surrounding mountains, pass through rivers and end at low-lying areas to form lakes. The major geomorphological features of these water systems and lakes are detailed in Chapters 4 and 7.

Fig. 3.11. Lakes in the Qaidam Basin. (a) Xiaoqaidam Lake; (b) Daqaidam Lake; (c) Tuosu Lake; (d) Qarhan Lake.

3.3.4 *Playas and polygonal landforms*

Qaidam Basin is China's largest "salt sea" with an area of ~1.0×10^4 km^2 and 1.0×10^3 km^2 for dried and undried salt lakes, respectively. The major playas include Gasikule, Dalangtan, Chahansilatu, Kunteyi, Yiliping, Mahai, and Qarhan (Fig. 3.10). They formed ~15,000 years ago due to an extremely dry climate beginning since 25,000 years B.P. (Chen and Bowler, 1986). Since the Cenozoic, the Qaidam Basin has been elevated to a considerable height and experienced several rises and falls due to a series of strong tectonically controlled transformations (Xia *et al.*, 2001). The basin depression was eventually separated into individual salt lakes over geologic time. The current dry environment of the basin is characterized by having an average annual rainfall of ~30 mm, but annual evaporation reaching up to ~3,000 mm (Kong *et al.*, 2013), resulting in a typical arid climate indicated by a striking desert landscape (Xiao *et al.*, 2017).

The thickness of salt deposits varies across the region from playa to playa (Chen and Bowler, 1986; Huang *et al.*, 1993). These playas were formed at different times and had distinct physical-chemical evolution histories. The western part of Qaidam Basin has a higher topography and is the earliest place to form playas, such as Dalangtan playa (Dang *et al.*, 2018). The western region has experienced the longest period of evaporation and deposition, resulting in a maximum salt thickness locally. The salt lakes in the eastern region are late to dry and have a minimum value of accumulated salt thickness. The compositional variation of salt minerals is dominated by a series of evaporite minerals including sulfate, carbonate, and halite components (Fig. 3.12). Their surfaces are characterized by polygonal networks as observed in the drying or dried playa deposits in the Qaidam Basin (Fig. 3.13). These typical patterned landforms can be considered as a combination and connection of an array of polygonal surface structures (PSSs), or polygons for simplicity.

Generally, the salt polygons in the Qaidam Playas have a size range from centimeters to hundreds of meters in diameter (Fig. 3.13). They display the forms of triangle, quadrangle, and complex-shaped polygons, depending on their edge/side number and arrangement pattern. Regular and irregular random orthogonal polygons (Neal *et al.*, 1968) are predominant in the Qaidam Playas. Commonly, large polygons contain small ones in their interior. The most prominent characteristic of these polygons is their positive top-cracked rims that formed from a subsurface process of lateral salt displacement (e.g. Artieda *et al.*, 2015; Christiansen,

Fig. 3.12. Evaporite minerals observed in the Qaidam Basin. (a) Chloride is crystallizing in Qarhan Playa; (b) Large amount of dark gray gypsum crystals are scattered in the surface soils; (c) Sulfate sedimentary layers exposed in the Dalangtan Playa, the light layer is mirabilite forming in the dry period, the dark layer is clay forming in a relatively wet period.

1963; Lowenstein and Hardie, 1985) due to crystallization pressures (Handford, 1991) or to thermal contraction and expansion (Dellwig, 1968; Handford, 1991; Tucker, 1981). In fact, polygonal ridges (uplifted rims) are thought to have formed through a combination of thermodynamic and geochemical mechanisms (e.g. Lowenstein and Hardie, 1985; Nield *et al.*, 2016). The durability of salt polygons is in most cases limited by seasonal rainfall or flooding events, which often destroy the surface morphology and then start a new cycle of the formation and evolution of salt polygons (Artieda *et al.*, 2015).

The appearances of polygonal terrains appear to be mainly controlled by a wide range of contributing factors such as regional and local topography, strength and thickness of the tabular-shaped sedimentary units, injection of foreign materials transported from other areas inside and/or outside the basin, historic water level change, temperature variation,

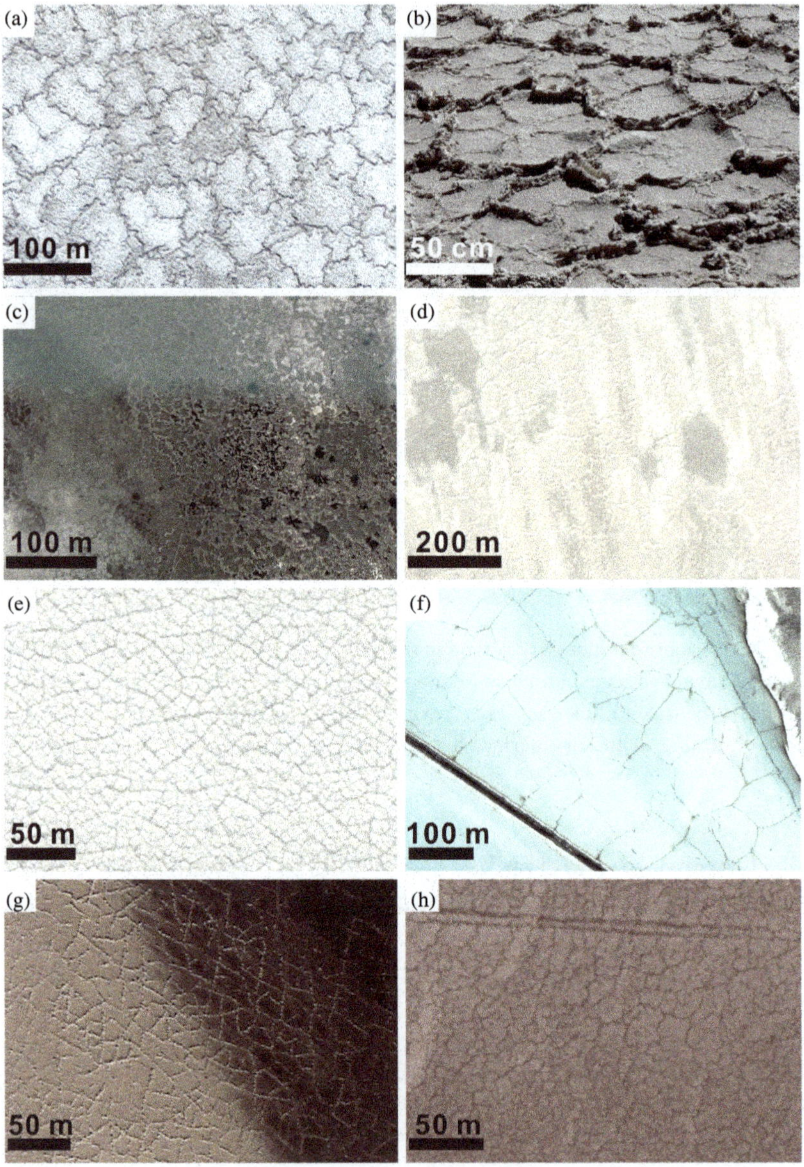

Fig. 3.13. Various types of polygons in different playas in the Qaidam Basin. (a)–(b) Dalangtan; (c) Gasikule; (d) Chahansilatu; (e) Kunteyi; (f) Yiliping; (g) Mahai; (h) Qarhan. Satellite image (credit: Google Earth); Photos (b) were taken during the field trip in Dalangtan Playa.

periodic evaporation, and brine activity (such as seasonal capillary upflow). Further, these factors acting upon each other add to the complexity of diversity of polygonal terrains. Understanding the role individual factors play in the formation of PSSs will help shed light on the historic environmental and climatic changes that once occurred on both Earth and Mars, since a series of salt polygons in Martian southern highland have also been observed (Dang *et al.*, 2018; El Maarry *et al.*, 2013). Please read Chapter 6 for more information.

3.4 Wind Erosion and Aeolian Landform

The arid climate and strong wind erosion have created special landforms in the Qaidam Basin, including Gobi Desert, yardangs and various sand dunes (Fig. 3.14). Under the long-term effect of the prevailing westerly winds, a large number of erosive landforms (e.g. various types of yardangs) formed in the western part of the basin. The loose sediments caused by erosion accumulated in front of the high mountains in the south of the basin, forming continuous sand seas. At the same time, the abrasion of the rocks also resulted in ventifacts with various shapes and colorful surface features.

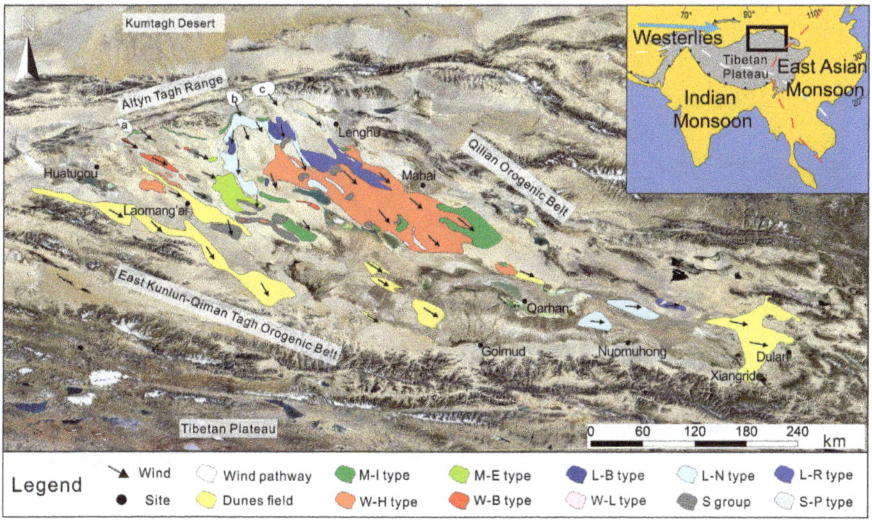

Fig. 3.14. Distribution of yardangs and dunes in the Qaidam Basin.

3.4.1 *Macroscale wind erosion features: Yardangs*

"Yardang" is a geological term that was introduced by Hedin in 1903, when he traveled the Lop Nur, northwestern China, to define the streamlined, wind-carved ridges that had been developed within lacustrine deposits. This term has now been extended to define wind-carved ridge-like features in general, whether highly streamlined or not (Halimov and Fezer, 1989; Laity, 2011). Typical yardangs mostly occur in large fields and arid to hyper-arid regions that have sparse vegetation and annual precipitation less than 50 mm (Niu *et al.*, 2011), which have been found in most of the deserts on Earth (McCauley *et al.*, 1977), including Central Asia, Southern Iran, Northern Saudi Arabia, Bahrain, Namibia, Egypt, southern California, Central Sahara, the high Andes of South America, and coastal Peru and Chile (Goudie, 2007) (Fig. 3.15).

As the most important yardang field, the climate in the northwest of China is relatively dry. Most of the yardang fields occur in the Qaidam Basin, the middle and lower Shule River, the Wuhe region of Inner Mongolia and the Xinjiang Province (Geng, 2009). Yardangs in the Qaidam Basin cover an area of about 3.88×10^4 km^2, nearly one-third of the basin, being the largest yardang field in China (Kapp *et al.*, 2011b; Li *et al.*, 2016).

As previously reported in our study, yardangs in the Qaidam Basin are mainly distributed in the central-eastern and northwestern parts

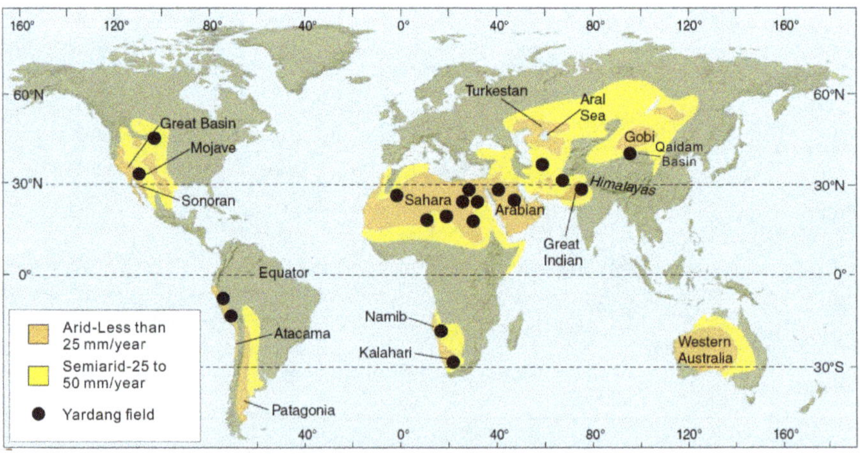

Fig. 3.15. Global distribution of yardangs (modified from Plummer *et al.*, 2016).

(Fig. 3.14). Generally, in the northern basin, their periphery is surrounded by modern piedmont alluvial-diluvial fans in the northern and northeastern parts, and sand fields in the southern part. While in the south of the northwestern part, they are bounded by playas, active saline lakes and rivers (Halimov and Fezer, 1989; Li *et al.*, 2016). In the central-eastern part, yardangs are mainly bounded by active rivers and alluvial-diluvial fans (Fig. 3.14).

Based on their morphologies and distributions, yardangs in the Qaidam Basin can be classified into 11 different types and sorted into 4 groups (Table 3.2) (see Figs. 3.16–3.18). Taking account of their

Table 3.2. Classification of the yardangs in the Qaidam Basin. Modified after Halimov and Fezer (1989) and Li *et al.* (2016).

Type	Characteristic	Group
Irregular mesa (M-I type)	Flat top, irregular in shape, cliffs, ambiguous orientations	Mesa (M group)
Elongated mesa (M-E type)	Flat top at the head, arranged to lines in one direction or two directions	
Broad-top long-ridge (L-B type)	Broad top, narrow troughs, gentle slope flanks, salt crust	Long-ridge (L group)
Narrow-top long-ridge (L-N type)	Narrow top, broad troughs, steep flanks	
Rough-top long-ridge (L-R type)	Broken top, ambiguous morphology	
Saw-toothed (S-T type)	Upwind zigzag-like shape, downwind areas are mostly unaltered	Saw-toothed (S group)
Pyramid (S-P type)	Large base with spire top	
Remnant cone (S-R type)	Large base with the rounded shape or small flat area on its top	
Whaleback (W-B type)	Classic form with a blunt head and tapered tail, salt crust on the surface	Whaleback (W group)
Hogback (W-H type)	Steep windward heads (>60°) with sharp crests on their backs, and some have undercut windward heads	
Low, streamlined whaleback (W-L type)	Low height, streamlined morphology, large spacing	

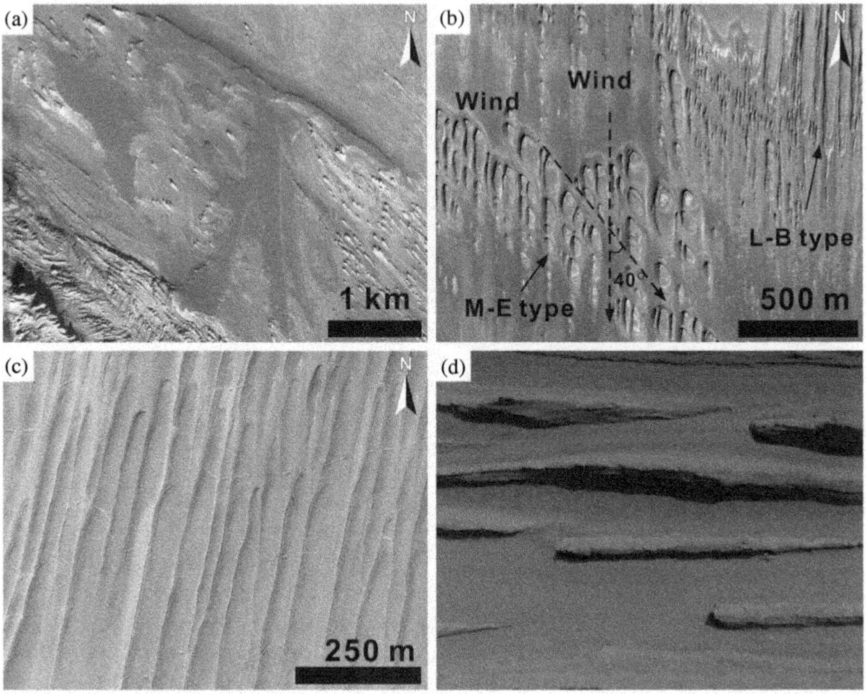

Fig. 3.16. (a) The M-I type yardangs in the southeast of Lenghu area (38.46°N, 93.45°E). (b) The M-E type yardangs in the bidirectional area (38.07°N, 92.14°E). (c) The L-B type yardangs in the western Qarhansilatu Playa (38.49°N, 92.26°E). (d) The L-N type yardangs in the southwest of Lenghu (38.52°N, 92.97°E).

morphological complexities and distributions in the field, our priority is to researches the morphological characteristics of the representative yardang in the Qaidam Basin (Table 3.2), for more details about yardangs in the Qaidam Basin, see Chapter 5.

3.4.2 *Meso-micro-scale wind erosional features: Ventifacts and Tafoni*

3.4.2.1 *Ventifacts*

Ventifacts are subaerially-exposed clasts or bedrock surfaces that have been abraded by the action of wind-carried particles (Greeley and Iversen, 1985), especially sands. Abrasion of the faces of these clasts and rock

Fig. 3.17. (a) The L-R type yardangs in Lenghu (38.65°N, 92.78°E). (b) The S-T type yardangs in the folded stratum (38.14°N, 91.87°E). (c) The S-P type yardangs in the Daxiongshan uplift (E'boliangYardang Geopark, image center: 38.35°N, 92.88°E). (d) The S-R type yardangs with rounded top and multiple layers (37.52°N, 92.30°E).

surfaces gives rise to distinctive and smooth plano-concave to plano-convex faces.

Surface textural features, such as pits, grooves, flutes, and scallops of sand-blasted ventifacts have been attributed to various erosional mechanisms and to variations in material hardness. Within the Qaidam Basin, especially the Gobi Desert regions, ventifacts are everywhere with different sizes and shapes. Actually, most exposed rock fragments have been abraded by wind and wind-carried particles and developed into ventifacts. They have normal single facet (large boulder or outcrops), multiple facets (several centimeters to tens of centimeters size boulders), or rounded (sub-centimeter to several centimeters) ventifacts (Fig. 3.19).

Salts are more easily eroded than rocks by wind erosion, usually shown as flutes and scallops (Fig. 3.20).

Fig. 3.18. (a) A typical small hogback (W-H type) yardang with a windward slope of ~60° (38.51°N, 92.91°E). (b) The whaleback (W-B type) yardangs in Nanyishan (38.41°N, 91.32°E). (c) The W-L type yardangs in the Nanbaxian area (37.61°N, 94.17°E). (d) The "great sphinx" yardang at Nanyishan.

3.4.2.2 *Tafoni*

Tafoni are large cavernous weathering features. They are characterized by having a flaking backwall and an overhanging lip in front. The opening of the tafoni is often near-circular. They are a common feature of deserts, semi-arid regions, and near coasts or anywhere salts can precipitate from dew or fog (Mustoe, 1982; Young, 1987). They occur in a wide range of rock types to be the result of flaking and granular disintegration, caused by hydration, salt crystallization and the chemical attack caused by saline solutions (Goudie, 2004).

Tafoni are often to be seen in the surrounding mountain ranges of the Qaidam Basin. In Hongya region, two types of tafoni co-exist. Tafoni with large pits are seen in thick-layered sandstone (Fig. 3.21(a)), while holey

Fig. 3.19. Ventifacts found in the Qaidam Basin. Pits and scallops were seen in these ventifacts. (a) *Dreikanter* (with three opposing edges, which together give rise to a pyramidal ventifact shape in plane view), (b) Ventifacts with plano-concave *faces*, (c) and (d) *Pitted ventifacts.* Width of images are 10 cm.

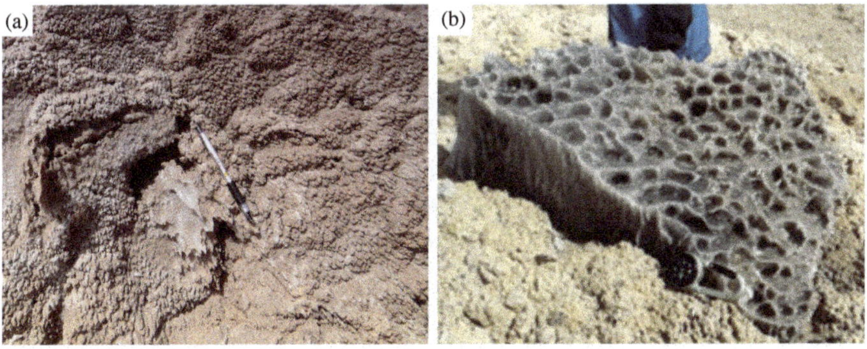

Fig. 3.20. Grooved, fluted (a), and alveolate (b) salt crust in the Qaidam Basin.

skeleton tafoni occurred in thin-layered brownish sandstone (Fig. 3.21(b) and 3.21(d)). The different morphology suggests that the former one was formed by the physical (wind force) process but the later one was dominated by chemical erosion.

Fig. 3.21. Tafoni near Hongya (Redcliff) region, Qaidam Basin. (a) Tafoni with large pits on thick layer sandstone; (b) Collapsed brownish thin-layered sandstones (holey) tafoni; (c) and (d) Fragments of brownish sandstone tafoni.

3.4.2.3 *Vermiculations*

There is also weathering formed "vermiculation" overprinted on some limestone ventifacts. They are featured with unique peculiar vermiculated surfaces, covered with tiny "rills". This was interpreted as specific carving among "micro solutional features". These stones are limestone (or dolomite) pebbles corroded by dew droplets that frequently form at night in many deserts, especially coastal-type ones such as the Atacama and the Namib. According to Viles and Bourke (2007), at sites sheltered from the wind, dew flows solely by gravity would cause rills with radial arrangements. On pebbles exposed to moderate wind, relatively constant in direction, the wind-driven drops create more or less parallel grooves. In areas of vortices or changing winds, the drawings are often more or less labyrinthiform.

Vermiculations are seen in front of Xitieshan Hill. Some limestone pebbles at the foot of the hill developed unique labyrinthiform (Fig. 3.22). The pattern of rills on the surface suggests that the wind direction is somehow vortices.

Fig. 3.22. Small vermiculated limestone cobbles in Xitieshan region, Qaidam Basin.

3.4.3 *Sand Dunes*

Sand dunes are a kind of accumulation of landforms where loose sand particles are deposited or reformed under the action of wind, which cover 1/4–1/3 areas of sand seas, being the major landscape in the desert. They act as moundy or ridged shapes that are several meters to tens of meters, or even hundreds of meters high. The sand grains are mainly weathered from rocks, and abraded by the wind. the main components of these sand grains are quartz and feldspar on Earth, while on Mars, they are mostly dark minerals formed by volcanic weathering.

The formation of sand dunes is closely related to the environment, such as climate change, hyacinth conditions, and the supply of sand sources. Their development processes are strongly restricted by the two-phase flow movement rules of wind and sand, which need two basic conditions: (1) a large amount of sand, (2) a sufficiently strong and durable wind to move the sand. If these two conditions are met, the sand can be moved hundreds or even thousands of kilometers away.

Due to the influence of wind conditions and sand sources, various forms of dunes have been formed. According to the relationship between the formation of the dunes and the wind direction, they can be divided into longitudinal dunes, transverse dunes, and multidirectional wind dunes. According to the vegetation cover on the dune surface, they can be

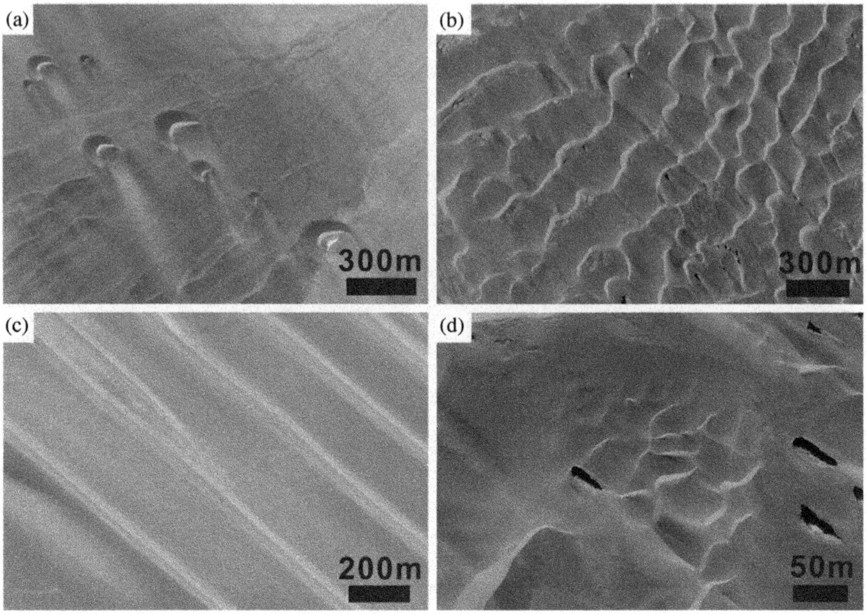

Fig. 3.23. (a) and (b) Barchan dunes and Barchan dune chains near the east Kunlun Mountains (37.66°N, 91.51°E) in the southern part of the basin. (c) Linear dunes in the Sanhu area (37.00°N, 94.03°E) in the middle of the basin. (d) Star dunes in the south of Nanyishan (38.22°N, 91.38°E) in the northwest of the basin.

divided into mobile dunes, semi-fixed dunes, and fixed dunes. According to the shape of the dunes, there are barchan dunes, parabolic dunes, trellis dunes, and pyramid dunes.

Dunes are distributed mainly in three regions, the southwestern piedmont area, the Sanhu area, and the southeastern area of the Qaidam Basin (Yu and Lai, 2012). In this region, dunes have been morphologically classified as barchan dunes, barchan dune chains, linear dunes, and star dunes (Liu *et al.*, 2008; Yanlin, 1995; Zhu *et al.*, 1980) (Fig. 3.23). Most of the dunes overlay well-bedded sediments.

3.5 Plants in Deserts

Affected by nature and climatic environment, the live species in the basin are sparse. In order to adapt to the basin's topography, soil, climate, water, salt, and other conditions, most of them have evolved and selected the dry life and dry life tendency. They should not only adapt to the drought of the soil and atmosphere in the basin and the physiological drought caused by soil salinization, but also adapt to the high temperature and low latitude and the low temperature caused by the Siberian cold air mass. This is consistent with the salt tolerance of some plants.

In order to adapt to drought, wind, and frost, the plants in the Qaidam Basin have the following characteristics: (1) Plants are short, large trees are few, and there are not many small trees. Weeds and shrubs are common. (2) mostly clustered; (3) dry form, many plants have narrowed or even completely degenerated; (4) deep root: they are under conditions of low rainfall, large evaporation, low humidity, and low groundwater level. Develop roots as much as possible to extend the roots to areas where they can absorb moisture from the soil and make the roots thicker to store nutrients and water.

There are many plants in the arid regions in the Qaidam Basin. Only representative populus, tamarisk, and medlar are displayed here (Fig. 3.24–3.26).

Fig. 3.24. Populus diversifolia. (a) Populus diversifolia field in summer; (b) Young populous diversifolia in autumn; (c) Old ones in autumn; (d) Dead but still standing trunk of Populus diversifolia, it can last for hundreds of years.

Fig. 3.25. Tamarisk is common in the desert area. The barks of old stems and branches are dark gray, the growing branches of the current year are red or orange-yellow, long and straight.

Fig. 3.26. Medlar is one of the major specialty products in the Qaidam Basin. It is widely distributed in Golmud, Dulan, Xiangride, Nuomuhong, and Delingha regions. The ripe fruits are mostly red in color, but black medlar is also common. Both of them are grown in fields, but more are cultivated at present.

Populus diversifolia often grows in the desert, it is resistant to cold, drought, salt and alkali, wind and sand, and has a strong vitality. It has a very long life and its trunk can stand without rot for hundreds of years. It is the only tree species growing in the desert, and it is very valuable, can be compared with the "living fossil plant" known as the ginkgo tree. In the Qaidam Basin, populus diversifolia are mostly grown in Tuolahai area, which is also the highest altitude in the world for populous diversifolia growth.

3.6 Climate Environment

The Qaidam Basin is mainly influenced by the westerlies (Fig. 3.14). Due to the shielding effect of the Tibetan Plateau and surrounding high mountain ranges, little moisture can arrive in this Eurasian Hinterland Basin after a long trip. The main moisture source is carried by the East Asian Monsoons, which can extend to the southeastern basin after crossing the Hengduan Mountains. Most of the rainfall comes from the convergence zone of these two flows, but both of them are very little (Wu *et al.*, 1985; Wu, 2009). Thus, their influences declines from east to west throughout the basin. The annual precipitations are ~150–~200 mm, ~40–~50 mm and <50–<20 mm in the southeastern, central, and northwestern basins, respectively. While their annual evaporations can be as high as ~2000–3600 mm (Wu, 2009; Zhang and Liu, 1985), making it one of the driest regions on Earth (Rohrmann *et al.*, 2013; Xiao *et al.*, 2017).

The prevailing wind direction in the western basin is NW-SE, and becomes nearly W-E in the eastern part. Except for some closed terrains (e.g. Daqaidam, Delingha), the average monthly wind velocity is >3 m/s, in the western Laomang'ai area it is ~5 m/s, while the maximum historical wind velocity is 40 m/s (Ren, 1996; WEI, 2013). Annual fresh gale conditions (wind velocity ≥17 m/s) last from 57 to 105 days, with an average of 84.2 days per year. Most of the gale days (about 50–70%) occur from March to June (Wu *et al.*, 1985; Wu, 2009).

During the past 50 years (1961–2010), air temperature records from eight meteorological stations (Mangya, Lenghu, Xiaozhaohuo, Dachaidan, Delinha, Golmud, Nuomuhong, and Dulan) show that the warming trend of the average air temperature, average maximum temperature, and average minimum temperature were respectively 0.54°C/10a, 0.43°C/10a, and 0.74°C/10a, much higher than the linear tendency of China and the whole world; the average minimum temperature increased most obviously and the winter temperature was the biggest contributor to the increased annual temperature; spatial analysis results show that the increment rate of temperature increased from the east to the west across the basin (E *et al.*, 2013). During the past 32 years (1980 ~ 2011), climate data collected from the meteorological station (Mangya, Dalangtan Playa) show that the temperature ranges from < −25°C (January) to >30°C (July), with an annual average temperature of 3.5°C (Kong *et al.*, 2018). The annual average air pressure recorded was 709 mbar (Kong, 2013), about two-thirds of the standard atmospheric pressure.

Acknowledgments

This study was supported by the Science and Technology Development Fund (FDCT) of Macau (Nos. 121/2017/A3), National Natural Science Foundation of China (Nos. 41830214 and 41773061), MOST Special Fund from the State Key Laboratory of Geological Processes and Mineral Resources, China University of Geosciences (No. MSFGPMR05). We would like to thank Dr. Zhiyong Xiao, Dr. Jun Huang, Dr. Ting Huang, Ms. Zhuqing Xue, Mr. Hansheng Liu, Mr. Jiawei Zhao, Mr. Siyuan Zhao, and Mr. Ning Zhang for fieldwork assistance.

References

Artieda, O., Davila, A., Wierzchos, J., Buhler, P., Rodríguez-Ochoa, R., Pueyo, J. and Ascaso, C. (2015). Surface evolution of salt-encrusted playas under extreme and continued dryness. *Earth Surf. Proc. Land.*, **40**, 1939–1950.

Chen, K. and Bowler, J. M. (1986). Late Pleistocene evolution of salt lakes in the Qaidam Basin, Qinghai Province, China. *Palaeogeography, Palaeoclimatology, Palaeoecology*, **54**, 87–104.

Christiansen, F. W. (1963). Polygonal fracture and fold systems in the salt crust, Great Salt Lake Desert, Utah. *Science*, **139**, 607–609.

Dang, Y., Xiao, L., Xu, Y., Zhang, F., Huang, J., Wang, J., Zhao, J., Komatsu, G. and Yue, Z. (2018). The polygonal surface structures in the Dalangtan Playa, Qaidam Basin, NW China: Controlling factors for their formation and implications for analogous martian landforms. *J. Geophys. Res. Planets*, **123**, 1910–1933.

Dellwig, L. F. (1968). Significant features of deposition in the Hutchinson Salt, Kansas, and their interpretation. *Geol. Soc. Am. Spec. Pap.*, **88**, 421–428.

E, C. Y., Hu, G. L., Wu, C. Y. W., Yang, P. Y. and Liu, W. W. (2013). The analysis on air temperature variation characteristics in Qaidamu basin (1961–2011) under the background of global warming. *J. Arid Land Reso. Environ.*, **27**, 94–99.

El Maarry, M. R., Pommerol, A. and Thomas, N. (2013). Analysis of polygonal cracking patterns in chloride-bearing terrains on Mars: Indicators of ancient playa settings. *J. Geophys. Res. Planets*, **118**, 2263–2278.

Fang, X., Zhang, W., Meng, Q., Gao, J., Wang, X., King, J., Song, C., Dai, S. and Miao, Y. (2007). High-resolution magnetostratigraphy of the Neogene Huaitoutala section in the eastern Qaidam Basin on the NE Tibetan Plateau, Qinghai Province, China and its implication on tectonic uplift of the NE Tibetan Plateau. *Earth Planet. Sci. Lett.*, **258**, 293–306.

Geng, Y. (2009). On Distribution of the Yardang in Xinjiang, 286–290.

Goudie, A. S. (2004). Tafoni. *Encyclopedia of Geomorphology*, **2**, 1034–1035.

Goudie, A. S. (2007). Mega-Yardangs: A global analysis. *Geogr. Compass*, **1**, 65–81.

Greeley, R. and Iversen, J. (1985). Wind as a geological process, no. 4, In *Cambridge Planetary Science Series*. Cambridge University Press, New York, NY 3, 3.2.

Halimov, M. and Fezer, F. (1989). Eight yardang types in Central-Asia. *Zeitschrift fur Geomorphologie*, **33**, 205–217.

Han, W., Fang, X., Ye, C., Teng, X. and Zhang, T. (2014). Tibet forcing Quaternary stepwise enhancement of westerly jet and central Asian aridification: Carbonate isotope records from deep drilling in the Qaidam salt playa, NE Tibet. *Glob. Planet. Change*, **116**, 68–75.

Handford, C. R. (1991). Chaper 1 Marginal Marine Halite: Sabkhas and Salinas, *Developments in Sedimentology*, Elsevier, pp. 1–66.

Huang, H., Huang, Q. and Ma, Y. (1996). *Geology of Qaidam Basin and Its Petroleum Prediction.* Geological Publishing House: Beijing.

Huang, Q., Ku, T. L. and Phillips, F. M. (1993). Evolutionary characteristics of lakes and paleoclimatic undulation in the Qaidam Basin, China. *Chin. J. Oceanol. Limnol.*, **11**, 34–45.

Jin, Z. and Zhang, B. (1997). Geological Map of the Qaidam Basin, (1:500000).

Kapp, P., Pelletier, J. D., Rohrmann, A., Heermance, R., Russell, J. and Ding, L. (2011a). Wind erosion in the Qaidam basin, central Asia: Implications for tectonics, paleoclimate, and the source of the Loess Plateau. *GsA Today*, **21**, 4–10.

Kapp, P., Pelletier, J. D., Rohrmann, A., Heermance, R., Russell, J. and Ding, L. (2011b). Wind erosion in the Qaidam basin, central Asia: Implications for tectonics, paleoclimate, and the source of the Loess Plateau. *GSA Today*, **21**, 4–10.

Kong, F. J., Kong, W. G., Hu, B. and Zheng, M. P. (2013). Meterological Data, Surface Temperature and Moisture Conditions at the Dalantan Mars Analogous Site, in Qinghaitibet Plateau, China, Lunar and Planetary Science Conference, p. 1743.

Kong, F. J. K., W. G.; Hu, B., and Zheng, M. P. (2013). Meterological Data, Surface Temperature and Moisture Conditions at the Dalantan Mars Analogous Site, in Qinghaitibet Plateau, China, 44th Lunar and Planetary Science Conference, 1743, 1719.

Laity, J. E. (2011). Wind erosion in drylands, *Arid Zone Geomorphology: Process, Form and Change in Drylands*, Third Edition, 539–568.

Li, J., Dong, Z., Qian, G., Zhang, Z., Luo, W., Lu, J. and Wang, M. (2016). Yardangs in the Qaidam Basin, northwestern China: Distribution and morphology. *Aeolian Res.*, **20**, 89–99.

Liu, H., Chai, H., Cheng, W., Zhong, D. and Zhou, C. (2008). A research of aeolian landform in northern China based on remote sensing imagery. *Geogr. Res.*, **27**, 108–118.

Lowenstein, T. K. and Hardie, L. A. (1985). Criteria for the recognition of salt-pan evaporites. *Sedimentology*, **32**, 627–644.

McCauley, J., Grolier, M. and Breed, C. (1977). Yardangs. *Geomorphology in Arid Regions*, 233–269.

Mustoe, G. (1982). The origin of honeycomb weathering. *Geolog. Soc. Am. Bull.* **93**, 108–115.

Neal, J. T., Langer, A. M. and Kerr, P. F. (1968). Giant desiccation polygons of Great Basin playas. *Geolog. Soc. America Bull.*, **79**, 69–90.

Nield, J. M., Wiggs, G. F., King, J., Bryant, R. G., Eckardt, F. D., Thomas, D. S. and Washington, R. (2016). Climate–surface–pore-water interactions on a salt crusted playa: Implications for crust pattern and surface roughness development measured using terrestrial laser scanning. *Earth Surf. Proc. Land.*, **41**, 738–753.

Niu, Q., Qu, J., Li, X., Dong, G., Zhang, K. and Han, Q. (2011). Review and prospect of Yardangs Landforms research. *Adv. Earth Sci.*, **26**(5), 516–527.

Plummer, C. C., Carlson, D. H., Hammersley, L. (2016). *Physical Geology*. New York, NY: McGraw-Hill/Education, Inc.

Ren, J. (1996). Physical geographical characteristics of Qaidam region. *J. Qinghai Environ.*, **21**, 149–150.

Rohrmann, A., Heermance, R., Kapp, P. and Cai, F. (2013). Wind as the primary driver of erosion in the Qaidam Basin, China. *Earth Planet. Sci. Lett.*, **374**, 1–10.

Song, T. and Wang, X. (1993). Structural styles and stratigraphic patterns of syndepositional faults in a contractional setting: Examples from Quaidam Basin, Northwestern China. *AAPG bulletin*, **77**, 102–117.

Tapponnier, P., Zhiqin, X., Roger, F., Meyer, B., Arnaud, N., Wittlinger, G. and Jingsui, Y. (2001). Oblique stepwise rise and growth of the Tibet Plateau. *Science*, **294**, 1671–1677.

Tucker, R. M. (1981). Giant polygons in the Triassic salt of Cheshire, England: A thermal contraction model for their origin. *J. Sediment. Res.*, **51**, 779–786.

Wang, E., Xu, F.-Y., Zhou, J.-X., Wan, J. and Burchfiel, B. C. (2006). Eastward migration of the Qaidam basin and its implications for Cenozoic evolution of the Altyn Tagh fault and associated river systems. *Geological Soc. America Bull.*, **118**, 349–365.

WEI, K.-H. (2013). Study on open Terme Rick Mine Landform Evolution Model and Morphological Differences. *Science Technology and Engineering*, **13**, 1671–1815.

Wu, G., Hu, S., Zhang, Z., Zhao, H. and Fang, X. (1985). The Qaidam Basin. *Journal of Lanzhou University(natural science)*, **1985**, 60–63.

Wu, Z. (2009). *Deserts and Desertification Control in China*, Beijing: Science Press, 714.

Wu, Z., Yang, Y., Barosh, P. J., Wu, Z. and Zhang, Y. (2014). Tectonics and topography of the Tibetan Plateau in Early Miocene. *Acta Geologica Sinica-*English Edition, **88**, 410–424.

Xia, W., Zhang, N., Yuan, X., Fan, L. and Zhang, B. (2001). Cenozoic Qaidam basin, China: A stronger tectonic inversed, extensional rifted basin. *AAPG Bull.*, **85**, 715–736.

Xiao, L., Wang, J., Dang, Y., Cheng, Z., Huang, T., Zhao, J., Xu, Y., Huang, J., Xiao, Z. and Komatsu, G. (2017). A new terrestrial analogue site for Mars research: The Qaidam Basin, Tibetan Plateau (NW China). *Earth-Sci. Rev.*, **164**, 84–101.

Xin, Y. (1995). Windblown Sand In Salt Lakes In Qaidam Basin. *J. Desert Res.*, **15**, 252–255.

Yang, J., Xu, Z., Zhang, J., Chu, C.-Y., Zhang, R., Liou, J.-G., Hendrix, M. and Davis, G. (2001). Tectonic significance of early Paleozoic high-pressure rocks in Altun-Qaidam-Qilian Mountains, northwest China. *Memoirs-Geol. Soc. Am.*, 151–170 doi: 10.1130/0-8137-1194-0.151.

Yin, A., Dang, Y.-Q., Zhang, M., Chen, X.-H. and McRivette, M. W. (2008). Cenozoic tectonic evolution of the Qaidam basin and its surrounding regions (Part 3): Structural geology, sedimentation, and regional tectonic reconstruction. *Geol. Soc. Am. Bull.*, **120**, 847–876.

Yin, A., Dang, Y., Zhang, M., McRivette, M. W., Burgess, W. P., and Chen, X. (2007). Cenozoic tectonic evolution of Qaidam basin and its surrounding regions (Part 2): Wedge tectonics in southern Qaidam basin and the Eastern Kunlun Range. *Geol. Soc. Am. Spec. Pap.*, **433**, 369–390.

Yin, A., Rumelhart, P., Butler, R., Cowgill, E., Harrison, T., Foster, D., Ingersoll, R., Qing, Z., Xian-Qiang, Z. and Xiao-Feng, W. (2002). Tectonic history of the Altyn Tagh fault system in northern Tibet inferred from Cenozoic sedimentation. *Geol. Soc. Am. Bull.*, **114**, 1257–1295.

Young, A. R. (1987). Salt as an agent in the development of cavernous weathering. *Geology*, **15**, 962–966.

Yu, L. and Lai, Z. (2012). OSL chronology and palaeoclimatic implications of aeolian sediments in the eastern Qaidam Basin of the northeastern Qinghai-Tibetan Plateau. *Palaeogeography, Palaeoclimatology, Palaeoecology*, **337**, 120–129.

Yu, X., Guo, Z., Zhang, D. and Zhang, Y. (2018). Quaternary shaking: Evidences of tectonic intensification from the Qaidam Basin and its influence on multi-spherical interaction. *Quatern. Sci.*, **38**(1), 39–53.

Zhang, J. and Liu, E. (1985). Hydrological characteristics of streams in Qaidam Basin. *J. Geograph. Sci.*, **52**(3), 242–255.

Zhang, P.-Z., Shen, Z., Wang, M., Gan, W., Bürgmann, R., Molnar, P., Wang, Q., Niu, Z., Sun, J. and Wu, J. (2004). Continuous deformation of the Tibetan Plateau from global positioning system data. *Geology*, **32**, 809–812.

Zhang, W., Fang, X., Song, C., Appel, E., Yan, M. and Wang, Y. (2013). Late Neogene magnetostratigraphy in the western Qaidam Basin (NE Tibetan Plateau) and its constraints on active tectonic uplift and progressive evolution of growth strata. *Tectonophysics*, **599**, 107–116.

Zhu, Z., Wu, Z., Liu, S. and Di, X. (1980). *The General Induction to Chinese Desert*. Science Press: Beijing.

https://doi.org/10.1142/9789811224027_0004

Chapter 4

Lakes

Jiannan Zhao[*] and Ting Huang[*,†,‡]

[*]*Planetary Science Institute, School of Earth Sciences, China University of Geosciences, Wuhan 430074, PR China*

[†]*State Key Laboratory of Lunar and Planetary Sciences, Macau University of Science and Technology, Macau, PR China*

[‡]*CNSA Macau Center for Space Exploration and Science, Macau, PR China*

4.1 Introduction

Lakes are water bodies formed from depressions on the surface of the Earth, accounting for 1.8% of the world's land area. The size and shape of lakes vary greatly. The largest lake on Earth is the Black Sea, with an area of 371,000 km^2, while the deepest one is the Lake Baikal, with a depth of 1,740 m.

Large lake basins are mostly formed by tectonic movements. As an example of the third pole of Earth, the strong uplift of the Qinghai–Tibet Plateau, with its towering mountains, has prevented the moisture carried by the southwest and southeast monsoons from invading the plateau, making the whole plateau cold and dry. This leads to the shrinking and disintegrating of the lakes and the decomposition of some large lakes into several small ones. Other lake basins formed from different dynamic geological processes have smaller sizes generally. Such as lake basins formed by ponding of crater water, lava or landslide (barrier lakes), rivers' geological processes (oxbow lakes), glacial processes (glacial lakes), wind process (aeolian lakes), groundwater geological process (dissolution lakes), and marine geology (lagoons).

Lake water mainly comes from atmospheric precipitation, surface water, and groundwater, followed by melting water from ice and snow and seawater. Different lakes have different sources of water, and they are mainly influenced by climate and location. The composition of lake water is mainly influenced by the climate and the lithology of the lake. Based on their salinity, lakes are divided into fresh and brackish lakes. Among them, the lakes that crystallize when the salt reaches supersaturation are called dry salt lakes. The chemical composition of the lake is not fixed all the time. With the change of tectonic movement, climate and other factors, especially the change of ratio between rainfall and evaporation, the fresh lake can be transformed into a brine lake. On the contrary, the brine lake can be diluted into a fresh lake. In the later stage of a lake's development, it dries up and becomes an anhydrous depression or lacustrine plain due to evaporation, reduction of river water injection, a burst of levee, or excessive filling of debris.

Lacustrine deposits include detrital, chemical, and biological deposits. Clastic sediments are mainly deposited in lakeside shoals and deltas, which generally have obvious stratification and incline to the center of the lake. Chemical deposits, which usually account for a large proportion of lake sediments, consist of phosphorus, silica, and magnesium. Carbonates, sulfates, and chlorides are often abundant in the lacustrine deposits of saltwater lakes in arid regions.

Various biological, physical, and chemical indicators of lake sediments and the evolution of lakes themselves are important data for the restoration of paleoclimate conditions, which also gives lake sediments irreplaceable advantages in the reconstruction of paleoclimate conditions, regional and global environmental comparison, and even the comparative study of paleoclimate conditions with Mars.

4.2 Paleolakes on Mars

Present-day Mars is a desert planet with a cold and dry climate. However, in the early history of Mars, the climate may have been different, and water could have played a significant role in shaping the Martian surface. A great number of landforms that may be related to water have been recognized, including outflow channels, valley networks, deltas, gullies, and paleolakes (Baker, 2006; McSween, 2006; McEwen *et al.*, 2007). This attracts scientists to study whether the planet was a habitable world and to search for signs of life. Paleolake is one of the hottest research fields as it contains information on the ancient aqueous environment of Mars and has been targeted for Martian exploration missions.

The identification and study of paleolakes on Mars began with the analysis of images from Mariner and Viking orbiters. Although the low-resolution images made it difficult to conduct a comprehensive and detailed investigation of paleolakes on the surface of Mars, paleolakes were still considered to exist according to the discovery of valley-like features and the study of Martian physical conditions and climate models (Sagan *et al.*, 1973; Toon *et al.*, 1980; Carr, 1980; Haberle and Jakosky, 1990; Baker *et al.*, 1991; McKay and Davis, 1991; Clifford, 1993). In addition, as impact craters are depressions on the Martian surface, they usually act as a catchment of local water and become lakes. According to this approach, early researchers identified 179 craters and paleolakes on the Martian surface (Goldspiel and Squyres, 1991; De Hon, 1992; Forsythe and Zimbelman, 1995; Forsythe and Blackwelder, 1998; Cabrol and Grin, 1999).

In recent years, with the successful launching of Martian exploration rovers as well as orbiters such as Mars Global Surveyor (MGS), Mars Odyssey, Mars Express, and Mars Reconnaissance Orbiter (MRO), a large number of high-resolution images and spectral data have been obtained, making it possible to make comprehensive investigations and detailed studies of paleolakes. Previous studies have been carried out on the morphology, distribution, mineral compositions, and ages of Martian paleolakes, and the existence of lacustrine environments on early Mars has been confirmed (Carr, 2006).

4.2.1 *Classification and distribution of Martian paleolakes*

Global identification of Martian paleolakes is mainly based on remote-sensing imagery and topographic data. Impact craters whose rims are breached by valleys are usually considered as paleolake basins (Forsythe and Blackwelder, 1998). Besides lacustrine deltas, lake terraces, stratified sediments, and exposed basin floor sediments are also important evidence for the confirmation of paleolakes on Mars. On this basis, Goudge *et al.* (2012a, 2015) identified 431 paleolakes on the Martian surface and our new investigation with high-resolution Context Camera (CTX) on board MRO obtained more than 900 paleolakes on Mars (Fig. 4.1).

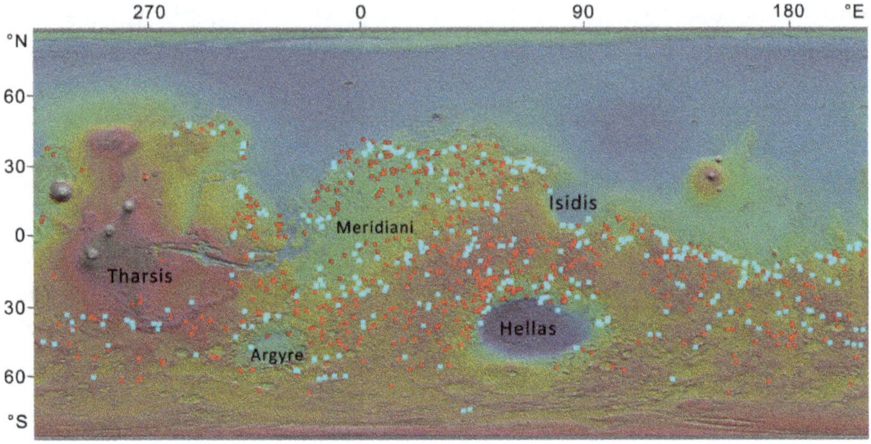

Fig. 4.1. Global distribution of Martian paleolakes. Blue and red dots denote open-basin and closed-basin lakes, respectively. The background is MOLA colorized topography map.

On Earth, the classification of lakes can be based on the physical, chemical, and biological processes, regional distribution, formation, and evolution of lakes (Allen and Collinsen, 1986; Dou *et al.*, 1996; Jiang *et al.*, 1998). However, our understanding of the physical and chemical environment of Mars during its geological history is limited, and there is no direct evidence of life. Therefore, researchers classified the paleolakes mainly based on geomorphology by analyzing the obtained remote-sensing data. For example, according to the formation mechanisms of lake basins, they can be divided into tectonic lake, volcanic lake, and impact crater lake (Tokano, 2005). In addition, classification based on the relative position of lake basins and valleys (De Hon, 1992) and hydrodynamic systems (Cabrol and Grin, 1999) have also been proposed. Among them, the most widely used classification scheme is based on the hydrologic system composed of lakes and their connecting valleys, which divides the paleolakes into three categories: (1) closed-basin lakes, (2) open-basin lakes, and (3) lake chain systems (Cabrol and Grin, 1999; Fassett and Head, 2008a) (Fig. 4.2). A closed-basin lake has only inlet valley while an open-basin lake has both the inlet and outlet valley. Lake chains consist of a series of paleolakes that are connected by one or more valleys to form a lake system, and the lake at the end of the system may be an open- or closed-basin lake. Goudge *et al.* (2015) further divided closed-basin lakes into two categories. The first category has short inlet valley (less than 20 km),

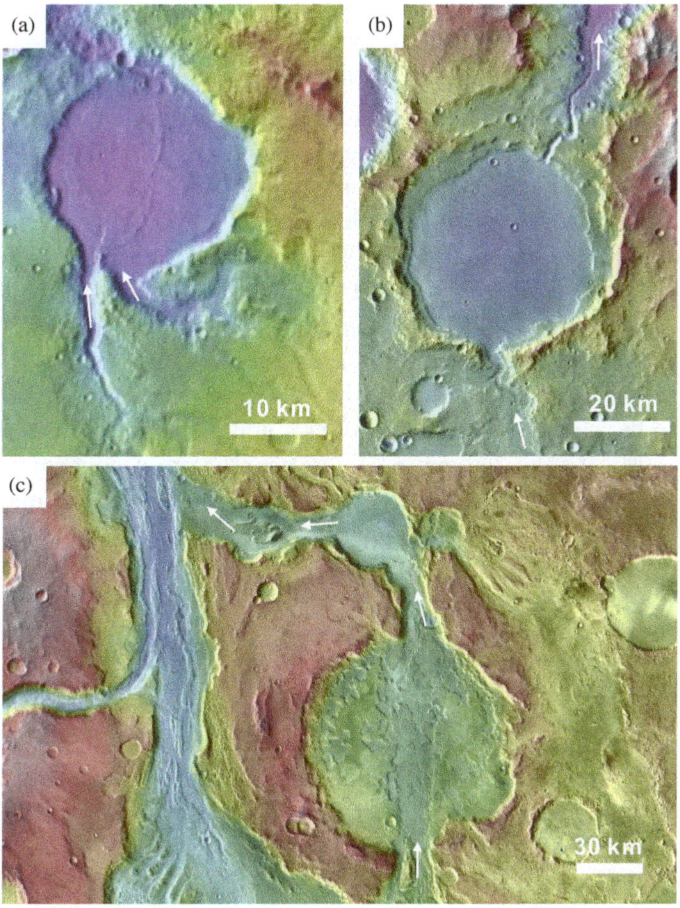

Fig. 4.2. Different types of lakes. (a) Closed-basin lake; (b) Open-basin lake; (c) Lake chain. The images are THEMIS daytime IR mosaics overlain by MOLA colorized topographic map.

and the second category includes paleolakes with inlet valleys longer than 20 km. This classification is based on the consideration that the variation of the length of feeding valleys may be related to different types of water sources.

At present, a total of 566 candidate closed-basin lakes and 356 open-basin lakes have been identified on the Martian surface (Fig. 4.1). By analyzing their distribution, it can be found that about 70% of the paleolakes were distributed between 30°S and 30°N. In addition, the distribution of the paleolakes was also closely related to the geological conditions of different surface units. For example, most paleolakes are located in the southern highlands of Mars, especially near the dichotomy boundary. This may be caused by the fact that a large number of ancient impact craters were preserved in the southern plateau, which provided favorable conditions for water storage. What's more, many valley networks developed near the dichotomy boundary, which provided sufficient water supply for the paleolakes (Cabrol and Grin, 1999). In addition, of the 566 closed-basin paleolakes, about 70% (394) have short inlet valleys, and only about 30% have long inlet valleys. Paleolakes with long and short inlet valleys have roughly the same distribution area and are also roughly consistent with the distribution of open-basin lakes. However, special regions were also observed, such as in the Arabia Terra and Xanthe Terra where paleolakes with short inlet valleys are significantly more concentrated than other types of paleolakes. Goudge *et al.* (2015) argue that this may be due to an intense activity of groundwater in these regions, but overland flows could be more likely to form the short inlet valleys.

4.2.2 *Martian paleolake morphology*

As Martian paleolakes are mostly developed in impact craters, they usually have circular shape and greatly varied diameters, ranging from several kilometers to hundreds of kilometers. There are also some sedimentary landforms developed in the lake basins which are similar to those in terrestrial lakes, including lacustrine deltas, terraces, layered deposits, exposed lake floor deposits, etc.

Lacustrine deltas (Fig. 4.3) are formed by deposition of detrital materials carried by water into the stable water body of the lake (Cabrol and Grin, 1999). Cabrol and Grin (2001) identified 75 possible deltas using Viking image data and classified them into three categories based on their shapes: fan-shaped, elongated, and lobate. The morphology of the delta can reflect the particle size of materials carried by the water, hydrodynamic conditions, and environmental conditions during the formation of the delta, which is of great significance for the identification of paleolakes and the discussion of their formation environment (Tokano, 2005). For example, some deltas on Mars have large surface areas (up to 250 km^2 at the end of the Ma'adim Vallis), indicating that the delta formed under relatively stable environmental conditions. On Earth, such large deltas take thousands of years to form. Thus, the presence of large deltas on Mars

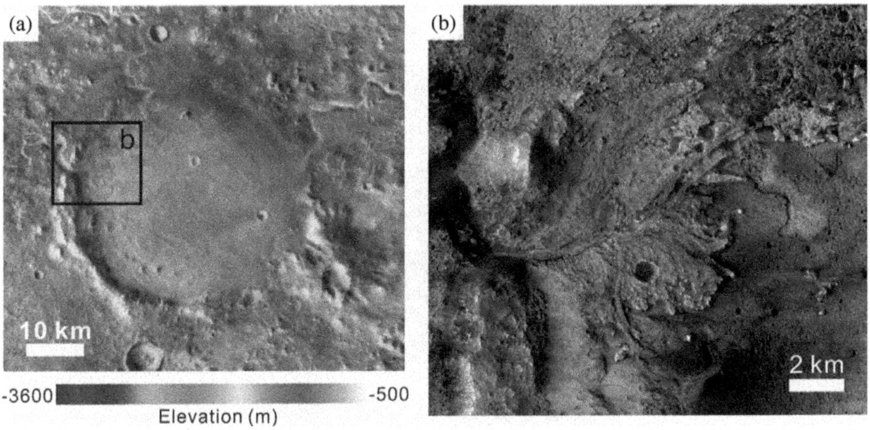

Fig. 4.3. Jezero Crater (a) THEMIS daytime IR mosaics overlain by MOLA colorized topographic map and a lacustrine delta in it; (b) CTX image ID: D14_032794_1989_XN_18N282W.

may indicate that large lakes were once stable on the surface. However, it should be noted that not all the fan deposits are formed by water, as they may also be formed by mass wasting and wind. It is necessary to carefully distinguish the morphological differences between these fan deposits and lacustrine deltas.

Terraces (Fig. 4.4(a)) generally occur near the edges of paleolakes, and may be formed by erosion of crater floor by water and the transportation of eroded materials to the edge of the basin (Carr, 2006). However, some terraces are formed by the collapse of the materials near the rim of craters rather than water activity (Tokano, 2005; Carr, 2006). Leverington and Maxwell (2004) suggested that the terrace in craters may also be formed by magma retreat in magma pools. Therefore, it is necessary to use high-resolution image data to analyze the terraces and their related geomorphological features in detail to distinguish the terraces from other similar landforms. In general, continuous and smooth annular terraces in impact craters may be related to the erosion of running water.

With the availability of high-resolution images, layered deposits on the surface of Mars have been widely found (Fig. 4.4(b)). The formation of layered deposits may be related to aeolian, glacial, volcanic, or fluvial activities (Tokano, 2005), thus it is necessary to study the morphology and distribution of these sediments in detail to make sure they have a lacustrine origin. In general, lacustrine layered deposits have the following

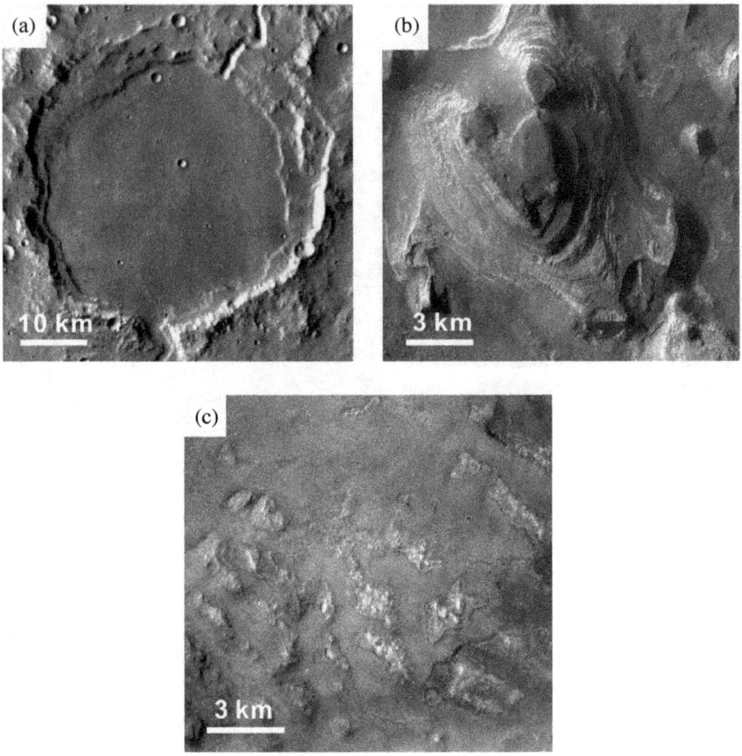

Fig. 4.4. Typical sedimentary landforms in paleolakes. (a) Terraces (174.1°W, 14.2°S, THEMIS Daytime IR mosaic); (b) Layered deposits (located in Terby Crater, 74.3°E, 27.5°S, CTX image: B02_010259_1522_XI_27S285W); (c) Exposed floor materials (1.8°W, 4.4°S, CTX image: P11_005291_1758_XN_ 04S001 W).

characteristic: firstly, these deposits should be located below the elevation of the inlet valley as well as other post-lacustrine stratigraphic units; secondly, the outcrops of layered deposits should be continuous throughout the whole paleolake basin; thirdly, the deposits should only be distributed within the basin (Goudge *et al.*, 2012a). By analyzing the morphological features and mineral compositions of the layered deposits, we can acquire the volume of sediments in the lake basin and estimate the water level during the active period of the lake. At the same time, the cyclicity of the activity of paleolakes can also be reflected, revealing the changes of paleoclimate and paleoenvironment (Cabrol and Grin, 2002). Gale Crater, for example, hosts layered deposits that have been considered as an evidence of paleolakes. The composition of the layered deposits exhibits changes from clay-rich strata to sulfate-rich strata, which could indicate a climate change from warm and wet to cold and dry around the boundary of Noachian and Hesperian (~3.7 Ga) (Milliken *et al.*, 2010). The *in situ* exploration of the layered deposits in Gale Crater by Curiosity Rover also indicates that there were paleolakes in the crater periodically, and the cycle of the lake activity ranged from 100 to 10,000 years (Grotzinger *et al.*, 2015).

Exposed basin floor deposits are light-colored, hilly, eroded materials at the bottom of a paleolake basin, and they are generally surrounded by or below the units where surface modification occurs (Goudge *et al.*, 2012a; Fig. 4.4(c)). The evidences for a lacustrine origin of these materials are as follows:

- These sediments are generally under and often surrounded by any other stratigraphic unit that has undergone later modification;
- They are generally bright to medium tone and may be sulfates formed by evaporation (Glotch *et al.*, 2006; Noe Dobrea *et al.*, 2006);
- These sediments were only distributed within the crater (Goudge *et al.*, 2012a).

However, care should be taken to distinguish the central peak or central peak ring materials in the impact crater to avoid confusion with basin floor deposits.

4.2.3 *Mineralogy of lake deposits*

The study of the mineral composition of paleolakes on Mars mainly depends on the analysis of remote-sensing spectral data. Currently available spectral data are mainly derived from the Thermal Emission Spectrometer (TES, 3 km/pixel) on board the MGS, the Visible and Infrared Mineralogical Mapping Spectrometer (OMEGA, 0.3–4.8 km/pixel) on board the Mars Express, the Thermal Emission Imaging System (THEMIS, 100 m/pixel) on board the Mars Odyssey, the Compact Reconnaissance Imaging Spectrometer for Mars (CRISM, 18 or 36 m/pixel) on board the Mars Reconnaissance Orbiter. The compositions obtained by these orbiters can be mutually verified with the results from *in situ* exploration landers and rovers.

Previous results identified hydrated minerals in only 20 paleolakes of more than 100 confirmed paleolakes with lake sediments (Goudge *et al.*, 2012a, 2015). The rest of the paleolakes have neither obvious differences in mineral composition with their surrounding areas nor special mineral compositions such as aqueous minerals. The possible reasons are as follows: first, high-spectral-resolution CRISM data only have limited coverage and many paleolakes do not have CRISM data; second, thin layers of dust or other mantling materials covered the underneath hydrated minerals and evaporates so that their spectral characteristics were not detected; third, lake deposits on Mars are composed of minerals different from those on Earth. Aqueous minerals that have been detected in lake deposits are mainly Fe/Mg-smectite, kaolinite, hydrated silica, and evaporates (Ehlmann *et al.*, 2008a, 2008b, 2009; Mustard *et al.*, 2008; Dehouck *et al.*, 2010; Ansan *et al.*, 2011; Zhao *et al.*, 2020). However, in general, the distribution of the minerals is relatively limited, and they could be detrital in origin rather than formed *in situ* (Ehlmann *et al.*, 2008a, 2008b). This may indicate Martian lakes were active for a relatively short period of time, making it difficult for carbonate and other evaporates forming in the paleolakes. Previous studies on the morphology of Martian paleolakes also suggested that these lakes existed for a relatively short time (e.g. Howard *et al.*, 2005; Irwin *et al.*, 2005). Another possibility is that special chemical properties of water in paleolakes may make it difficult to form common hydrated minerals and evaporates *in situ*, but this hypothesis is difficult to verify by current available data.

In recent years, Mars rovers such as Opportunity, Spirit, and Curiosity have provided valuable *in situ* data support for the study of paleolakes. Opportunity landed in Meridiani Planum, and discovered hydrogenic sediments, evaporates such as sulfates and chlorides, and "blueberry-like" materials that are rich in hematite (Christensen *et al.*, 2004; Squyres *et al.*, 2006). Besides, the rover identified phyllosilicates (mainly Fe-smectite) in Endeavor Crater, confirming the existence of water in the exploration region (Arvidson *et al.*, 2014). Spirit landed in Gusev Crater, a typical closed-basin lake on Mars (Grin and Cabrol, 1997), and found sulfates, chlorides, hematite, and goethite in the Columbia Hills, indicating that aqueous alteration process had occurred in the crater. Compared with results from CRISM data, Spirit found more hydrated minerals, indicating that the Martian crustal materials underwent more widespread aqueous alteration (Arvidson *et al.*, 2008; Morris *et al.*, 2008). Gale Crater is also a closed-basin lake (Cabrol *et al.*, 1999) and layered deposits containing hematite, clay minerals, and evaporates were detected in the crater by remote sensing (Thomson *et al.*, 2011; Fraeman *et al.*, 2013; Ehlmann and Buz, 2015). The *in situ* exploration of curiosity rover found lacustrine mudstone, sulfates, goethite, Fe/Mg-smectite, and a variety of sedimentary landforms, which confirmed the remote-sensing detection results (Grotzinger *et al.*, 2014, 2015; Vaniman *et al.*, 2014; Treiman *et al.*, 2016). In general, *in situ* exploration acquires more diverse and precise mineral detection results while remote-sensing detection has larger coverage and can obtain major minerals with high concentrations.

4.2.4 *Age of Martian paleolakes*

The chronology of paleolakes is of great significance for understanding the evolution of aqueous environments on Mars. Cabrol and Grin (1999, 2001) used stratigraphic methods and crater statistics to obtain the approximate age range of Martian paleolakes. Their stratigraphic analysis showed that most of the craters were formed between the middle and late Noachian (about 3.9–3.7 Ga.), which is the upper age limit of the paleolakes. According to the statistics of newly formed impact craters on the floor of paleolakes, it is found that most of the paleolakes underwent post-lacustrine modification events (such as the modification of volcanism and glaciation) from the Hesperian to the early Amazonian Period (about 3.7–1.4 Ga.). However, due to the limitation of the dating methods and lack of high-resolution image data, their results are not accurate.

Fassett and Head (2008b) used the high-resolution image data from High Resolution Stereo Camera (HRSC) on board Mars Express and THEMIS to study the age of valleys connected to the paleolakes with the buffered crater counting method, and found that the mean age of highland valleys is 3.53 Ga (according to Hartmann's dating methods) or 3.75 Ga (according to Neukum's dating method), indicating that Martian paleolakes are active around the Noachian–Hesperian boundary (Fig. 4.5).

In addition, Fassett and Head (2008a) also obtained the time when paleolakes underwent modification processes by counting the craters on the floor of the lake basins. The time spans from Hesperian to the border of early/middle Amazonian period, which was consistent with the results suggested by Cabrol and Grin (2001). Goudge *et al.* (2012b) dated 30 paleolakes that have undergone volcanic modification and found the modification events happened from the Noachian–Hesperian boundary to the early Amazonian period, but most of the paleolakes were resurfaced

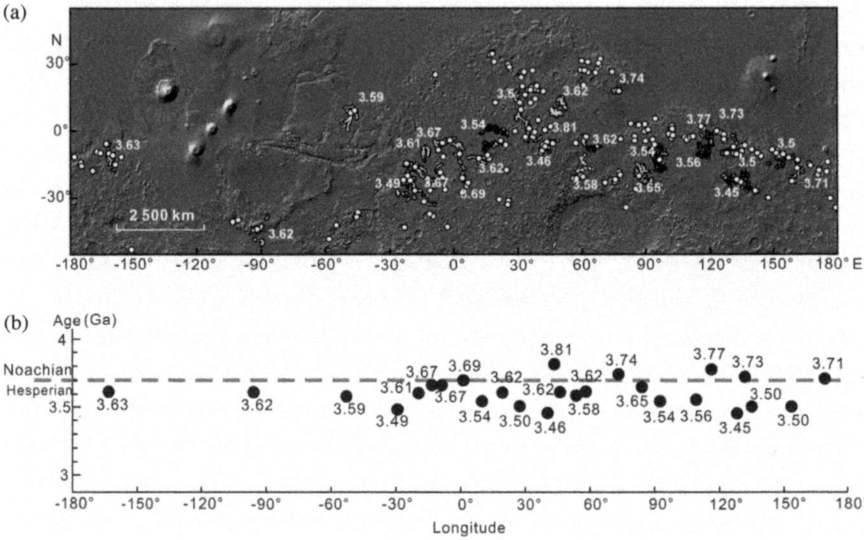

Fig. 4.5. (a) Model ages (Ga) of some paleolakes derived from crater counting results of connecting valleys (Fassett and Head, 2008a). White dots indicate paleolakes. Yellow regions indicate locations of the valley systems examined by crater counting. (b) Age distribution of paleolakes at different longitudes.

in the Hesperian period, which was also the formation period of large-scale volcanic plains on Mars (Scott and Tanaka, 1986; Greeley and Guest, 1987; Head *et al.*, 2002).

Although most of the paleolakes have ages from the Noachian to the early Hesperian, there could be still some paleolakes active in the Amazonian according to studies with high-resolution remote-sensing data in recent years (Adeli *et al.*, 2016; Wilson *et al.*, 2016). Howard and Moore (2011) proposed that Newton and Gorgonum basins on Mars might have hosted lakes during the boundary of Hesperian and Amazonian (~3 Ga.). Wilson *et al.* (2016) discovered nearly 10 paleolakes in the Arabia Terra that may have been active in the Amazonian.

4.2.5 *Origin of Martian paleolakes*

The origin and development of Martian paleolakes were controlled by the water supply. A variety of water sources have been proposed in previous studies, such as precipitation, snowmelt water, and groundwater (Clifford and Parker, 2001; Hanna and Phillips, 2005; Fassett and Head,

2008a; Goudge *et al.*, 2016, 2018). Fassett and Head (2008a) suggested that the water source of paleolakes is related to the elevation of the lake. Surface runoff formed by atmospheric precipitation provides water for the paleolakes with higher elevations, while groundwater could be the water source of the paleolakes with lower elevations. Different water sources can be reflected by the inlet valley morphology of paleolakes. For example, paleolakes supplied by groundwater generally have short but deep-incised inlet valleys with alcove-like terminations (Malin and Carr, 1999; Harrison and Grimm, 2005; Lapotre and Lamb, 2018). Goudge *et al.* (2015) found that paleolakes with short inlet valleys are significantly more concentrated than other types of paleolakes in the Arabia Terra and Xanthe Terra, suggesting that there could be strong groundwater activities in the region as short inlet valleys are indicators of groundwater sapping. In addition, Late Noachian to Early Hesperian age supervolcanoes identified by Michalski and Bleacher (2013) in the Arabia Terra could be triggers for the release of groundwater that fed the paleolakes. Wilson *et al.* (2016) studied the inlet valleys of the recognized Amazonian paleolakes and found that these valleys are relatively narrow, shallow, and relatively late in formation time. This indicated that snowmelt may have provided water source for the above-mentioned young paleolakes.

4.3 Lakes and Playas in the Qaidam Basin

The Qaidam Basin in northwest China is very dry today, with annual precipitation ranging from 16 to 190 mm and annual evaporation varying between 1974 and 3183 mm (Jin *et al.*, 2013). However, the basin developed a paleolake in the Tertiary Period. From the early middle Cenozoic to the late Pliocene (23.03–2.58 Ma), the lakes gradually expanded and showed a tendency of spreading from northwest to southeast. During this period, the deepest lake reached 125 m and the area covered 59,000 km^2. Subsequently, due to the dual effects of climate change and tectonics, the ancient lakes gradually dried up and split into many small lakes, and over time the water-containing lakes eventually became dried salt lakes, also known as salt playas (Kezao and Bowler, 1986).

In the Qaidam Basin, loose lacustrine sediments, lake evaporation products (e.g. salt crusts), and water bodies occupy 15.17%, 3.18%, and 0.78% of the land cover area, respectively (Ma *et al.*, 2008). At present, there are 26 water-containing lakes (one of which is a fresh lake, the others are brine lakes) and 7 large salt playas: Dalangtan, Chahanslatu, Khuntei, Mahai, Huahaizi, Yiliping, and Qarhan.

4.3.1 *Lakes in the Qaidam Basin*

4.3.1.1 *Distribution and morphology of lakes in the Qaidam Basin*

The Qaidam Basin is one of the most high-plateau-lake intensive-distribution regions in China. The neotectonic movement from the late Mesozoic to Cenozoic raised the Qilian and Kunlun Mountain systems strongly, and the middle zone settled and formed many lakes due to low-lying land.

Most of the lakes in the basin are ancient lakes left over from the Tertiary Period, which has created favorable conditions for the formation of salt lakes in the closed natural geographical environment in the inland for a long time. With an elevation of 2,675–3,171 m, this area is one of the highest salt-lake regions on Earth. Most of the lake water is very shallow, generally tens of centimeters, which is conducive to evaporation. As a result, the water has very high salinity. The average salt content is more than 300 g/l, nearly 10 times than the salt content of seawater. Salt playas with no surface water or less surface water are widely distributed in the Qaidam Basin, with a total area of 10,000 km^2, accounting for about one-tenth of the basin bottom area (Lu, 2014).

The trend and arrangement of lakes in the basin are consistent with the tectonic direction of the basin. A series of east–west direction large lakes are formed in the eastern subsidence area, such as Huobuxun Lake, Dabuxun Lake, and Taijinaier Lake. In the north, a series of small inter-mountain basins are distributed with lakes of different sizes, such as Sugan Lake, Daqaidam Lake, Keluke Lake, and Tuosu Lake. Due to the flowing factors: dry climate, vigorous evaporation, lake's shrinking, and salt concentration, many brackish lakes gradually evolved into brine lakes, finally developed to be salt playas. While the lake surface is decreasing, there is a very common migration phenomenon: the brine lakes in the middle of the basin move to the settlement center (Dabuxun Lake), and those at the edge of the basin move to the interior.

There are 26 lakes in the Qaidam Basin (listed in Fig. 4.6 and Table 4.1). Some of the representative lakes are introduced as follows.

(1) East and West Taijinaier Lakes: The East and West Taijinaier Lakes (Fig. 4.7(a)) are located in the middle of the Qaidam Basin. The two lakes

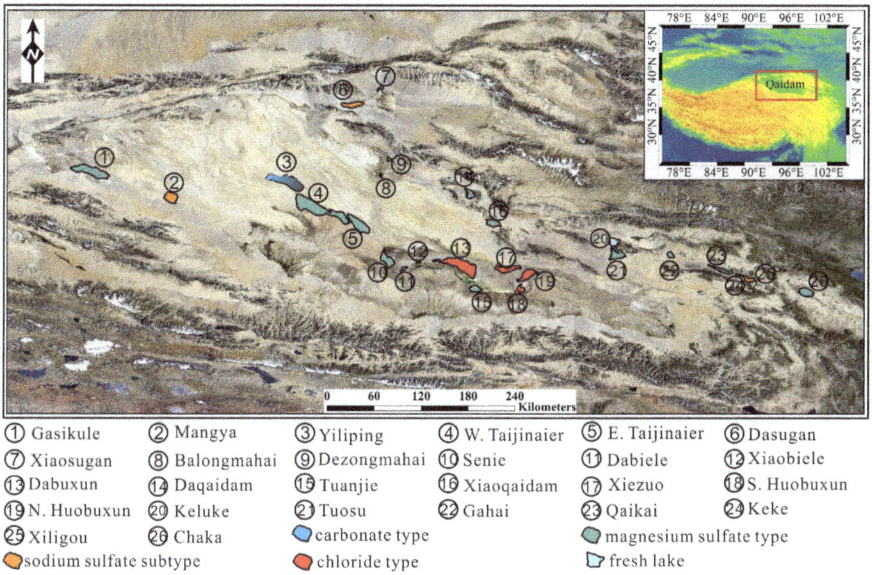

Fig. 4.6. The lakes in the Qaidam Basin (Xiao *et al.*, 2017).

are adjacent to each other and were originally the same lake. However, they were separated into independent lakes after the lake retreated. The periphery of the lake basin is composed of Quaternary alluvial, aeolian, and lacustrine clay, as well as chemical deposits. Most of the lakesides are deposited with lacustrine and chemical sediments, and saline-alkali lands are widely distributed. The eastern part of E. Taijinaier Lake is a sandy salt playa of approximately 50.0 km^2, with a thickness of 15.0–20.0 m, and a lake bed sediment thickness of 2.0–3.0 m. The water level is 2,681.00 m, the length is 24.0 km, the maximum width is 8.7 km, the average width is 5.05 km, and the area is 121.3 km^2. The catchment area is 38,500.0 km^2, and the recharge coefficient is 156.0 (including W. Taijinaier Lake). The lake water mainly relies on the supply of the E. Taijinaier which is derived from the Nalenggele River and the Hongshui River, and originates from the Kunlun Mountain in the south.

In addition, yardangs developed in the E. Taijinaier lake (Fig. 4.7(b)). According to the study by Mao *et al.* (2018), the strata of semi-submerged yardang landforms in the study area belongs to the Shangyoushashan Formation, and the ostracod assemblages indicate that the formation age is late Miocene. The analysis of sedimentary facies and sporo-pollen

Table 4.1. The general characteristics of salt lakes in the Qaidam Basin (Xiao *et al.*, 2017).

No.	Name	Center coordinate	Area (km^2)	Altitude (m)	Hydrochemical type	Degree of mineralization (g/l)	pH
1	Gasikule	90°55′E, 38°04′N	103	2853	Magnesium sulfate	333.2	7.6
2	Mangya	91°50′E, 37°46′N	2	2886	Sodium sulfate	41.4	8.7
3	Yiliping	93°09′E, 38°59′N	360	2863	Magnesium sulfate	327.2	7.3
4*	W. Taijinaier	93°22′E, 38°43′N	82	2678	Magnesium sulfate	336.3	7.7
5*	E. Taijinaier	93°55′E, 37°30′N	116	2681	Magnesium sulfate	331.5	7.7
6*	Dasugan	93°55′E, 37°30′N	104	2850	Sodium sulfate	25-32	9.7
7*	Xiaosugan	94°12′E, 39°01′N	12	2691	Carbonate	1.2	9.5
8	Balongmahai	94°13′E, 38°02′N	5	2743	Chloride	313.9	7.4
9	Dezongmahai	94°19′E, 38°14′N	11	2740	Magnesium sulfate	355.2	7.4
10	Senie	94°19E, 37°03N	69–85	2676	Magnesium sulfate	332.2	7.1
11	Dabiele	94°31E, 36°59N	7	2677	Magnesium sulfate	362.9	7.0
12	Xiaobiele	94°39′E, 37°02′N	6	2677	Chloride	386.9	6.2
13	Dabuxun	95°06E, 36°58N	184–334	2765	Chloride	307.6	7
14	Daqaidam	95°10′E, 37°50′N	23–36	3148	Magnesium sulfate	274.4	7.9
15	Tuanjie	95°21′E, 36°48′N	6	2673	Magnesium sulfate	425.3	5.4
16	Xiaoqaidam	95°30′E, 37°31′N	36–40	3147	Magnesium sulfate	131.08	8.83
17	Xiezuo	95°39E, 37°00N	17	2696	Chloride	358.5	5.5
18*	S. Huobuxun	95°47E, 36°44N	33	2670	Chloride	346.1	6.7
19*	N. Huobuxun	95°55E, 36°54N	51–90	2676	Chloride	306.8	6.7
20	Keluke	96°55E, 37°18N	58	2812	/	0.7	8.8
21	Tuosu	96°57E, 37°04N	166	2808	Magnesium sulfate	29.1	8.8
22	Gahai	97°33E, 37°08N	37	2851	Magnesium sulfate	66.58	8.35
23	Qaikai	98°01E, 37°01N	18	2940	Magnesium sulfate	324.1	6.9
24	Keke	98°14E, 36°58N	4	3010	Magnesium sulfate	326.4	6.8
25	Xiligou	98°27E, 36°50N	20	2938	Sodium sulfate	213.3	7.6
26	Chaka	99°07E, 36°33N	104	3059	Magnesium sulfate	337.86	7.56

Notes: *Some modifications of lakes' names have been made with respect to (Xiao *et al.*, 2017). (4) W. Taijinaier = West Taijinaier, (50) E. Taijinaier = East Taijinaier, (6) Dasugan, (70) Xiaosugan, (18) S. Huobuxun = South Huobuxun, (19) N. Huobuxun = North Huobuxun.

Fig. 4.7. (a) Remote-sensing images of the East and West Taijinaier Lakes (Google Earth); (b) Semi-submerged yardang field within the E. Taijinaier Lake.

assemblages reveal that this area was warm and arid climate in late Miocene, and it was covered by paleolake and accepted lacustrine and delta front facies sediments. In the Pliocene, the paleolake shrank and disappeared under the dual role of regional tectonic uplift and arid climate. In the extreme drought and strong monsoon environments, the upper strata of the Shangyoushashan Formation were lifted up to the surface, formed yardang landforms under the action of wind erosion-dominated external forces. From late Pleistocene to early Holocene, due to the arrival of the last interglacial period, the climate turned warm and moist, and the lake invaded to form two lakes. In recent years, the distribution pattern of lakes has been changed under the action of human activities.

(2) Qarhan Lake: Qarhan Lake is a seasonal salt lake located at the end of Golmud River. The lake surface varies with seasons between dry and wet. The area varies from 184 to 1001 km^2. The water depth is generally 1 to 0.4 m, and the maximum depth is 1.7 m. Due to continuous development, the original salt lake has been gradually replaced by man-made salt ponds, forming a famous potash production base in China (Fig. 4.8).

(3) Gasikule Lake and its water source: The Gasikule Lake basin (Fig. 4.9) is a closed inner flow basin, which receives the recharge from

Fig. 4.8. (a) Large brine pool shown on satellite image (central coordinates: 95.28°E, 36.84°N; Google Earth); (b) The dam that separates the salt ponds is several kilometers long; (c) and (d) halite and mirabilite crystal flowers in the salt lake. Adapted from: http://zsq219391.blog.163.com/blog/static/21006810120129129459460/.

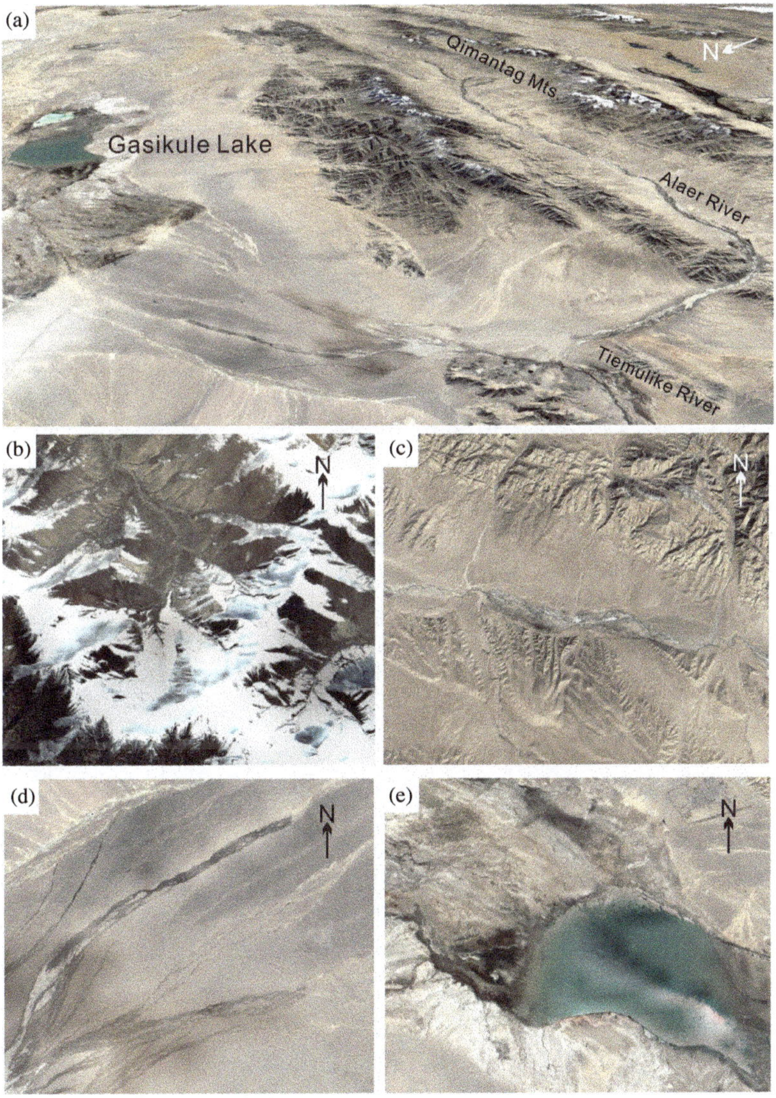

Fig. 4.9. (a) The Gasikule Lake and its water source (Qimantag Mountains) (panel width 80 km). (b) Qimantag Mountains, the water source of the Gasikule Lake (~5500 m, center at 37°26'14.85"N, 90°18'34.03"E, panel width 7.5 km). (c) Seasonal meltwater and rainfall form the upstream water and flow to the west (panel width ~20 km). (d) Middle-downstream, the water flows into the Qaidam Basin, slowly spreads northwest and finally fills the Gasikule Lake. (e) Panel width 30 km (Google Earth).

the Alar River and the Tiemulik River in the western region of the Kumkule Basin. The Gasikule Lake is a superlarge halite lake deposit with the coexistence of solid and liquid phases.

The source of lake Gasikule is seasonal meltwater and rainfall from the Qimantag Mountains in the south (elevation 5,500 m). The western part of the mountains ranges flows northward into the Qaidam Basin and turns northwest, forming the Gasikule Lake. In addition, water from the Gasikule Lake also feeds the Jade Lake (Fig. 4.10), which is located to the east of Gasikule Lake.

(4) Dahallerten River-Sugan Lake and its water source: Located at the foot of the Altyn-Tagh Mountains, Sugan Lake consists of two lakes with a total area of 116 square kilometers (Fig. 4.11). The Sugan Lake is a mountain fault basin with an altitude of 2700–2800 m. The water source mainly comes from the reservoir of surface water in the eastern Halten Basin (Danghe Nanshan and Tuergendaban Mountains). The surface

Fig. 4.10. The Jade Lake, formed by diversion of water from Gasikule Lake.

Fig. 4.11. In the north, the east–west direction is the Altyn-Tagh Mountains, and the right-north-west-south direction is the Danghe Nanshan Mountain in the Qilian Mountains. Both of them provide water for the Sugan Lakes. The two lakes are interconnected. Panel width 200 km.

water is first injected into the Xiao Sugan Lake, and then the Xiao Sugan Lake is injected into the Da Sugan Lake. Therefore, the Xiao Sugan Lake is less salty than Da Sugan Lake.

(5) Daqaidam and Xiaoqaidam Lakes: The water source of Daqaidam and Xiaoqaidam Lakes (Fig. 4.12) comes from the north side of the Zongwulongshan and the Tuergendaban Mountains, which are mainly supplied by seasonal rainfall. The two large lakes are formed in the catchment area. Among them, the river that flows into Xiaoqaidam is longer than that which enters Daqaidam. The large amount of incoming water is the meltwater of the Torgenda snow-capped mountains with a very large catchment area.

The Daqaidam Lake and the Xiaoqaidam Lake belonged to the same big lake in the Quaternary period. Later, due to the dry climate, the lakes retreated. The alluvial fan formed by the foothills of the Korek Mountain emerged from the dew surface and evolved into independent lakes.

(6) Keluke and Tuosu Lakes: Keluke Lake and Tuosu Lake (Fig. 4.13 and 4.14) are collectively known as Talian Lake. “Keluke” is Mongolian,

Fig. 4.12. (a) Daqaidam and Xiaoqaidam Lakes and their water source region (Tuergendaban snow mountain) (Google Earth). Panel width 30 km. (b) Xiaoqaidam Lake is 9 km wide from north to south, 13 km from east to west, and covers an area of 69 km^2. (c) The close view of Xiaoqaidam Lake.

meaning “a place with many grassy beaches and beautiful water plants.” Located in the eastern part of the Qaidam Basin, it is a freshwater lake with clear waters, calm lakes, and beautiful scenery.

Tuosu is Mongolian, which means “ghee lake”. Tuosu Lake is a typical inland saltwater lake surrounded by the vast Gobi Desert. With the high temperature and great evaporation rate of the lake water, the salinity in the water is increased. There are few aquatic animals and planktons.

4.3.1.2 *Hydrochemical characteristics of the lakes*

Among the 26 lakes in the Qaidam Basin, 25 are brine lakes, and only one is a freshwater lake. The differentiation of brine lakes from freshwater

Fig. 4.13. The interconnected lakes Keluke (top) and Tuosu (bottom). Keluke Lake is 2817 m and Tuosu Lake is 2808 m above sea level, respectively. Panel width 25 km.

lakes is adopted as the salinity of 0.5‰ or the mineralization of 0.5 g/l, i.e. freshwater lakes have salinity/mineralization <0.5‰ or <0.5 g/l (Hammer, 1986; Zhao *et al.*, 2010). The mineralization of the lakes in the Qaidam Basin increases from west to east, and the average mineralization is 321.3g/l.

Generally, the contents of Na^+, K^+, Ca^{2+}, Mg^{2+}, SO_4^{2+}, Cl^-, and HCO_3^- in the brine of salt lakes in the basin are relatively high, and the contents of other components are lower than those mentioned above (Table 4.2).

Depending on hydrochemical properties, i.e. using Kurnakov–Valyashko classification, the lakes were identified as different types, they are chloride type, sulfate type (including sodium sulfate and magnesium sulfate subtypes), and carbonate type (Fig. 4.6) (Zheng and Liu, 2009).

Magnesium sulfate type brine lakes are the most widely distributed ones from northwest to southeast. Along the edge of the basin, Mangya and Dasugan Lake near Altyn-Tagh and Kunlun Mountains are sodium

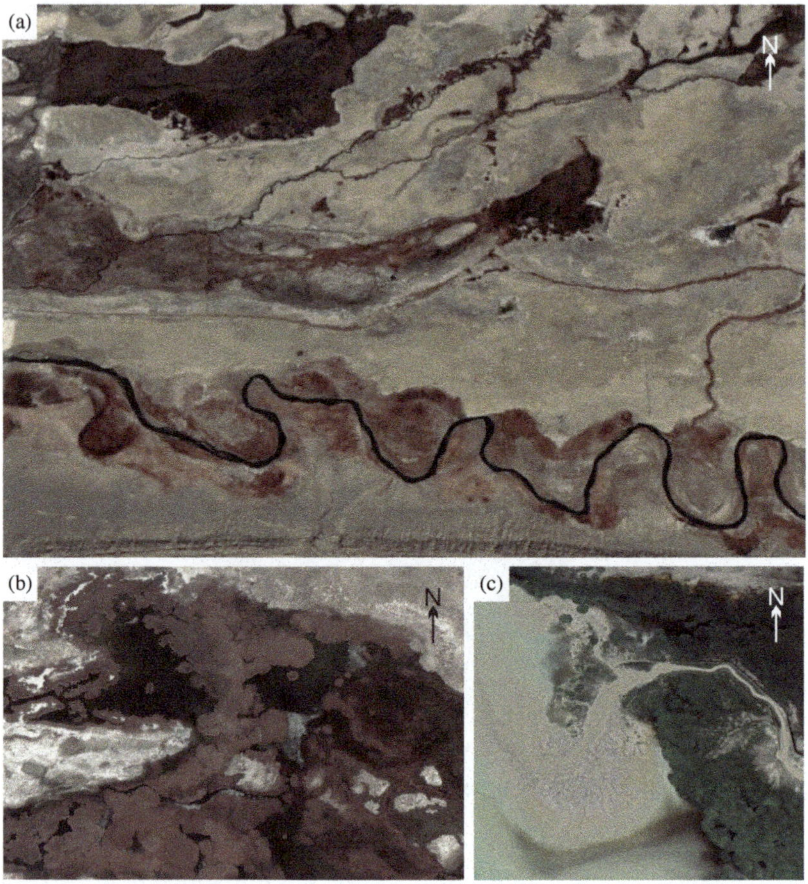

Fig. 4.14. (a) Drainage network flowing into the Keluke Lake (panel width 5 km). (b) Marsh in the winter of the year of 2015 (panel width 1.6 km). (c) A river-dominated delta in the Keluke Lake in summer of 2019 (panel width 1500 m) (Google Earth).

sulfate types, while Xiao Sugan is carbonate type. Chloride type lakes distribute mainly in the middle area of the basin, including Dabuxun, Xiezuo, S. Huobuxun, and N. Huobuxun (Hu, 1990). In general, the mineralization of lake water decreases from chloride to magnesium sulfate, to sodium sulfate and then to carbonate type.

The pH values of brine lakes in the basin are related to both hydrochemical parameters and water mineralization. They tend to decrease as lakes change from carbonate type to sodium sulfate subtype, to magnesium sulfate subtype and to chloride type. While the pH values of the

Table 4.2. Examples of geochemical parameters of lakes in the Qaidam Basin (Yang *et al.*, 2013; Liu *et al.*, 2018).

Lake	Keluke	Erhai	Qinghai	Tuosu	Gahai	Xiaoqaidam	Dabuxun	Chaka
pH	8.8	9.4	9.1	8.8	8.4–8.9	8.4	7.0	7.8
Sulfate (mg L^{-1})	BDL	0.1	BDL	BDL	0.1	0.1	0.2	0.0
DOC* (mg L^{-1})		179.1	29.4	26.7	25.5–31.2	169.7	—	92.0
Concentration of major ions (mg L^{-1})								
K^+	6.5	9.0	269.2	326.9	452.9–613.0	920.0	2163.0	2089.3
Na^+	135.2	218.8	3993.0	8087.0	9384.0–26770.0	54691.0	77950.0	107460.3
Ca^{2+}	38.5	20.0	17.7	32.4	22.8	625.0	1711.0	823.2
Mg^{2+}	53.6	68.5	824.1	2107.0	1467.0–4097.0	2588.7	24700.0	9470.8
SO_4^{2-}	140.4	117.9	2188.0	6454.0	6058.0–10790.0	28735.2	4462.0	17099.8
Cl^-	206.2	231.1	5625.7	13129.5	13034.7–41059.4	72063.4	196231.9	187633.2
CO_3^{2-}	0.0	46.6	405.0	416.0	112.9–515.4	366.1	BDL	BDL
HCO_3^{1-}	149.8	268.5	854.1	767.9	312.2–824.1	380.1	359.6	153.1
NO^{3-}	0.5	0.2	0.4	0.4	0.5–0.6	0.5	1.1	1.4
NH_4^+	1.0	1.2	1.0	0.8	0.5–0.6	0.6	0.4	0.4

DOC: Dissolved Organic Carbon.

lake are generally inversely related to mineralization (Fig. 4.15) (Zheng and Liu, 2009). During the process of concentration and evaporation of lakes' brine water, pH values change in a certain rule. When chloride and magnesium sulfate subtype lakes evaporate, pH value gradually decreases, while when alkali carbonate or borate accounts for a large proportion of brine, pH value gradually increases with evaporating (Wu, *et al.*, 2012).

The formation of brine lakes in the Qaidam Basin is closely related to two kinds of water bodies in the basin: surface river water and deep water. Surface river water and deep water are mixed and doped in different proportions, and the water chemical types and salt-producing ways of water are different, which have a good correspondence with the different brine lake types and salt-producing ways in the basin (Liu *et al.*, 2002). In the short term (e.g. the last 40 years), the changes in the area of modern lakes are mainly affected by the precipitation and runoff in mountainous areas, and other influences include some human

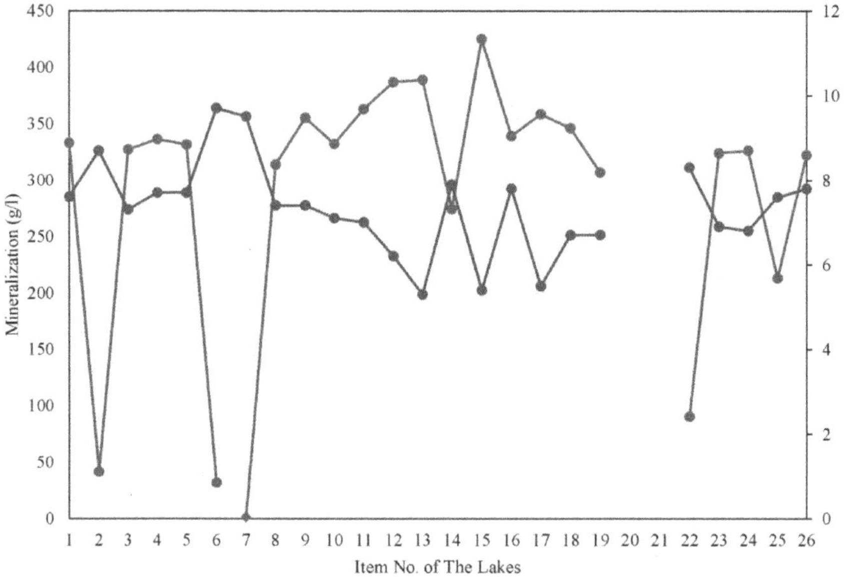

Fig. 4.15. Mineralization and pH values of the lakes have an inverse correlation where lower pH values can be observed with higher mineralization of the water. Data of mineralization and pH values are collected from (Hou *et al.*, 2010a; Xiao *et al.*, 2017; Zheng *et al.*, 2002).

activities, such as social and economic water consumption, artificial barriers between rivers and lakes, and brine mining. A new lake was formed in the southern part of Dabuxun, the area of Tuanjie Lake was significantly expanded, the East and West Taijinaier Lakes gradually shrank and dried up, and the seasonal Yiliping Lake became a large lake in the basin after it was filled with water in the former dry salt playas (Duan, 2018).

The Dasugan Lake is mainly fed by snow and ice meltwater flowing from mountains around with an altitude of over 4800 m. Groundwater in this region is buried at a depth of fewer than 5 m and is subject to strong evaporation. The serious soil salinization is resulted from higher mineralization of the Dasugan Lake than that of the Xiao Sugan Lake whose water source is fed by rainfall. The temperature of the recharge area of Xiao Sugan Lake is higher than that of Dasugan (Hou *et al.*, 2010b). Little destruction is subjected to these salt deposits as the source water is saturated before recharging to the lakes (Hu, 1990).

Salts are important mineral resources which have significance to a nation's economy and agriculture. According to the different types of salt minerals and their application value in salt resources, China's saline lake minerals can be divided into four types: potassium and magnesium type, special type (containing lithium, boron, and potassium), nitrate and halite–mirabilite type, and ordinary type (containing salt, trona, and mirabilite). The percentage of each type was 1.3%, 9.0%, 88.9%, and 0.8%, respectively (Zheng, 2010).

The Qaidam Basin is rich in salt lake deposits. The solid deposits include halite, mirabilite, trona, gypsum, carnallite, sylvite, etc. The saline lakes are known as comprehensive "liquid deposits" because of the high concentration of economic minerals in the brine water (Table 4.3). Potassium, boron, lithium, rubidium, cesium, and other useful chemical elements are relatively low in the Earth's crust. However, they are very rich and closely symbiotic in brines compared to those in the seawater.

The integrative exploration of salts resources benefits regional prosperity immensely. Saline lake is a kind of dynamic resource closely related to the changes in the surrounding climate and environment. Only by accumulating the basic data of the natural changes of the lake area, deepening the understanding of the changes of the lakes, and mastering their change rules can we achieve the goal of rational development, utilization, and protection of the salt lake resources and environment. Besides monitoring hydrology and meteorological system of the lakes, dynamic monitoring system using remote sensing is also fully applied to provide the important scientific basis for rational exploitation and protection.

4.3.2 *Playas in the Qaidam Basin*

In the late stage of brine lake evolution, the concentrated brine is gradually evaporating and precipitating. Because of the different solubility of salts in the water, carbonate which has the least solubility is deposited first, followed by sulfate, and finally chloride (Fan *et al*., 2010). As the Qaidam Basin is a diamond-shaped concentric catchment center, the precipitation in the surrounding mountains provides water source and metallogenic materials for the basin.

The climate of the basin is arid desert climate, with large daily temperature differences, strong wind force and strong UV radiation.

Table 4.3. A comparison of hydrochemical characteristic coefficient values between the Qaidam Basin water and seawater (Yu, 1981).

Water body \ Coefficient value	$\frac{B_2O_3 \cdot 10^3}{\Sigma}$	$\frac{B_2O_3 \cdot 10^3}{K}$	$\frac{B_2O_3 \cdot 10^3}{Mg}$	$\frac{Li \cdot 10^3}{\Sigma}$	$\frac{Li \cdot 10^3}{Mg}$	$\frac{K \cdot 10^3}{\Sigma}$	$\frac{K \cdot 10^3}{Cl}$	$\frac{Rb \cdot 10^3}{\Sigma}$	$\frac{Cs \cdot 10^3}{\Sigma}$
Brine lake in the Qaidam Basin	3	206	34	0.32	3.62	14.65	29.68	0.034	0.000106
Brine lake in the northern Tibetan Plateau	8.6	262.6	374.6	1.58	68.79	32.73	72.01	0.0259	0.0082
Sea water	0.42	38.23	11.42	0.004	0.116	11.07	20.03	0.0057	0.0000106

The Qaidam is an inland environment with strong evaporation rate, and abundant salt deposits have been formed since the Quaternary Period (Wei and Jiang, 1993). The evaporite deposition model is proposed as a "great salty basin encrusted small saline lake" (Jin and Zha, 2000). "Great salty basin" refers to the widely distributed shallow-water environment dominated by calcareous mudstone, marlstone, and siltstone deposition. The "small saline lake" refers to the deep-water environment where most gypsum, halite, and marlstone are deposited. At the same time, it is believed that most of the salts sourced from deep brine enter the basin through active faults, and form evaporite through evaporation, while a small amount of surface water is dissolved and injected. This is the genesis and sedimentary model of the basin from the Paleogene to Neogene period.

4.3.2.1 *Spatial and temporal distribution of dry salt playas in the basin*

There are seven large salt playas distributed in the Qaidam Basin, among which Dalangtan, Chahanslatu, Kunteyi, Mahai, and Huazhaizi are located in the northwest, Yiliping in the central region, and Qarhan in the southwest (Fig. 4.16). At present, in addition to Dalangtan, Chahanslatu, and Kunteyi, there are still remaining water-containing brine lakes in the other salt playas.

According to the characteristics of geological structures described in (Yang, 1988), three sediment zones in the Qaidam Basin have been classified: the salt deposits zone in the intermountain basins at north (Huahaizi), the Middle Cenozoic sedimentation with salt deposits zone (Kunteyi-Mahai, Chahansilatu-Dalangtan, Yiliping-E. Taijinaier-Qarhan), downfaulted basin salt deposits in southwestern region and southern salt deposits zone, as well as the mineral boron ore deposit zone in the southern Qilian intermontane valley area.

There are great differences in the salt depositions in different areas, which are mainly reflected in the time of salt formation, the hydrochemical type of brine lake, the symbiotic combination and thickness of salt deposits, and the enrichment degree of beneficial elements. Salt depositing process in the Qaidam Basin started in the Oligocene Epoch of Tertiary Period. It can be divided into four stages: Late Oligocene, Late Pliocene, Middle-Early Pleistocene, and Early Pleistocene-Holocene

(Yang, 1988). According to Wang *et al.* (2013), the earliest salt-forming regions are Dalangtan, Yiliping, Chahansilatu, and Kunteyi successively with a diagenesis time between 3.92 Ma and 1.16 Ma. Whereas the diagenesis time for Mahai and Qarhan regions is between 358 ka and 24 ka. These differences are mainly influenced by the following: the tectonic movements, the development and evolution of the ancient lake basin, and the supply of material sources. As the second reason, the gradually dry and cold climate is also responsible for the long-term salt deposition.

The spatial distribution of salt playas is closely related to the evolution of brine lakes (Yang, 1988). It has the following pattern: old ones lying in the west and younger ones in the east. For example, the salt deposits in the Late Oligocene are mainly distributed in the Shizigou area (Fig. 4.16) in the west, with a distribution range of only dozens of square kilometers. In the Late Pliocene, salt deposits extended eastward to Xiaoliangshan, Nanyishan, and Youdunzi areas (Fig. 4.16), covering thousands of square kilometers. The range of salt deposits in the Middle and Lower Pleistocene was basically the same as those in the Upper Pliocene,

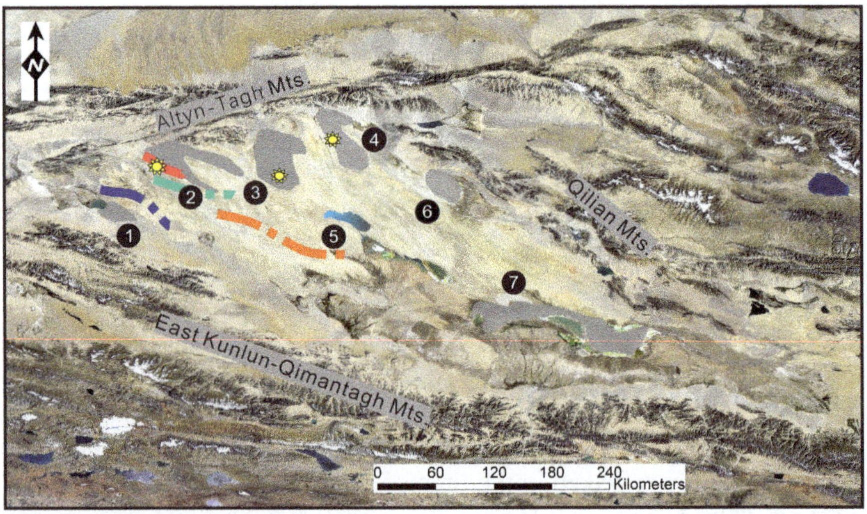

Fig. 4.16. Playas in the Qaidam Basin, related regions mentioned in the context (dotted lines) and boreholes (yellow stars) are marked on the map. Playas: (1) Gasikule; (2) Dalangtan; (3) Qahansilatu; (4) Kunteyi; (5) Yiliping; (6) Mahai; (7) Qarhan. Regions: red-Xiaoliangshan, green-Nanyishan, blue-Shizigou, and orange-Youdunzi. Boreholes from left to right: SG-1, ZK02 and ZK3208. Base Image Credit: Google Earth.

but the thickness of the salt layers increased. By Holocene, salt deposits had covered the whole basin. But during Late Pleistocene, the salt deposition was mainly confined to the east.

The salt deposition is also closely related to Neotectonic movement. Neotectonic movement is very well developed in the basin, which controls the development and evolution of the basin, as well as the distribution and thickness of salt deposits. In general, since the Cenozoic Era, all the places with salt deposition have been in the tectonic settlement. The larger the settlement, the thicker the salt deposition (Yang, 1988). Tertiary lacustrine tectonic evolution in the Qaidam Basin has an obvious controlling effect on sedimentary evolution (Zhao *et al.*, 2004). Due to the influence of tectonic movement and ancient topography, the sedimentary center moved regularly from south to north and from west to east.

4.3.2.2 *Mineral types of salt playas in the Qaidam Basin*

The salt minerals and the thickness of salt deposits vary between the seven playas. In general, halite and mirabilite are widely distributed in the Qaidam, especially halite which is present at almost every salt playa, while mirabilite is commonly found in deposits of sulfate type salt playas. Sulfate-rich evaporites, such as gypsum, are enriched in the western region of Gaskule, Kunteyi, Chahanslatu, and Dalangtan. Halite and small amounts of sulfate are found in the eastern Mahai, Yiliping, and Qarhan salt playas (Kezao and Bowler, 1986).

The salt deposits of Chahanslatu and Yiliping-E began to form at the end of Pliocene, but the salt layers were in a small range and thin (Qi *et al.*, 1993). The salt deposition thickness of Yiliping salt playa is over 100 m in the west and 20–40 m in the east,with a maximum thickness of 73 m.

The Dalangtan playa in the northwest of the Qaidam Basin is about 44 km long and 6–15 km wide, which was the main sedimentary center in Pliocene (Shen, 1993). Its surface is covered with Neogene–Quaternary sediments, and the evaporite crusts are composed of halite and gypsum. Evaporites with up to 100 m thickness were deposited in Dalangtan area (Wang *et al.*, 1993; Zheng *et al.*, 1993). Halite deposits occurred approximately 4 million years ago, which were also mixed with slight gypsum and mirabilite. Bloedite and polyhalite were found in the lower Pleistocene (Wei and Jiang, 1993). Specific mineral types of each salt playa are shown in Table 4.4.

Table 4.4. Large dry salt playas and their mineral types in Qaidam Basin (Yang, 1988).

Salt playa	Minerals
Dalangtan	Halite, Gypsum, Kaluszite, Carnallite, Picromerite, Langbeinite, Kainite, Polyhalite, Epsomite, Leonhardtite, Hexahydrite, Bloedite, Mirabilite, Thenardite, and Glauberite
Chahansilatu	Halite, Mirabilite, Thenardite, Glauberite, Gypsum, Carnallite, Ulexite, and Borax
Kunteyi	Gypsum, Hemihydrate gypsum, Mirabilite, Thenardite, Glauberite, Bloedite, Vanthoffite, Epsomite, Hexahydrite, Leonhardtite, Polyhalite; Halite, Sylvite, Carnallite, Bischofite, Koenenite, Antarcticite
Yiliping	Halite (dominated), Mirabilite, Gypsum, Bloedite, and Sylvite
Mahai	Halite, Gypsum, Thenardite, Glauberite; Carnallite, Bischofite, Vanthoffite; few Polyhalite, Sylvite, Langbeinite, Kainite
Qarhan	Gypsum, Kaluszite, Polyhalite, Kainite, Halite, Hydrohalite, Carnallite, Bischofite, Antarcticite
Huahaizi	Halite and Mirabilite

According to the core of SG-1 (38°24'35.3"N, 92°30'32.7"E, 938.5 m, Fig. 4.16), which is located in the Chahansilatu playa, the evaporites in the upper 400 m are mainly chlorite salts and sulfates, specifically stone salts, gypsum, anhydrous gypsum, mirabilite, thenardite, calcium mirabilite, eugsterite, bloedite, and polyhalite (Li *et al.*, 2010). The carbonate clay layer is mainly composed of clastic minerals and carbonates, trace sulfate, and borate. Clastic minerals include quartz, plagioclase, potassium feldspar, kaolinite, chlorite, mica, talc, montmorillonite, dolomite, ferrodolomite, calcite, aragonite, amphibole, analcite, and cronstedtite. The salt minerals in the carbonated clay layer are mainly halite and gypsum, followed by a small amount of glauberite and thenardite, with traces of carbonate, barite, and paternoite.

The evaporite layer in the borehole alternates with the carbonate clay layer. The thickness of the single salt layer is 5–385 m. The salt layer minerals are mainly halite and gypsum, which are compact and with particle sizes of several millimeters to centimeters. Although sulfate minerals are widely present in boreholes, they are only major minerals in individual layers (Li *et al.*, 2010). While in the lower 538.5 m, the mineral composition includes halite, gypsum, anhydrite, gaylussite, calcite, aragonite, ankerite, dolomite, and an unnamed mineral ($Mg_{0.92}Ca_{0.08}CO_3 \cdot 3H_2O$), with trace amounts of eugsterite. Salt minerals occur in thin layers of centimeters or in fragmentary layers (Li *et al.*, 2013). Carbonate consists of calcite, aragonite, and dolomite.

The borehole ZK02 (38°33'N, 91°21'E, 1000 m) is located in the sedimentary center of the Dalangtan Playa. The core samples were recovered up to 95%. According to Qin *et al.* (2012), the age of the bottom is 2.6 Myr. The lithology of this core includes clastic minerals and evaporites. Clastic minerals are mainly silty clay and mudstone, and a small amount of sandstone is developed at the bottom of the borehole. Evaporites include a complete set of halite, gypsum, polyhalite, mirabilite, bloedite, sylvite, and carnallite. The evaporites content tends to increase from bottom to top.

4.3.2.3 *Paleoclimatic characteristics of salt playas*

The chemical properties of Quaternary saline lakes are closely related to climate change. Saline lakes are evolved from fresh lakes under certain conditions. Saline lakes eventually become salt flats through continuous drying. The change of lake environment and climate is recorded in lakes' sediments. Therefore, there is a close relationship between sediments and paleoenvironment as well as paleoclimate.

The types of evaporites and associated sediments accumulated are sensitive to climate-related factors such as inflow water volumes, temperatures, regional storm tracks, wind patterns, and evaporation rates (Lowenstein *et al.*, 1999). The areas where chlorite, sulfate, and nitrate deposited in the western Qaidam Basin often suffered from the Quaternary drought. Since the Pleistocene, the average thickness of salts and the content of highly soluble salts increased from the bottom up. Among the evaporites, layered mirabilite ($Na_2SO_4 \cdot 10H_2O$) indicates cold climate, layered mirabilite (Na_2SO_4) indicates warm and dry climate, and layered blodite [$Na_2Mg\ (SO_4)_2 \cdot 4H_2O$] indicates warm climate. In addition, basalt is treated as a cold phase mineral (Zheng *et al.*, 2016).

Kunteyi Playa: The oldest salt layers in the Kunteyi Playa formed in the late early Pleistocene. Since then, there had been eight events of big fluctuations which changed between wet/dry and warm/cold climate. A large number of salts began to deposit since around 300 ka. The salt layers were formed in the dry, cold, and acidic oxidation environment, while the mudstone layers were formed in the wet, warm, and alkaline reduction environment. According to the study of the drill hole ZK3208 (Fig. 4.16)

(Han *et al*., 1995), the salt formation stage is recorded from 336 m underground to the surface. The mineral assemblage of salt layers in Kunteyi Playa is mainly composed of halite and mirabilite. The solubility of mirabilite in water has a positive relation with temperature, thus the reduction of temperature is conducive to precipitation of mirabilite from water. This indicates that the halite and mirabilite layers might form under dry and cold climates. The mineral assemblages in the argillaceous strata of Kunteyi Playa are composed of illite and chlorite, which are generally considered as clay mineral assemblages presenting in glacial areas. Perhaps because there were moraines in the material source area of Kunteyi, the clay from the moraines in the source area is transported to and deposited in the lake by the melted water during the warming seasons. This, in turn, suggests a warm and humid climate for argillaceous shales' formation. Meanwhile, the illite, carbonate, and chlorite minerals in mudstone are indicators of the alkaline environment. The occurrence of organic matter and sulfide in mudstone indicates a reductive environment. At last, the presence of high valence sulfate minerals in the salt layer, such as mirabilite, epsomite, and bloedite, indicates that the salt layer was formed in an acidic oxidation environment.

The climate evolution of the Kunteyi region is characterized by the changing between wet/dry and cold/warm climate, which corresponds to the salt forming sequence (Han *et al*., 1995). Since 2 million years ago, there have been ten major climatic changes and salt-forming cycles in Kunteyi Playa. There is at least one layer of halite and mudstone in each salt-forming period, where from bottom to top, Ca^{2+}, Al^{3+}, and SO_4^{2-} decreased, while Mg^{2+} increased. The ancient salinity increased from small to large, and the dry and cold climate increased upward, while the warm and wet climate gradually weakened.

Chahansilatu Playa: The main factors that control minerals are compositions of mother brine and climate (Li *et al*., 2013). Past climatic information can be obtained through the analysis of deposited minerals.

According to borehole SG-1 (938.5 m) in Chahansilatu (Li *et al*., 2010), the upper 400 m is characterized as variations of salt layers (>45 cm) and carbonated clay layers (>300 cm). This part is mainly divided into eighteen evaporates-forming stages. The alternations between salts and carbonated clay layers indicated that the region where the drilling was conducted had an alternate dry and wet climate since 1 Ma. The arid climate appeared in 0. 97–0. 03 Ma.

The lower 538.5 m sediments are composed of salt and clay alternations (Li *et al.*, 2013). It can be divided into six sections, which have recorded the brine evolution as well as climatic oscillation between drought and relative humidity ranging from 2.77 Ma to 1.222 Ma. Since 2.77 Ma, the evaporation rate had been higher than the precipitation rate. During 2.77 Ma–2.221 Ma and 2.053 Ma–1.222 Ma, Chahansilatu area presented as a shallow brine lake. The deposited salt minerals were mainly Ca-carbonate and Ca-sulfate.

Based on the assemblage of identified five lithofacies: (1) gray-black laminated mudstone, (2) gray mudstone and siltstone, (3) bedded halite, (4) loose muddy halite, and (5) brownish-yellowish salt-bearing siltstone and mudstone along core SG-1, the stratigraphic sequence could be divided as nine groups of phases (Fig. 4.17). The succession of main sedimentary environments reveals an evolution of a fresh to semi-brackish water lake into a playa mudflat plain. And it thus demonstrates a general trend of long-term drying of the Asian inland from a relatively humid to sub-humid climate, and to an extremely arid climate with acceleration of drying phases.

To study the paleoclimate of saline lake areas, it is better not only to pay attention to lacustrine deposition, but also to monitor the dynamic changes of modern lake environments. Modern lakes are natural sedimentary laboratories, which are helpful to understand the diagenesis and mineralization of ancient sediments. Long-term and continuous monitoring will provide useful information for studying the paleoclimate in the saline lake area. Meanwhile, geology of the playas has recorded the range, depth, hydrochemistry and even paleoclimate of this area. The combined study of modern lakes and ancient deposits is the way to better understand the evolution of lacustrine history of the Qaidam Basin.

4.4 Summary

Lakes provide important records of geologic and climatic history. Hundreds of paleolakes have been identified on the Martian surface. They are mainly distributed in the southern highlands and mostly have ages of Noachian to Early Hesperian. They have been classified into open-basin lakes, closed-basin lakes, and lake chains based on their hydrological characteristics. Different sedimentary landforms such as deltas, layered deposits, and exposed floor materials have been identified in the paleolakes. Aqueous minerals including phyllosilicates, evaporites, and iron oxides were detected by orbiters and rovers, indicating a different

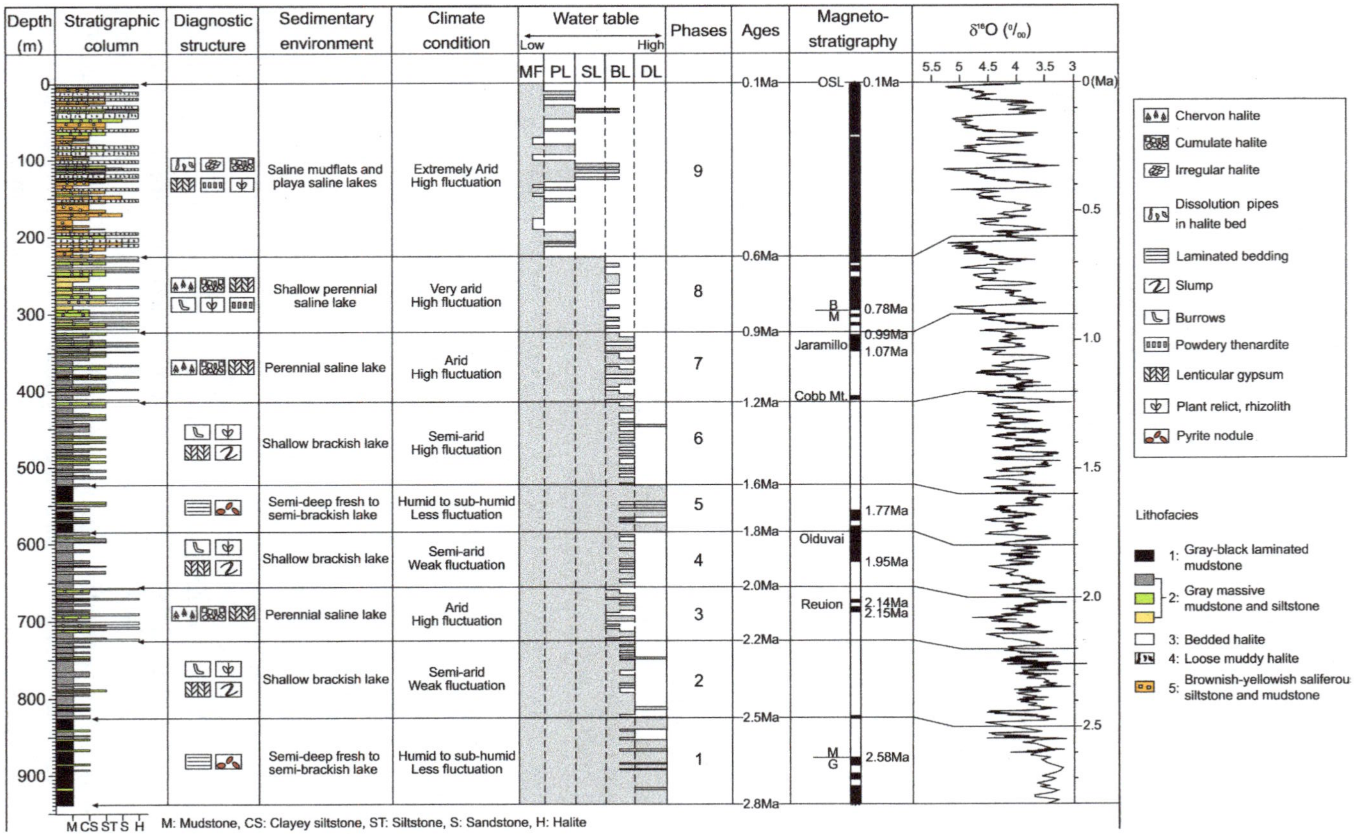

Fig. 4.17. The stratigraphic sequence, lithology, and lithofacies of core SG-1, which shows the sedimentary environment and climate changes since 2.77 Ma (Wang *et al.*, 2012).

environment in the early Martian history from present-day Mars. The water source of the paleolakes could be surface runoff, groundwater, or melted snow, which is still a matter of controversy.

Lakes and playas are widely distributed in the Qaidam Basin. The 26 lakes are identified as chloride type, sodium sulfate subtype, magnesium sulfate subtype, and carbonate type. Among these, magnesium sulfate type is most widely distributed. In general, the mineralization of lake water decreases from chloride to magnesium sulfate, to sodium sulfate, and then to carbonate type. The pH values of brine lakes in the basin tend to decrease from carbonate type to sodium sulfate subtype, to magnesium sulfate subtype and to chloride type, while being inversely related to lakes' mineralization.

The exploration of playas allows us to enter a time machine where the past can be deciphered. The seven playas (Dalangtan, Chahanslatu, Kunteyi, Mahai, Huazhaizi, Qarhan, and Yiliping) are the products of the late evolution of the paleolakes in the Qaidam Basin. Four of them still remain watery brine lakes. Salt depositing process of the playas started in the Oligocene Epoch of the Tertiary Period. The earliest salt-forming regions are Dalangtan, Yiliping, Chahansilatu, and Kunteyi successively with a diagenesis time between 3.92 and 1.16 Ma. Whereas the diagenesis time for Mahai and Qarhan region is between 358–24 ka. These differences are mainly influenced by tectonic movements, the development of the ancient lake basin, the supply of material sources, and the climate changing. The spatial distribution of the playas is closely related to the evolution of brine lakes: old ones lying in the west and younger ones in the east. Halite and mirabilite are widely distributed in the Qaidam. The former is present at almost every salt playa, while the latter is commonly found in the deposits of sulfate type salt playas. The climate evolution corresponds to the salt forming sequence of the playas.

The study of the paleolakes on Earth, which are counterparts to those on Mars, can help to understand the evolution process of paleolakes on Mars and the source of lake water. The presence of liquid water on ancient Mars is the main reason for considering that there might be traces of life on the planet. The identification and analysis of paleolakes on Mars will provide clues for this search for evidence of life on Mars.

Acknowledgments

This study is supported by the Natural Science Foundation of China (42002305, 41830214), China Postdoctoral Science Foundation

(2018M630889), and the Pre-research Project on Civil Aerospace Technologies (D020101) of CNSA. We thank Long Xiao for his suggestions in improving the manuscript, and Jiang Wang for his assistance in data collection.

References

Adeli, S., Hauber, E., Kleinhans, M., Le Deit, L., Platz, T., Fawdon, P. and Jaumann, R. (2016). Amazonian-aged fluvial system and associated ice-related features in Terra Cimmeria, Mars. *Icarus*, **277**, 286–299, doi:10.1016/j.icarus.2016.05.020.

Allen, P. and Collinson, J. (1986). Lakes. *Sedimentary Environments and Facies*, **2**, 63–94, UK: Blackwells.

Ansan, V., Loizeau, D., Mangold, N., Le Mouélic, S., Carter, J., Poulet, F., Dromart, G., Lucas, A., Bibring, J. P., Gendrin, A., Gondet, B., Langevin, Y., Masson, P., Murchie, S., Mustard, J. F., and Neukum, G. (2011). Stratigraphy, mineralogy, and origin of layered deposits inside Terby crater, Mars. *Icarus*, **211**, 273–304, doi:10.1016/j.icarus.2010.09.011.

Arvidson, R. E., Ruff, S. W., Morris, R. V., Ming, D. W., Crumpler, L. S., Yen, A. S., Squyres, S. W., Sullivan, R. J., Bell, J. F., Cabrol, N. A., Clark, B. C., Farrand, W. H., Gellert, R., Greenberger, R., Grant, J. A., Guinness, E. A., Herkenhoff, K. E., Hurowitz, J. A., Johnson, J. R., Klingelhöfer, G., Lewis, K. W., Li, R., McCoy, T. J., Moersch, J., McSween, H. Y., Murchie, S. L., Schmidt, M., Schröder, C., Wang, A., Wiseman, S., Madsen, M. B., Goetz, W. and McLennan, S. M. (2008). Spirit Mars Rover Mission to the Columbia Hills, Gusev Crater: Mission overview and selected results from the Cumberland Ridge to Home Plate. *J. Geophys. Res.*, **113**, E12S33, doi:10.1029/2008je003183.

Arvidson, R. E., Squyres, S. W., Bell, J. F., 3rd, Catalano, J. G., Clark, B. C., Crumpler, L. S., de Souza, P. A., Jr., Fairen, A. G., Farrand, W. H., Fox, V. K., Gellert, R., Ghosh, A., Golombek, M. P., Grotzinger, J. P., Guinness, E. A., Herkenhoff, K. E., Jolliff, B. L., Knoll, A. H., Li, R., McLennan, S. M., Ming, D. W., Mittlefehldt, D. W., Moore, J. M., Morris, R. V., Murchie, S. L., Parker, T. J., Paulsen, G., Rice, J. W., Ruff, S. W., Smith, M. D. and Wolff, M. J. (2014). Ancient aqueous environments at Endeavour crater, Mars. *Science*, **343**, 1248097, doi:10.1126/science.1248097.

Baker, V., Strom, R., Gulick, V., Kargel, J., Komatsu, G. and Kale, V. (1991). Ancient oceans, ice sheets and the hydrological cycle on Mars. *Nature*, **352**, 589–594, doi:10.1038/352589a0.

Baker, V. R. (2006). Geomorphological evidence for water on Mars. *Elements*, **2**, 139–143, doi:10.2113/gselements.2.3.139.

Cabrol, N. and Grin, E. (2001). The evolution of lacustrine environments on Mars: Is Mars only hydrologically dormant? *Icarus*, **149**, 291–328, doi:10.1006/icar.2000.6530.

Cabrol, N. A. and Grin, E. A. (1999). Distribution, classification, and ages of Martian impact crater lakes. *Icarus*, **142**, 160–172, doi:10.1006/icar.2000.6530.

Cabrol, N. A., Grin, E. A., Newsom, H. E., Landheim, R. and McKay, C. P. (1999). Hydrogeologic evolution of Gale crater and its relevance to the exobiological exploration of Mars. *Icarus*, **139**, 235–245.

Carr, M. (1980). The morphology of the Martian surface. *Space Sci. Rev.*, **25**, 231–284, doi:10.1007/bf00221929.

Christensen, P., Wyatt, M., Glotch, T., Rogers, A., Anwar, S., Arvidson, R., Bandfield, J., Blaney, D., Budney, C. and Calvin, W. (2004). Mineralogy at Meridiani Planum from the Mini-TES experiment on the Opportunity Rover. *Science*, **306**, 1733–1739, doi:10.1126/science.1104909.

Clifford, S. M. (1993). A model for the hydrologic and climatic behavior of water on Mars. *J. Geophys. Res. Planets*, **98**, 10973–11016, doi:10.1029/93JE00225.

Clifford, S. M. and Parker, T. J. (2001). The evolution of the Martian hydrosphere: Implications for the fate of a primordial ocean and the current state of the northern plains. *Icarus*, **154**, 40–79.

De Hon, R. (1992). Martian lake basins and lacustrine plains. *Earth, Moon, and Planets*, **56**, 95–122.

Dehouck, E., Mangold, N., Le Mouélic, S., Ansan, V. and Poulet, F. (2010). Ismenius Cavus, Mars: A deep paleolake with phyllosilicate deposits. *Planet. and Space Sci.*, **58**, 941–946, doi:10.1016/j.pss.2010.02.005.

Dou, H., Wang, S., Jiang, J. and Wang, X. (1996). On the principles, scale division and procedures of comprehensive classification of Chinese Lakes. *J. Lake Sci.*, **8**, 173–178.

Duan, Y. Q. (2018). Lake evolution in the Qaidam Basin during 1976–2015 and their changes in response to climate and anthropogenic factors. *J. Lake Sci.*, **30**, 256–265.

Ehlmann, B., Mustard, J. and Team, C. (2009). Regional carbonate-and kaolinite-bearing rock units and how precursor lithologies control alteration products on Mars: An example from the Nili Fossae region. *LPI Contributions*, **1502**, 13.

Ehlmann, B. L. and Buz, J. (2015). Mineralogy and fluvial history of the watersheds of Gale, Knobel, and Sharp craters: A regional context for the Mars Science Laboratory Curiosity's exploration. *Geophys. Res. Lett.*, **42**, 264–273, doi:10.1002/2014gl062553.

Ehlmann, B. L., Mustard, J. F., Fassett, C. I., Schon, S. C., Head Iii, J. W., Des Marais, D. J., Grant, J. A. and Murchie, S. L. (2008a). Clay minerals in delta deposits and organic preservation potential on Mars. *Nat. Geosci.*, **1**, 355–358, doi:10.1038/ngeo207.

Ehlmann, B. L., Mustard, J. F., Murchie, S. L., Poulet, F., Bishop, J. L., Brown, A. J., Calvin, W. M., Clark, R. N., Des Marais, D. J. and Milliken, R. E. (2008b). Orbital identification of carbonate-bearing rocks on Mars. *Science*, **322**, 1828–1832.

Fan, Q., Lai, Z., Long, H., Sun, Y. and Liu, X. (2010). OSL chronology for lacustrine sediments recording high stands of Gahai Lake in Qaidam Basin, northeastern Qinghai–Tibetan Plateau. *Quat. Geochronol.*, **5**, 223–227.

Fassett, C. I. and Head, J. W. (2008a). The timing of martian valley network activity: Constraints from buffered crater counting. *Icarus*, **195**, 61–89, doi:10.1016/j.icarus.2007.12.009.

Fassett, C. I. and Head, J. W. (2008b). Valley network-fed, open-basin lakes on Mars: Distribution and implications for Noachian surface and subsurface hydrology. *Icarus*, **198**, 37–56, doi:10.1016/j.icarus.2008.06.016.

Forsythe, R. D. and Blackwelder, C. R. (1998). Closed drainage crater basins of the Martian highlands: Constraints on the early Martian hydrologic cycle. *J. Geophys. Res.*, **103**, 31421–31431, doi:10.1029/98je01966.

Forsythe, R. D. and Zimbelman, J. R. (1995). A case for ancient evaporite basins on Mars. *J. Geophys. Res. Planets*, *(1991–2012)*, **100**, 5553–5563, doi:10.1029/95JE00325.

Fraeman, A. A., Arvidson, R. E., Catalano, J. G., Grotzinger, J. P., Morris, R. V., Murchie, S. L., Stack, K. M., Humm, D. C., McGovern, J. A., Seelos, F. P., Seelos, K. D. and Viviano, C. E. (2013). A hematite-bearing layer in Gale Crater, Mars: Mapping and implications for past aqueous conditions. *Geology.*, **41**, 1103–1106, doi:10.1130/g34613.1.

Glotch, T. D., Bandfield, J. L., Christensen, P. R., Calvin, W. M., McLennan, S. M., Clark, B. C., Rogers, A. D. and Squyres, S. W. (2006). Mineralogy of the light-toned outcrop at Meridiani Planum as seen by the Miniature Thermal Emission Spectrometer and implications for its formation. *J. Geophys. Res.*, **111**, E12S03, doi:10.1029/2005je002672.

Goldspiel, J. M. and Squyres, S. W. (1991). Ancient aqueous sedimentation on Mars. *Icarus*, **89**, 392–410, doi:10.1016/0019-1035(91)90186-W.

Goudge, T. A., Aureli, K. L., Head, J. W., Fassett, C. I. and Mustard, J. F. (2015). Classification and analysis of candidate impact crater-hosted closed-basin lakes on Mars. *Icarus*, **260**, 346–367, doi:10.1016/j.icarus.2015.07.026.

Goudge, T. A. and Fassett, C. I. (2018). Incision of Licus Vallis, Mars, From Multiple Lake Overflow Floods. *J. Geophys. Res. Planets*, **123**, 405–420, doi:10.1002/2017je005438.

Goudge, T. A., Fassett, C. I., Head, J. W., Mustard, J. F. and Aureli, K. L. (2016). Insights into surface runoff on early Mars from paleolake basin morphology and stratigraphy. *Geology*, **44**, 419–422, doi:10.1130/g37734.1.

Goudge, T. A., Head, J. W., Mustard, J. F. and Fassett, C. I. (2012a). An analysis of open-basin lake deposits on Mars: Evidence for the nature of associated lacustrine deposits and post-lacustrine modification processes. *Icarus*, **219**, 211–229, doi:10.1016/j.icarus.2012.02.027.

Goudge, T. A., Mustard, J. F., Head, J. W. and Fassett, C. I. (2012b). Constraints on the history of open-basin lakes on Mars from the composition and timing of volcanic resurfacing. *J. Geophys. Res. Planets*, **117**, E00J21, doi:10.1029/2012je004115.

Greeley, R. and Guest, J. (1987). Geologic map of the eastern equatorial region of Mars, US Geological Survey.

Grin, E. A. and Cabrol, N. A. (1997). Limnologic analysis of Gusev crater paleolake, Mars. *Icarus*, 130, 461–474, doi:10.1006/icar.1997.5817.

Grotzinger, J. P., Gupta, S., Malin, M. C., Rubin, D. M., Schieber, J., Siebach, K., Sumner, D. Y., Stack, K. M., Vasavada, A. R., Arvidson, R. E., Calef, F., 3rd, Edgar, L., Fischer, W. F., Grant, J. A., Griffes, J., Kah, L. C., Lamb, M. P., Lewis, K. W., Mangold, N., Minitti, M. E., Palucis, M., Rice, M., Williams, R. M., Yingst, R. A., Blake, D., Blaney, D., Conrad, P., Crisp, J., Dietrich, W. E., Dromart, G., Edgett, K. S., Ewing, R. C., Gellert, R., Hurowitz, J. A., Kocurek, G., Mahaffy, P., McBride, M. J., McLennan, S. M., Mischna, M., Ming, D., Milliken, R., Newsom, H., Oehler, D., Parker, T. J., Vaniman, D., Wiens, R. C. and Wilson, S. A. (2015). Deposition, exhumation, and paleoclimate of an ancient lake deposit, Gale crater, Mars. *Science*, **350**, aac7575, doi:10.1126/science.aac7575.

Grotzinger, J. P., Sumner, D. Y., Kah, L. C., Stack, K., Gupta, S., Edgar, L., Rubin, D., Lewis, K., Schieber, J., Mangold, N., Milliken, R., Conrad, P. G., DesMarais, D., Farmer, J., Siebach, K., Calef, F., 3rd, Hurowitz, J., McLennan, S. M., Ming, D., Vaniman, D., Crisp, J., Vasavada, A., Edgett, K. S., Malin, M., Blake, D., Gellert, R., Mahaffy, P., Wiens, R. C., Maurice, S., Grant, J. A., Wilson, S., Anderson, R. C., Beegle, L., Arvidson, R., Hallet, B., Sletten, R. S., Rice, M., Bell, J., 3rd, Griffes, J., Ehlmann, B., Anderson, R. B., Bristow, T. F., Dietrich, W. E., Dromart, G., Eigenbrode, J., Fraeman, A., Hardgrove, C., Herkenhoff, K., Jandura, L., Kocurek, G., Lee, S., Leshin, L.A., Leveille, R., Limonadi, D., Maki, J., McCloskey, S., Meyer, M., Minitti, M., Newsom, H., Oehler, D., Okon, A., Palucis, M., Parker, T., Rowland, S., Schmidt, M., Squyres, S., Steele, A., Stolper, E., Summons, R., Treiman, A., Williams, R., Yingst, A. and Team, M. S. L. S. (2014). A habitable fluvio-lacustrine environment at Yellowknife Bay, Gale crater, Mars. *Science*, **343**, 1242777, doi:10.1126/science.1242777.

Haberle, R. M. and Jakosky, B. M. (1990). Sublimation and transport of water from the north residual polar cap on Mars. *J. Geophys. Res. Solid Earth*, **95**, 1423–1437, doi:10.1029/JB095iB02p01423.

Hammer, U. T. (1986). *Saline Lake Ecosystems of the World*. Netherlands: Springer Science & Business Media.

Han, F. Q., Huang, Q., Wang, K. J., Wang, H. A. A. and Yuan, L. (1995). Study of geochemical evolution and palaeoclimatic fluctuation of Kunteyi salt lake in the Qaidam Basin, Qinghai. *Oceanologia et Limnologia Sinica*, **26**, 502–508.

Hanna, J. C. and Phillips, R. J. (2005). Hydrological modeling of the Martian crust with application to the pressurization of aquifers. *J. Geophys. Res. Planets (1991–2012)*, **110**, E01004, doi:10.1029/2004JE002330.

Harrison, K. P. and Grimm, R. (2005). Groundwater-controlled valley networks and the decline of surface runoff on early Mars. *J. Geophys. Res.*, **110**, E12S16, doi:10.1029/2005je002455.

Head, J. W. (2002). Northern lowlands of Mars: Evidence for widespread volcanic flooding and tectonic deformation in the Hesperian Period. *J. Geophys. Res.*, **107**, 3-1–3-29, doi:10.1029/2000je001445.

Hou, X., Zheng, M., Zhang, C., Shi, L. and Wang, Y. (2010a). Sedimentary characteristics and paleoenvironmental of Dalangtan Salt Lake in Western Qaidam Basin, Since 140 ka BP. *Acta Geol. China*, **84**, 1623–1630.

Hou, Y. J., Wang, J. H. and Zhu, J. L. (2010b). Research on water supplying resources of big and small Sugan Lake in Sugan Lake Basin by using hydrogen and oxygen isotope. *Gansu Geology*, **19**, 66–69.

Howard, A. D. and Moore, J. M. (2011). Late Hesperian to early Amazonian midlatitude Martian valleys: Evidence from Newton and Gorgonum basins. *J. Geophys. Res.*, **116**, E05003, doi:10.1029/2010je003782.

Howard, A. D., Moore, J. M. and Irwin, R. P. (2005). An intense terminal epoch of widespread fluvial activity on early Mars: 1, Valley network incision and associated deposits. *J. Geophys. Res.*, **110**, E12S14, doi:10.1029/2005je002459.

Hu, D. (1990). Geomorphology of Qaihan Salt Lakes. *Scientia Limnologica Sinica*, **2**, 37r43.

Irwin, R. P., Howard, A. D., Craddock, R. A. and Moore, J. M. (2005). An intense terminal epoch of widespread fluvial activity on early Mars: 2, Increased runoff and paleolake development. *J. Geophys. Res.*, **110**, E12S15, doi:10.1029/2005je002460.

Jiang, J. and Wang, S. (1998). Study on the classified system of Chinese lakes. *Adv. in Water Sci.*, **9**, 170–175.

Jin, Q. and Zha, M. (2000). Co-sedimentation of Tertiary evaporites and oil source rocks in the western Qaidam Basin. *Scientia Geologica Sinica*, **4**, 465–473.

Jin, X., Guo, R. and Xia, W. (2013). Distribution of actual evapotranspiration over Qaidam Basin, an arid area in China. *Remote Sensing*, **5**, 6976–6996.

Kezao, C. and Bowler, J. (1986). Late Pleistocene evolution of salt lakes in the Qaidam basin, Qinghai Province, China. *Palaeogeography, Palaeoclimatology, Palaeoecology*, **54**, 87–104.

Lapotre, M. G. and Lamb, M. P. (2018). Substrate controls on valley formation by groundwater on Earth and Mars. *Geology*, **46**, 531–534.

Leverington, D. and Maxwell, T. (2004). An igneous origin for features of a candidate crater-lake system in western Memnonia, Mars. *J. Geophys. Res.*, **109**, E06006, doi:10.1029/2004je002237.

Li, M., Fang, X., Wang, J., Song, Y., Yang, Y., Zhang, W. and Liu, X. (2013). Evaporite minerals of the lower 538.5 m sediments in a long core from the Western Qaidam Basin, Tibet. *Quat. Int.*, **298**, 123–133.

Li, M., Fang, X., Yi, C., Gao, S., Zhang, W. and Galy, A. (2010). Evaporite minerals and geochemistry of the upper 400m sediments in a core from the Western Qaidam Basin, Tibet. *Quat. Int.*, **218**, 176–189, doi:http://dx.doi.org/10.1016/j.quaint.2009.12.013.

Liu, W., Jiang, H., Yang, J. and Wu, G. (2018). Salinity and DOC influence the distribution of free-living and particle-attached aerobic anoxygenic phototrophic bacteria in the Qinghai–Tibetan Lakes. *Geomicrobiol. J.*, **35**, 247–254.

Liu, X. Q., Cai, K. Q. and Yu, S. S. (2002). Geochemical simulation of formation and evolution of salt lakes and their water sources in Qaidam Basin: Application of Pitzer's model. *Geochemica*, **31**, 501–507.

Lowenstein, T. K., Li, J., Brown, C., Roberts, S. M., Ku, T.-L., Luo, S. and Yang, W. (1999). 200 ky paleoclimate record from Death Valley salt core. *Geology*, **27**, 3–6.

Lu, N. (2014). Changes of lake area in Qaidam basin and the influence factors. *J. Arid Land Reso. and Environ.*, **8**, 015.

Ma, J. F., Lv, C. H. A. and Wang, Q. (2008). Land-cover classification of the Qaidam Basin based on MODIS NDVI time series data. *Arid Land Geog.*, **31**, 442–448.

Malin, M. C. and Carr, M. H. (1999). Groundwater formation of Martian valleys. *Nature*, **397**, 589–591.

Mao, X. C., Liu, X., Dong, Y., Bai, Y. S., Xu, Y. D. A. and Song, B. W. (2018). Research on the Genesis of Semi-submerged Yardang Landform in the Duck Lake Area of Qaidam Basin. *Geol. Rev.*, **64**, 1505–1518.

McEwen, A. S., Hansen, C. J., Delamere, W. A., Eliason, E. M., Herkenhoff, K. E., Keszthelyi, L., Gulick, V. C., Kirk, R. L., Mellon, M. T., Grant, J. A., Thomas, N., Weitz, C. M., Squyres, S. W., Bridges, N. T., Murchie, S. L., Seelos, F., Seelos, K., Okubo, C. H., Milazzo, M. P., Tornabene, L. L., Jaeger, W. L., Byrne, S., Russell, P. S., Griffes, J. L., Martinez-Alonso, S., Davatzes, A., Chuang, F. C., Thomson, B. J., Fishbaugh, K. E., Dundas, C. M., Kolb, K. J., Banks, M. E. and Wray, J. J. (2007). A closer look at water-related geologic activity on Mars. *Science*, **317**, 1706–1709, doi:10.1126/science.1143987.

McKay, C. P. and Davis, W. L. (1991). Duration of liquid water habitats on early Mars. *Icarus*, **90**, 214–221, doi:10.1016/0019-1035(91)90102-Y.

McSween, H. Y. (2006). Water on Mars. *Elements*, **2**, 135–137, doi:10.2113/gselements.2.3.135.

Michalski, J. R. and Bleacher, J. E. (2013). Supervolcanoes within an ancient volcanic province in Arabia Terra, Mars. *Nature*, **502**, 47–52, doi:10.1038/nature12482.

Milliken, R. E., Grotzinger, J. P. and Thomson, B. J. (2010). Paleoclimate of Mars as captured by the stratigraphic record in Gale Crater. *Geophys. Res. Lett.*, **37**, L04201, doi:10.1029/2009gl041870.

Morris, R. V., Klingelhöfer, G., Schröder, C., Fleischer, I., Ming, D. W., Yen, A. S., Gellert, R., Arvidson, R. E., Rodionov, D. S., Crumpler, L. S., Clark, B. C., Cohen, B. A., McCoy, T. J., Mittlefehldt, D. W., Schmidt, M. E., de Souza, P. A. and Squyres, S. W. (2008). Iron mineralogy and aqueous alteration from Husband Hill through Home Plate at Gusev Crater, Mars: Results from the Mössbauer instrument on the Spirit Mars Exploration Rover. *J. Geophys. Res.*, **113**, E12S42, doi:10.1029/2008je003201.

Mustard, J. F., Murchie, S., Pelkey, S., Ehlmann, B., Milliken, R., Grant, J., Bibring, J.-P., Poulet, F., Bishop, J. and Dobrea, E. N. (2008). Hydrated silicate minerals on Mars observed by the Mars Reconnaissance Orbiter CRISM instrument. *Nature*, **454**, 305–309, doi:10.1038/nature07097.

Noe Dobrea, E., Poulet, F. and Malin, M. (2006). OMEGA analysis of light-toned outcrops in the chaotic terrain of the eastern Valles Marineris region, 37th Annual Lunar and Planetary Science Conference, 2068.

Qi, H., Ku, T.-L. and Phillips, F. M. (1993). Evolutionary characteristics of lakes and palaeoclimatic undulation in the Qaidam Basin, China. *Chin. J. Oceanol. Limnol.,* **11**, 34–45.

Qin, Y., Hou, X., Zheng, M., Yang, Z., Li, H. and Shi, L. (2012). Magnetostratigraphy of Liang-ZK02 borehole in Dalangtan, Qaidam Basin and its paleoenvironmental significance. *Geol. Rev.*, **58**, 553–564.

Sagan, C., Toon, O. B. and Gierasch, P. J. (1973). Climatic change on Mars. *Science*, **181**, 1045–1049, doi:10.1126/science.181.4104.1045.

Scott, D. H. and Tanaka, K. L. (1986). Geologic map of the western equatorial region of Mars, US: Geological Survey.

Shen, Z. (1993). *The Division and Sedimentary Environment of Quaternary Salt-Bearing Strata in Qaidam Basin*, Beijing: China Land Press.

Squyres, S. W., Arvidson, R. E., Bollen, D., Bell, J. F., Brückner, J., Cabrol, N. A., Calvin, W. M., Carr, M. H., Christensen, P. R., Clark, B. C., Crumpler, L., Des Marais, D. J., d'Uston, C., Economou, T., Farmer, J., Farrand, W. H., Folkner, W., Gellert, R., Glotch, T. D., Golombek, M., Gorevan, S., Grant, J. A., Greeley, R., Grotzinger, J., Herkenhoff, K. E., Hviid, S., Johnson, J. R., Klingelhöfer, G., Knoll, A. H., Landis, G., Lemmon, M., Li, R., Madsen, M. B., Malin, M. C., McLennan, S. M., McSween, H. Y., Ming, D. W., Moersch, J., Morris, R. V., Parker, T., Rice, J. W., Richter, L., Rieder, R., Schröder, C., Sims, M., Smith, M., Smith, P., Soderblom, L. A., Sullivan, R., Tosca, N. J., Wänke, H., Wdowiak, T., Wolff, M., Yen, A., (2006). Overview of the

Opportunity Mars Exploration Rover Mission to Meridiani Planum: Eagle Crater to Purgatory Ripple. *J. Geophys. Res.*, **111**, E12S12, doi:10.1029/2006je002771.

Thomson, B. J., Bridges, N. T., Milliken, R., Baldridge, A., Hook, S. J., Crowley, J. K., Marion, G. M., de Souza Filho, C. R., Brown, A. J. and Weitz, C. M. (2011). Constraints on the origin and evolution of the layered mound in Gale Crater, Mars using Mars Reconnaissance Orbiter data. *Icarus*, **214**, 413–432, doi:10.1016/j.icarus.2011.05.002.

Tokano, T. (2005). Water on Mars and life. Netherlands: Springer Science & Business Media.

Toon, O. B., Pollack, J. B., Ward, W., Burns, J. A. and Bilski, K. (1980). The astronomical theory of climatic change on Mars. *Icarus*, **44**, 552–607, doi:10.1016/0019-1035(80)90130-x.

Treiman, A. H., Bish, D. L., Vaniman, D. T., Chipera, S. J., Blake, D. F., Ming, D. W., Morris, R. V., Bristow, T. F., Morrison, S. M. and Baker, M. B. (2016). Mineralogy, provenance, and diagenesis of a potassic basaltic sandstone on Mars: CheMin X-ray diffraction of the Windjana sample (Kimberley area, Gale Crater). *J. Geophys. Res. Planets*, **121**, 75–106.

Vaniman, D., Bish, D., Ming, D., Bristow, T., Morris, R., Blake, D., Chipera, S., Morrison, S., Treiman, A. and Rampe, E. (2014). Mineralogy of a mudstone at Yellowknife Bay, Gale crater, Mars. *Science*, **343**, 1243480.

Wang, J., Fang, X., Appel, E. and Song, C. (2012). Pliocene–Pleistocene climate change at the NE Tibetan Plateau deduced from lithofacies variation in the drill core SG-1, western Qaidam Basin, China. *J. Sediment. Res.*, **82**, 933–952.

Wang, J., Fang, X., Appel, E. and Zhang, W. (2013). Magnetostratigraphic and radiometric constraints on salt formation in the Qaidam Basin, NE Tibetan Plateau. *Quat. Sci. Rev.*, **78**, 53–64.

Wang, M., Wang, Y., Liu, C. and Chen, Y. (1993). The features and genesis of the Dalangtan salt deposit in the Qaidam Basin. *Diqiu Xuebao = Acta Geoscientia Sinica*, **26**, 97–114.

Wei, X. J. and Jiang, J. (1993). The evolution of the Quaternary salt lake in the Qaidam Basin. *Acta Geol. Sin.*, **67**, 255–265.

Wilson, S. A., Howard, A. D., Moore, J. M. and Grant, J. A. (2016). A cold-wet middle-latitude environment on Mars during the Hesperian-Amazonian transition: Evidence from northern Arabia valleys and paleolakes. *J. Geophys. Res. Planets*, **121**, 1667–1694.

Wu, Z., Cui, X. A. and Zheng, M. (2012). pH value change trends in salt brine evaporation. *Chinese J. Inorg. Chem.*, **28**, 297–301.

Xiao, L., Wang, J., Dang, Y., Cheng, Z., Huang, T., Zhao, J., Xu, Y., Huang, J., Xiao, Z. and Komatsu, G. (2017). A new terrestrial analogue site for Mars research: The Qaidam Basin, Tibetan Plateau (NW China). *Earth-Sci. Rev.*, **164**, 84–101.

Yang, J., Jiang, H., Dong, H., Wang, H., Wu, G., Hou, W., Liu, W., Zhang, C., Sun, Y. and Lai, Z. (2013). amoA-encoding archaea and thaumarchaeol in the lakes on the northeastern Qinghai-Tibetan Plateau, China. *Frontiers in Microbiology*, **4**, 329, doi:10.3389/fmicb.2013.00329.

Yang, Q. (1988). The salt mineral resources and their spatial distribution in chaidamu basin, Qinghai. *Earth Sci. J. China Univ. Geosci.*, **13**, 49–58.

Yu, S. (1981). The hydrochemical characteristics of the saline lakes on the Qinghai-Xizang Plateau. *Oceanologia et Limnologia Sinica*, **1**, 498–511.

Zhao, J., Chen, X., and Du, Y. (2004). The Tertiary sedimentary evolution of the Qaidam Basin, Northwest China. *Petroleum Exploration and Development*, **31**, 2-0.

Zhao, J., Xiao, L. and Glotch, T. D. (2020). Paleolakes in the Northwest Hellas Region, Mars: Implications for the Regional Geologic History and Paleoclimate. *J. Geophys. Res. Planets*, 125, 2019JE006196, doi:10.1029/2019je006196.

Zhao, W., Yin, X. and Zhang, P. (2010). *Saline Lake Biology of China*, Beijing: Science Press.

Zheng, M. and Liu, X. (2009). Hydrochemistry of Salt Lakes of the Qinghai-Tibet Plateau. China. *Aquat. Geochem.*, **15**, 293–320.

Zheng, M., Tang, J., Liu, J. and Zhang, F. (1993). Chinese saline lakes. *Hydrobiologia*, 267, 2336.

Zheng, M., Zhang, Y., Liu, X., Qi, W., Kong, F., Nie, Z., Pu, L., Hou, X., Wang, H.A. and Zhang, Z. (2016). Progress and prospects of salt lake research in China. *Acta Geol. Sin.-English Edition*, **90**, 1195–1235.

Zheng, M. P. (2010). Salt lake resources and eco-environment in China. *Acta Geol. Sin.*, **84**, 1613–1622.

Zheng, X. Y., Zhang, M. G., Li, B. X. and Xu, C. (2002). *Salt Lakes of China*. Beijing: Science Press.

https://doi.org/10.1142/9789811224027_0005

Chapter 5

Aeolian Landforms

Jiang Wang, Jun Huang and Long Xiao
Planetary Science Institute, School of Earth Sciences, China University of Geosciences, Wuhan 430074, PR China

5.1 Introduction

Aeolian landform is a feature of the Earth's surface produced by either the erosive or constructive action of the wind that covers approximately 20% of the Earth's land surface. Aeolian processes play a significant role not only in the dynamics of deserts on Earth but also contribute to surface landforms on several bodies in our solar system (Bourke *et al.*, 2019). At least three other bodies have sufficient enough atmosphere to sustain winds that can transport sediments, including Mars, Titan, and Venus. Their atmospheres interact with geological processes and influence the morphology and composition of surfaces (Grotzinger *et al.*, 2013).

As we all know, Mars is the neighbor of Earth. The surface average temperature of Mars is about −123°C to −33°C (Kieffer *et al.*, 1977), and the daily highest temperature can be up to 27°C at local areas of the mid-southern latitude in the summer season (Carr, 2007). The current Martian atmosphere is much thinner, with an average surface atmospheric pressure of about 0.5–1% compared with Earth, comprising mostly of carbon dioxide (about 95%) with a small amount of nitrogen and argon (Read and Lewis, 2004). The surface wind velocities are quite different in the separate areas. For example, usually, the wind velocity is about 20 m/s to 30 m/s at the Viking landing site, and the highest wind velocity can reach 150 m/s (Chamberlain *et al.*, 1976). While wind velocity obtained from the Rover Environmental Monitoring Station (REMS) on board MSL ranges from 0 to 20 m/s, concentrated mostly between 2 m/s to 6 m/s (Viúdez-Moreiras *et al.*, 2017). Long-term observations show that dust storms are activated in nearly half of the Martian year, being the most important external agent to shape the surface of Mars. Remote sensing and *in situ* explorations show that the surface of Mars is abundant of aeolian landforms, such as ventifacts, yardangs, wind streaks, sand dunes, and so on. In this chapter, we will focus on yardangs and sand dunes.

5.2 Yardangs

5.2.1 *Yardangs on Mars*

We focus our study to the Medusae Fossae Formation (MFF) region and Gale Crater, which are also abundant and highly complex, but are similar to features within the Qaidam Basin.

According to the characteristics of yardangs in the Qaidam Basin (Table 3.2), we examined the yardangs of the MFF region and Gale Crater on Mars and sorted them into different types, for the purpose of comparative study.

In the MFF exposed region, we have found a large amount of dendritic shape mesa yardangs with flat broad tops in the Aeolis Dorsa (Fig. 5.1(a)). They can be classified into M-I type compared with those in the Qaidam Basin. Streamlined yardangs are developed both on their broad tops and surrounding depressions (Fig. 5.1(a)). We also have identified a group of M-E type yardangs coexisting with long-ridge yardangs in the MFF area at Zephrria Planum. These yardangs exhibit in NE-SW and N-S directions (Fig. 5.1(d)), with some of them eroded into streamlined shapes (e.g. W-B type).

Long ridge yardangs are widely spread in the MFF region. The representative L-B type yardang's analog is in Lucus Planum (Fig.5.2(a)). They have broad tops, steep flanks, and are parallel to each other in NW-SE direction over 30 km long. Their width ranges from tens of meters to ~1 km with height from tens of meters to ~290 m (Fig. 5.2(b)). Some of them are eroded into narrow-top yardangs (L-N type) with double tapered ends. The northwestern part of L-B type yardangs are eroded and change into M-E or W-H type yardangs. Some L-B type paleo-yardangs are observed in the north Zephyria Planum (Fig. 5.16(e)). They extended in W-E direction over 80 km and were overlaid by NE-SW directional

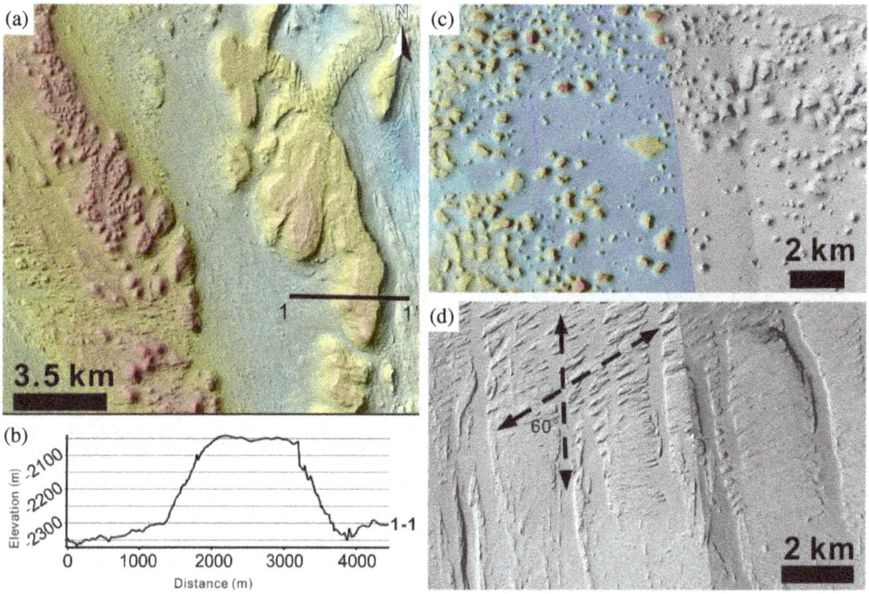

Fig. 5.1. M group yardangs on Mars. (a) The M-I type yardangs in Aeolis Dorsa of the Mars, CTX image with CTX DTM (f23_044686_1762_xi_03s209w/f19_043143_1761_xi_03s209w) overlay. Blue areas: lower elevations, red areas: higher elevations. (b) Elevation profile along 1-1' in figure (a). (c) The M-I type yardangs in Aeolis Planum on Mars (b18_016744_1804_xi_00s216w/b22_018089_1764_xi_03s215w). (d) The M-E type yardangs in Zephyria Planum on Mars (black dashed arrows indicate two wind directions, CTX image, p20_008766_1803_xi_00n206w).

yardangs with an included angle of about 45 (degrees). Their widths are similar to those in Lucus Planum, while their heights are much lower (Fig. 5.16(g)).

We find the L-N type yardangs in Aeolis Dorsa, presenting streamlined shapes with blunt heads and tapered tails. Faceted surfaces and curvilinear crestlines can be observed. Some of them are coexisting with the W-H type yardangs (Fig. 5.3(a)). The L-R type yardangs are extended in N-S directions in the Aeolis Dorsa (Fig. 5.3(b)). Their tops are quite tattered with faceted surface and curvilinear crestlines, and separated by broad troughs, where sand sheets are deposited in dark areas.

The S group yardangs are relatively common at the edge of layered deposits (Fig. 5.4(a) and 5.4(b)). The yardangs developed in the eastern part of Gale Crater are good analogs for the S-T type. Their stratified

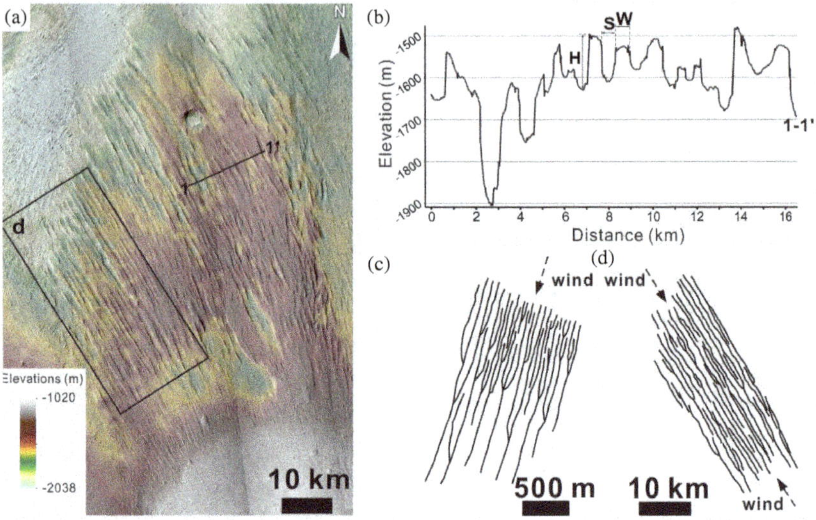

Fig. 5.2. (a) The L-B type yardangs in the MFF area at Lucus Planum of Mars, CTX images (p21_009200_1742_xi_05s176w/p16_007262_1747_xn_05s176w) with a MOLA DTM overlay. (b) Elevation profile (CTX DTMs, p14_006550_1726_xn_07s175w/ b17_016215_1728_xi_07s175w) along 1-1' in figure (a). (c) and (d) The schematic diagrams of the long axis of the L-B type yardangs in Fig. 4(a) and Fig. 5.9(a).

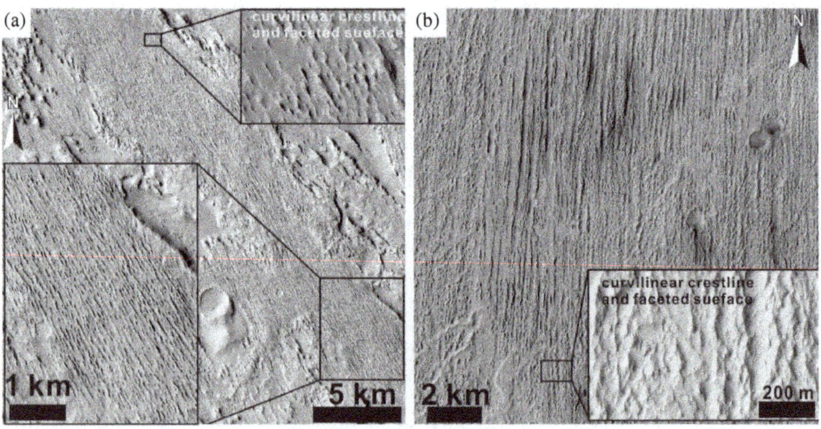

Fig. 5.3. (a) The L-N type yardangs in Aeolis Serpens (CTX image, f03_036800_1762_ xn_03s210w), inset figure of the upper right corner is part of HiRISE image (esp_036800_1750_red). (b) The L-R type yardangs in Aeolis Dorsa (CTX image, b16_016058_1739_xn_06s208w), inset figure of the bottom right corner is partial of HiRISE image (esp_016058_1740_red). The dark areas are sand sheets.

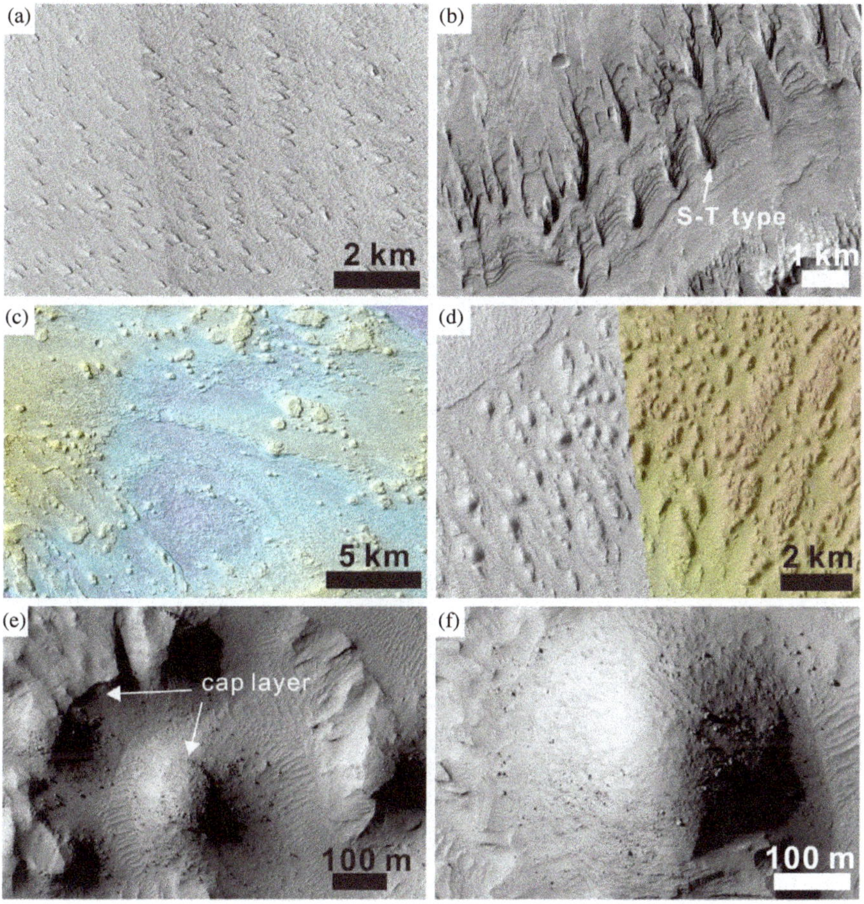

Fig. 5.4. S group yardangs in AZP area. (a) S-T type yardangs in north Zephyria Planum, CTX image, d04_028796_1835_xn_03n206w; (b) S-T type yardangs in Gale crater, CTX image, f20_043750_1747_xn_05s221w); (c) S-R type yardangs, CTX images (b18_016744_1804_xi_00n216w/b22_018089_1764_xi_03s215w); (d) S-R type yardangs, CTX images (p02_001857_1750_xn_05s209w/g04_019803_1748_xn_05s209w); (e) and (f) S-R type yardangs, HiRISE image, esp_036655_1735_red.

structures are noticeable (Fig. 5.4(b)). These yardangs are developed in the fine layers and orientated to the downslope. Most of them have been segmented from the zigzags forming W-B or W-H type yardangs. The S-R type yardangs found at the Aeolis Planum are characterized by rounded shapes (Fig. 5.4(d) and 5.4(e)), with some cap layers on the top (Fig5.4(f)).

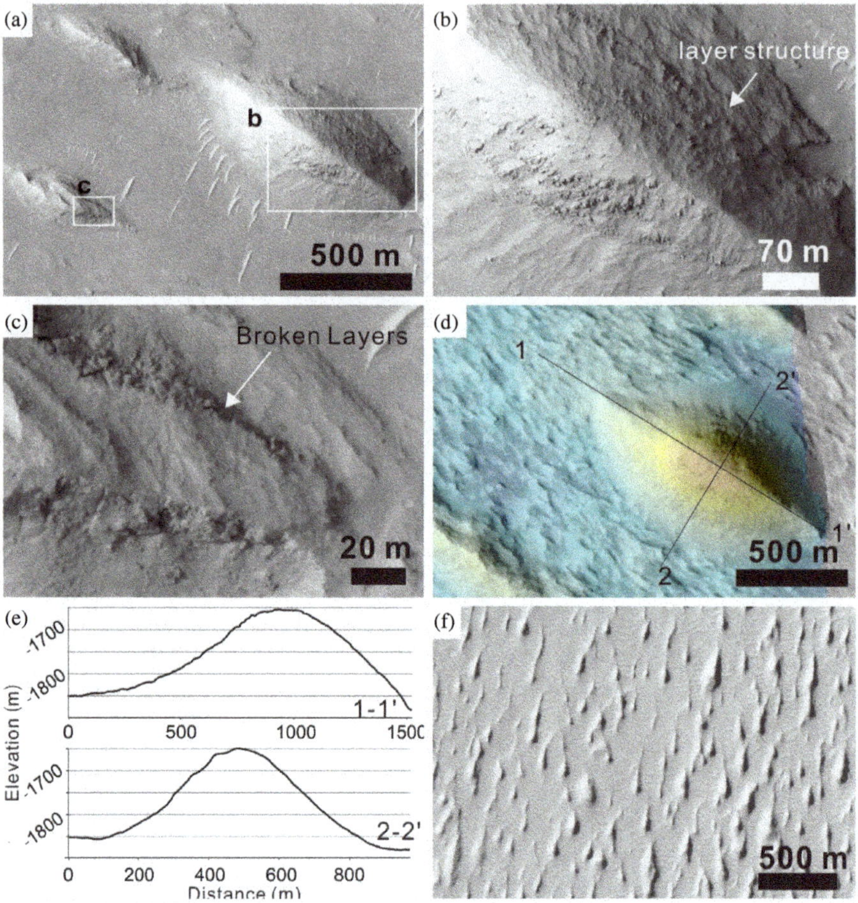

Fig. 5.5. The W-B type yardans in AZP area. (a), (b) and (c) The W-B type yardangs, HiRISE image, psp_010731_1820_red), (d) W-B-01 type yardangs, CTX image (d14_032673_1814_xn_01n216w/j05_046585_1825_xn_02n216w); (e) The profile of W-B type yardang in figure d; (f) The W-B type yardangs, CTX image, d21_035389_1738_xn_06s206w.

The W-B type yardangs are perfectly streamlined with blunt heads, symmetrical wings, and tapered tails, and some with slim drags (Fig. 5.5(f), Fig. 5.5(d) and 5.5(e)). They have layered structures (Fig. 5.5(b)), interlayered with some broken layers (Fig. 5.5(c)) and rough surfaces. The spaces between yardangs are relatively large, and the surface looks much rougher (Fig. 5.6(a)). Some of them show crestlines.

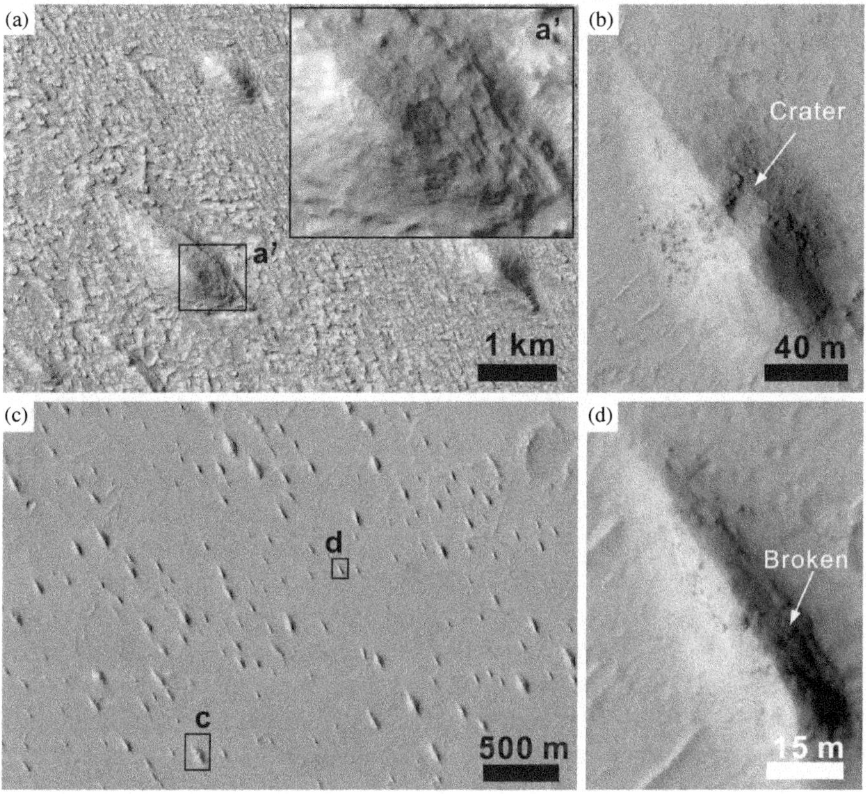

Fig. 5.6. The W-B type yardangs in Aeolis Planum on Mars (CTX image, d20_035231_1813_xn_01n215w). Inset a' shows the layered deposits. (b) The W-H type yardangs in Zephyria Planum (HiRISE image, psp_066828_1820_red). (c) and (d) Large and small size W-H type yardangs shown in (b). The W-H type yardang was partially destroyed by impact crater in (c).

The W-H type yardangs are widely distributed in the Aeolis-Zephyria Planum (AZP). They extend in SE-NW direction with streamlined shape and sharp crestlines. The multilayer structures of the basement of the W-H type yardangs in the northern Zephyria Planum are obvious. Their surfaces are characterized by cracks and blocks. Dark streaks are obvious at their wings (Fig. 5.6(d)). The spacings are large, and the bottoms of these troughs are flat wide with some polygons, and covered by sand ripples and dust. They can be divided into two groups in size, which are coexisting with each other (Fig. 5.6(b)). Sand ripples are obvious around the yardangs, indicating the current wind direction is parallel to the elongation of yardangs. There is no report yet about the W-L type yardangs in the MFF region due to their small sizes and data resolutions.

5.2.2 *Yardangs in the Qaidam Basin*

According to the classification in Table 3.2, we will describe the yardangs of the Qaidam Basin in more detail in the following section.

5.2.2.1 *Mesa group yardangs*

Mesa group (M group) yardangs include two different types, irregular (M-I type) and elongated mesa (M-E type) yardangs, both of which are elevated steep-sided and flat-topped hills (Fig. 5.7 and 5.8). The M-I type yardangs develop in piedmont layered alluvial-diluvial clinoplains (Fig. 5.7(a), 5.7(b), and 5.7(c)), and occur sporadically inside the western basin. The typical area is between Mahai and Xiaoqaidam Lake, and the southern margin of the Altyn-Tagh Mountains. They are large in size (up to ~60 km long and 1 km wide), and exhibit various shapes, ranging from irregular (Fig. 5.7(c) and 5.8(a)) to dendritic (Fig. 5.7(a)). Aeolian sediments and slope debris accumulate at the foot of these mesas (Fig. 5.7(d)). We find some gravel-enriched polygons (~1 m in diameter) on the M-I type yardangs' surface and gullies at their rim (Fig. 5.7(d)). Some M-I type yardangs are developed at the western margin of the playa. Light wind streaks (salt) are obvious.

The elongated mesa (M-E type) yardangs are usually in rhombus shape with a flat top, surrounding cliffs, and a relatively gentle slope at the leeward end, which mainly occurs in the bidirectional winds region, downwind of the L group yardangs. The typical area is between Dafengshan and Tulinbao. They are more streamlined than the M-I type yardangs. Generally, the leeward surfaces are covered by salt crusts exhibiting polygons, while the windward surfaces expose the original horizontal lacustrine sediments (Fig. 5.8(d)), sometimes with moats surrounding them (Fig. 5.8(b)). Overall, these M-E type yardangs are usually arranged in lines with one direction or two directions (Fig. 5.8(a)).

5.2.2.2 *Long-ridge group yardangs*

Morphologically, the long-ridge group (L group) yardangs contain three different types, i.e. broad-top long-ridge (L-B type, Fig. 5.9(a)),

Fig. 5.7. The irregular mesa (M-I type) yardangs in the Qaidam Basin. (a) The M-I type yardangs between Mahai and Xiaoqaidam (black arrows mark the position of some M-I type yardangs, image center: 37.70°N, 94.69°E). (b) The M-I type yardangs in the north of Dabuxun Lake (37.25°N, 95.10°E). Some M-I type yardangs have transformed into W-H type yardangs. (c) The M-I type yardangs in the southeast of Lenghu area (38.46°N, 93.45°E). (d) The M-I type yardangs in the southeast of Lenghu (38.46°N, 93.43°E). Sand ripples indicate the wind direction. Gullies and polygonal features are found on the M-I type yardangs.

narrow-top long-ridge (L-N type, Fig. 5.10(a)), and rough-top long-ridge (L-R type, Fig. 5.10(c)) yardangs. They are mainly distributed in the areas surrounding the Qahansilatu Playa, Kunteyi Playa, and the north area of Nuomuhong. The L-B and L-N type yardangs in the western basin occur mostly in Early-Middle Pleistocene strata consisting of clay, mudstone, gypsum, and saline lacustrine deposits (Li *et al.*, 2010; Wang *et al.*, 2012). The L-N type yardangs in the north of Nuomuhong formed on the Holocene flood plain of the Qaidam River. The L-R type yardangs mainly

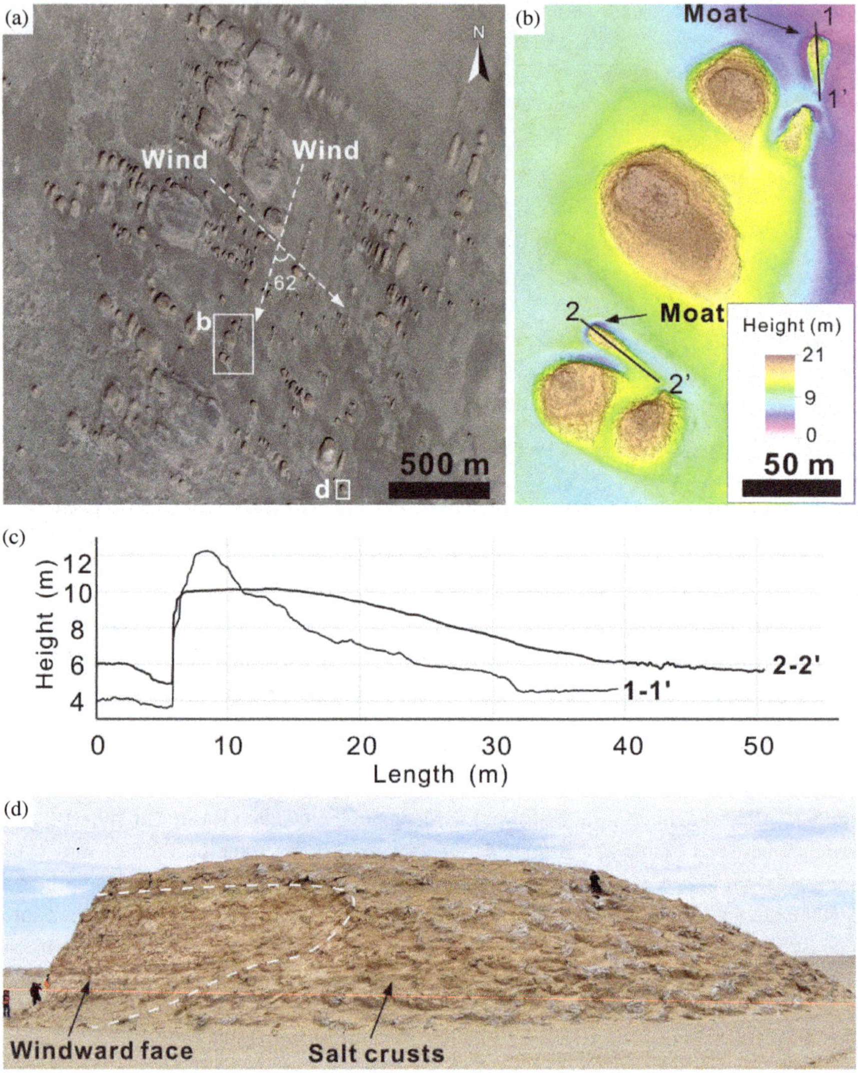

Fig. 5.8. The elongated mesa (M-E type) yardangs in the Qaidam Basin. (a) The M-E type yardangs in the bidirectional area (white dashed arrows indicate two wind directions, image center: 38.07°N, 92.14°E). (b) DTM superposed on drone images of the M-E type yardangs shown in (a). Clearly visible are moats surrounding the upwind heads of the yardangs. (c) The profiles of yardang 1 and yardang 2 in (b). Black lines mark the location of the profiles. The moat in site 2 is deeper than that in site 1. (d) The M-E type yardangs with an eroded windward face (white dashed line marks the boundary) shown in (a).

occur between E'boliang and Nanbaxian in Neogene strata comprising mudstone, siltstone, interbedded with gypsum, nodular limestone, and rock salt (Li *et al.*, 2010).

The L-B type yardangs are all parallel to each other and extend continuously for nearly 4 km at maximum. They have a broad top, two gentle convex flanks, and narrow troughs in-between them (Fig. 5.9(a)). On the upwind side, the broad tops are eroded into small ridges with tapered heads (Fig. 5.9(b)), while their downwind areas are still unaltered. The flanks of the new heads become much steeper. In some places, remnant cap layers occur on the top surface (Fig. 5.9(b)). Large polygons developed both on the surface of the yardangs and in the troughs (Fig. 5.9(c)).

The L-N type yardangs are also orientated parallel to each other but are less continuous. They have narrow tops, steep flanks and wide troughs in-between them (Fig. 5.10(a)). They are much shorter and narrower than

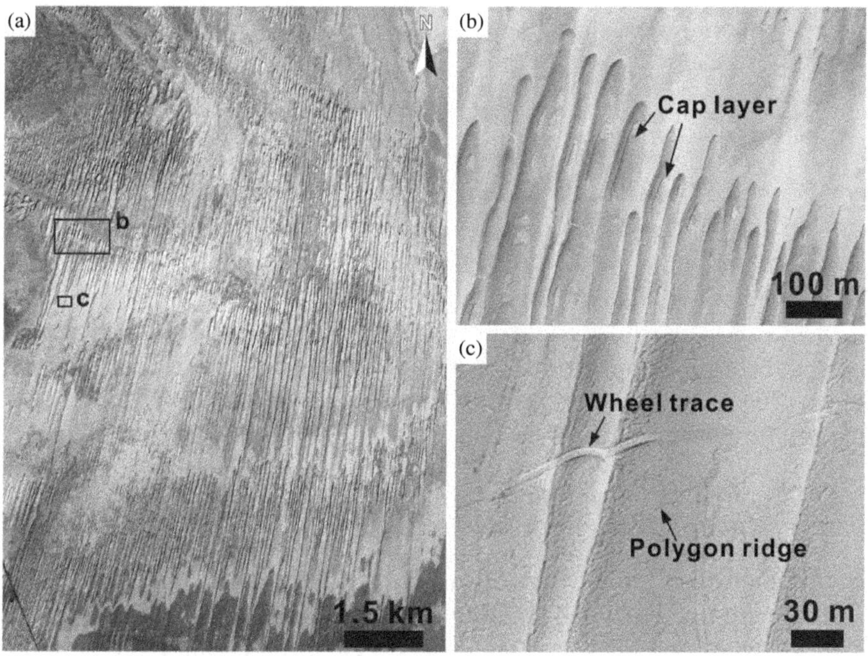

Fig. 5.9. The broad-top long-ridge (L-B type) yardangs in the Qaidam Basin. (a) L-B type yardangs in the western Qarhansilatu Playa (38.49°N, 92.26°E). (b) The heads of the L-B type yardangs are eroded into small ridges, some remnant cap rocks are on the surface. (c) Large polygons developed on the surface of yardangs.

Fig. 5.10. The narrow-top long-ridge (L-N type) and rough-top long-ridge (L-R type) yardangs in the Qaidam Basin. (a) The L-N type yardangs in the southwest of Lenghu (38.52°N, 92.97°E). (b) Multilayered L-N type yardangs (38.66°N, 93.27°E). The greenish layer is enriched with gypsum. (c) L-R type yardangs in Lenghu (38.65°N, 92.78°E).

the L-B type yardangs and even consist of only a wall (~1 m thick) with nearly vertical flanks. Mass wasting derived slope debris covers the lower parts of the yardangs. Colorful multilayered lacustrine sediments have been observed in the field, with remnant cap layers on the top surface, but no obvious salt crusts are found on the L-N type yardangs (Fig. 5.10(b)).

The L-R type yardangs are ambiguous in morphology, but still can be recognized from the parallel troughs and extensively broken tops (Fig. 5.10(c)). Original broad tops are eroded into fragments or small ridges with faceted surfaces, which seem rather rough in satellite images.

5.2.2.3 *Saw-toothed group yardangs*

Saw-toothed (S-T type), pyramid (S-P type), remnant cone (S-R type) yardangs are usually found together in the Qaidam Basin, so we consider them as one group, and here we describe each representative type individually. They usually develop in folded strata (Fan, 1962), which have similar dip directions with prevailing winds (Fig. 5.11(b)). Most of them are located in the southeastern limbs of anticlinal folded structures. They usually occur in Miocene–Middle Pleistocene mudstones. The best

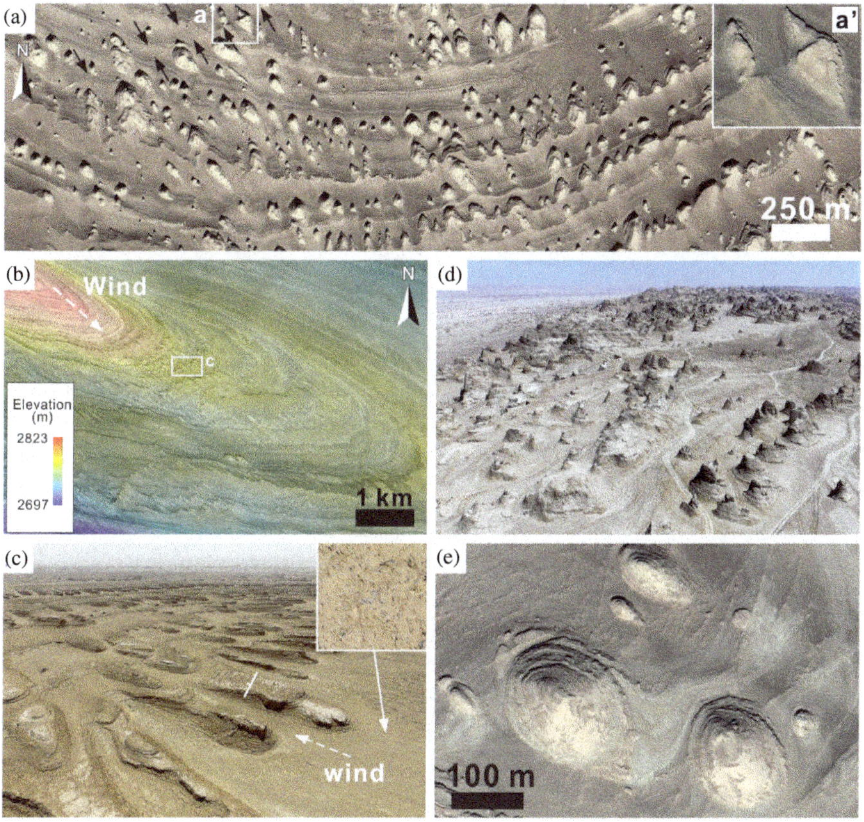

Fig. 5.11. Saw-toothed group (S group) yardangs in the Qaidam Basin. (a) Group of the S-T type, S-P type, and S-R type (remnant cone) yardangs (yellow arrows mark the fractures, image center: 37.62°N, 92.23°E). Inset a' shows that the yardangs are cut by fractures. (b) Google Terrain (SRTM-DEM) superposed on google images of the saw-toothed (S-T type) yardangs in the folded stratum (38.14°N, 91.87°E). (c) Perspective view using drone looking from north of the S-T type yardangs in (b), inset figure is the surface of the floor with gypsum. (d) The pyramid (S-P type) yardangs in the Daxiongshan uplift (E'boliang Yardang Geopark, image center: 38.35°N, 92.88°E). (e) The S-R type yardangs with rounded tops and multiple layers (37.52°N, 92.30°E).

developed S-P type yardangs are found in the Daxiongshan uplift (E'boliang Yardang Geopark) (Fig. 5.11(d)). The S-T type yardangs show zigzag-like morphologies (Fig. 5.11(a) and 5.11(b)). Layers exposed at their windward sides are eroded into small ridges (Fig. 5.11(c)). Most of their downwind sides are still unaltered, while some of their downwind sides are eroded to be divorced from the zigzag forming W-B or W-H type yardangs (Fig. 5.11(a)). In the field, we find salt crusts and small polygons on the top surfaces and gypsum-enriched layer on the floor (Fig. 5.11(c)). The S-P type yardangs look much more angular, with a large pedestal and a spire top (Fig. 5.11(d)) with no distinct coverings. The S-R type yardangs look like truncated cones. They usually have rounded shapes (Fig. 5.11(e)) or small flat tops.

5.2.2.4 *Whaleback group yardangs*

Whaleback (W-B type), hogback (W-H type), and low, streamlined whaleback (W-L type) yardangs often occur in groups mixed together in the field, which makes it difficult to separate individual types from each other. Considering their similar morphologies, we treat them as one group. They are widely distributed throughout the Qaidam Basin. They were mapped according to the locally most abundant type. The W-B type yardangs typically occur in the Dalangtan area consisting of Pliocene and Early–Middle Pleistocene mudstones. The W-H type yardangs mainly develop in Early–Middle Pleistocene mudstones which are distributed in the large area of the northwestern basin. The W-L type yardangs only appear in small areas in the basin. One representative area is in the north of the East Taijinaier Lake. They occur in Upper Pleistocene strata, which are characterized by evaporites.

The W-B type yardangs are archetypical yardang forms with blunt heads, convex flanks, and tapered tails. Viewed from the side, they show whaleback morphologies (Fig. 5.12(a)). They commonly have a thick (~50 cm), strong, salt crust covering horizontally bedded lacustrine sediments (Fig. 5.12(b)). In the field, there are also some small gullies carved by rainfalls on the flanks of these yardangs.

The W-H type yardangs have steep windward slopes (mostly >60°) and tapered tails (Fig. 5.12(c), 5.12(d), and 5.12(e)), and some have undercut windward heads (Fig. 5.15(a)). They are similar in morphology to the W-B type yardangs but are relatively less streamlined and larger. They usually have sharp crestlines on their backs (Fig. 5.12(c)). Some of their windward faces show evidence of collapses (Fig. 5.12(c)). Besides, their flanks are distinct from the W-B type yardangs. Some small W-H type yardangs exhibit slightly convex flanks, while some larger W-H type yardangs are characterized by convex flanks at the lower parts and concave flanks on their upper parts (Fig. 5.12(c) and 5.12(d)). Their lower

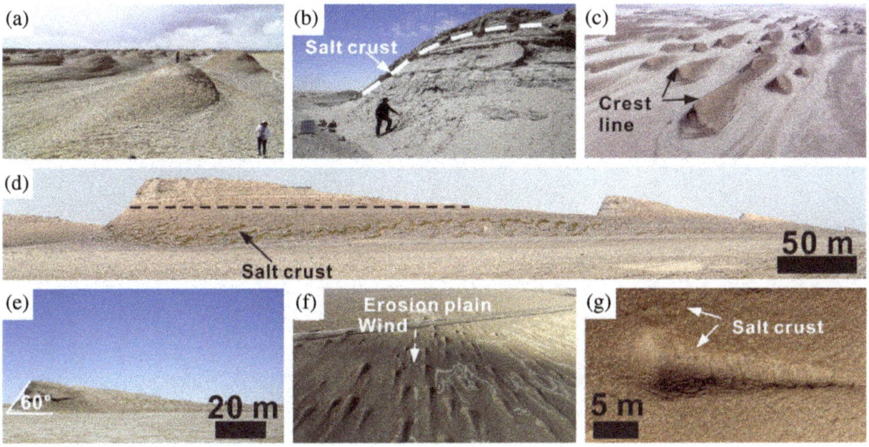

Fig. 5.12. The whaleback group (W group) yardangs in the Qaidam Basin. (a) The whaleback (W-B type) yardangs in Nanyishan (38.41°N, 91.32°E). Their surfaces are covered by salt crusts. (b) A well-exposed profile of the W-B type yardang (38.24°N, 92.10°E). The salt crust on its top is 40–50cm thick. Bird's-eye view (c) and side-view (d) of large W-H type yardangs with sharp crestlines (38.38°N, 93.54°E). Their flanks are convex at the lower slopes but concave at their upper parts (black dashed line indicates the boundary in (d). (e) A typical small hogback (W-H type) yardang with a windward slope of ~60° (38.51°N, 92.91°E). (f) and (g) The low, streamlined whaleback (W-L type) yardangs in the north of East Taijinaier Lake (37.61°N, 94.18°E). The upwind area is an erosion plain. Salt crusts cover both of the surfaces and the troughs of the yardangs.

parts are usually covered by salt crusts but their upper parts consist of exposed bedrocks (Fig. 5.12(d)). In contrast, the W-L type yardangs are relatively aerodynamically shaped and display perfect "teardrop"-like morphologies in plane view (Fig. 5.12(f) and 5.12(g)). They are smaller in size than the W-B type and W-H type yardangs, and usually have much larger spacing (Fig. 5.12(f)). Their surfaces are covered by polygonal fractured salt crusts (Fig. 5.12(g)).

5.2.3 *Evolution of yardangs in the Qaidam Basin*

Based on former studies (Brookes, 2001; Embabi, 1999; Dong *et al.*, 2012; Halimov and Fezer, 1989) and our work, we propose a four-step (embryonic, adolescent, mature, and receding) evolution model to describe the yardangs' formation and degradation in the Qaidam Basin (Fig. 5.13).

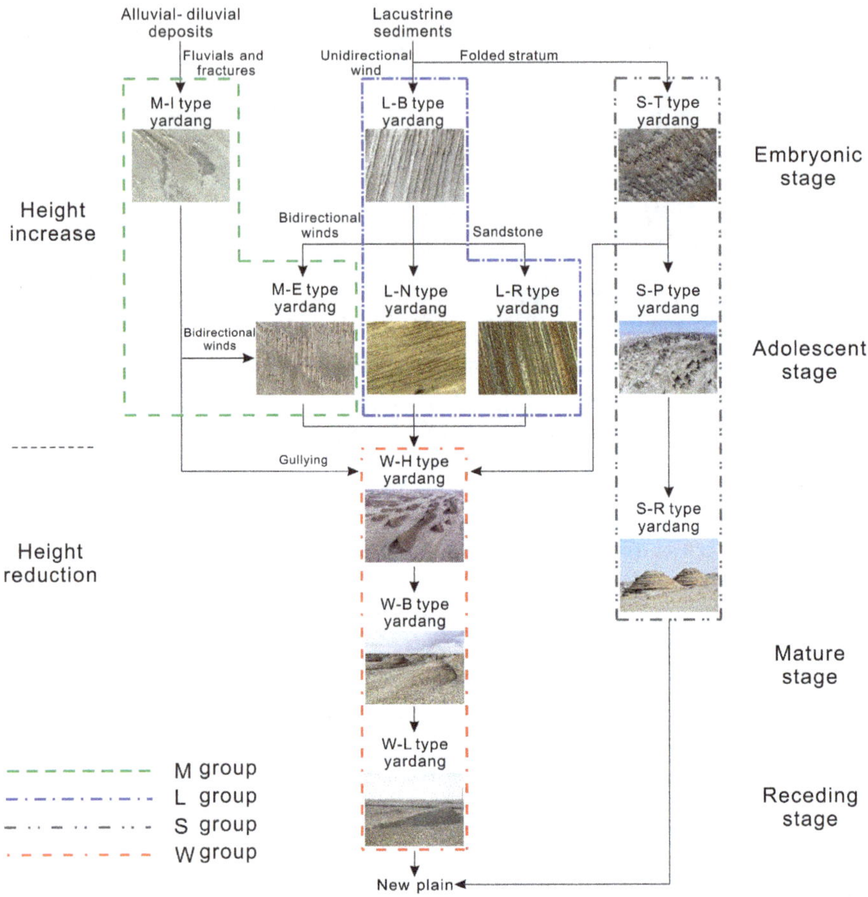

Fig. 5.13. The evolution model for yardangs in the Qaidam Basin (Wang *et al.*, 2018).

5.2.3.1 *Embryonic stage*

During the embryonic stage, local topography low areas, e.g. tectonic fractures, depressions, or surface runoff channels, provide initial passages for wind, where wind speed is strengthened, forming the turbulence when the speed >1 m/s (Bagnold, 1941), causing these passages to deepen, widen, and elongate along the dominant wind direction. The L group yardangs mainly form in horizontal lacustrine strata inside the basin surrounding the playas, they are characterized by relatively narrow troughs and broad tops (the L-B type) in the embryonic stage. While in the

Piedmont areas, alluvial-diluvial sediments are segmented into pieces by wind and fluvial processes forming large M-I type yardangs. Some M-I type yardangs inside the basin are controlled by pre-existing tectonic fractures. The dendritic shape M-I type yardangs are remnant steam channels, showing as positive terrains. They are protected by gravel-enriched layers, which are more resistant to wind erosion. The S-T type yardangs are developed in the downwind of the folded strata (Fig. 5.11(b) and 5.11(c)). Oblique topography and stratum accelerate the erosion. They form in relatively coherent sediments, suggesting that rock property plays an important role in their formation (Halimov and Fezer, 1989; Xiao *et al*., 2017). Their distributions are consistent with folded stratigraphy with blunt heads orientating to the upwind direction. The length of the sawtooth becomes shortened to the upwind side. In this period, the erosion rate of the troughs is greater than the yardang bodies. Thus, during this stage most of the erosion occurs in the troughs, the height of yardangs continuously increases.

5.2.3.2 *Adolescent stage*

During the early adolescent stage, the broadheads of the L-B type yardangs are eroded into small ridges with large spacings forming the L-N type yardangs. While some of the L-B type yardangs developed in the anticline uplifts consisting of sandstone interbedded with calcareous mudstones, and marls, their broad-top are eroded into broken tops. The M-E type yardangs usually transformed from the L-B type yardangs (Fig. 5.18(e)) or the M-I type yardangs (Fig. 5.7(b)) under the bidirectional winds inside the basin. It is obvious that the north-south orientated L-B type yardangs were segmented by the northwest wind from wind gap areas (Fig. 3.15(b) and Fig. 5.8(a)). In this period, the sawtooth length continues to shorten. When the downwind areas of the S-T type yardangs are segmented by fractures (Fig. 5.11(a)), rainfalls, or wind, the windward remnant parts form S-P type, W-H type, or S-R type yardangs. In the early adolescent stage, the height of the yardangs is still increasing. Meanwhile, fluvial erosion and mass wasting carve gullies surround the edge of yardangs, which provide passages for further erosion. As wind erosion continues, the height of yardangs begins to reduce, and the spacing between yardangs becomes larger. With leeward upper surface erosion accelerating, the L group yardangs change into many isolated W-H type yardangs that are less streamlined in shape and exhibit rough tops. In the field, the

W-H type yardangs surround the M group yardangs (Fig. 5.7(a), 5.7(b), and 5.7(c)), which are believed to be derived from the M-I type yardangs.

5.2.3.3 *Mature stage*

Mature stage yardangs are characterized by a streamlined shape and large spacing. During this stage, troughs usually appear with flat floors and large spacings because of more resistant underlayers, and yardangs are isolated occurrings, which have minimized the effects of the interference among yardang bodies in the incoming wind profile (Kerber, 2016). With erosion accelerating on the leeward upper surface and leeward corners, the W-H type yardangs are shortened in length, lowered in height, and become more streamlined, forming the W-B type yardangs. Statistics show that most of their aspect ratios are between 2:1 and 5:1, with an average value of about 3:1 (Wang *et al.*, 2018).

5.2.3.4 *Receding stage*

During the receding stage, the W-B type yardangs continually reduce in size, forming the W-L type yardangs. They are similar in morphology, but the latter ones are much smaller than the W-H and W-B type yardangs. As wind erosion continues, their heights decrease, the spacings increase, and finally a new erosion plain is formed before the next cycle of landscape evolution is initiated.

Brookes (2001) proposed a theoretical possibility that more advanced stages occur in upwind areas. The evolution of yardangs in the Qaidam Basin is cyclic. Thus, this sequence should be cyclic and varied spatially. According to previous studies (Halimov and Fezer, 1989; Li *et al.*, 2016) and our observations, the W group yardangs were evolved from the L group yardangs. Thus, the W group yardangs should occur at the upwind of the L group yardangs. However, the L group yardangs are distributed in the upwind area in contrast to the W group yardangs in the Qaidam Basin. This situation may indicate that the L group yardang represents the beginning of the next cycle. Thus, the yardang types show us a suite of cyclic yardang evolution stages.

In addition, we must be note that the evolution processes may be interrupted at some stages by the change of environment, e.g. rock competent, abrupt climate changes, tectonic movements, or infilling sediments. In the Qaidam Basin, competent layers in sediments play an important role in yardangs' evolution. These layers are more resistant to wind erosion, usually exhibit as "cap layers" (Fig. 5.10(b)), erosion floors (troughs) (Fig. 5.10(a)), or erosion plains (Fig. 5.12(f)), showing flat occurrence in morphology. The next cycle may begin after the breaking of the erosion floor (e.g. planation surface). The multi-stage evolution of yardangs in the Qaidam Basin is also evident by the occurrence of overlapped yardangs (Fig. 5.18(e)). As shown in Fig. 5.18(e), the approximately N-S oriented early stage yardangs are segmented by subsequent NW-SE winds on their top layers, and the new forming troughs are shallower than the former ones.

5.2.4 *Control factors of yardang morphology*

The formation and development of the Yardangs are related to many factors. Based on the previous morphological studies of the Qaidam Basin and Martian yardangs, in this section, we will discuss the role of lithology, wind, and water in the evolution of yardangs.

5.2.4.1 *Lithology*

Yardangs can be carved both in hard and soft materials under strong, unidirectional winds, with rock competent influencing their shape, scale, and rate of formation (Laity, 2011; McCauley *et al.*, 1977). The main characteristics of rocks that influence the evolution of yardangs are their compositions and textures, which can affect their anti-erosion resistance to wind (Dong *et al.*, 2012). The most common and widespread formation of yardangs on Earth occurs in softer materials, notably lacustrine sediments.

In the Qaidam Basin, most of the yardangs are developed in mudstones that are easily eroded. Some of them (e.g. the L-N type) have "cap layers" on their surface. These layers are usually rich in gypsum crystals and appear green in color (Fig. 5.10(b)). The size of these gypsum crystals can be as large as ~20 cm. These "green layers" exhibit as flat tops or surface layer (Fig. 5.11(c)), indicating more resistance to wind erosion that protects the underlayers from erosion. In addition, the gravel-enriched polygon layers (Fig. 5.7(d)) and salt crusts (Fig. 5.8(d), 5.9, and 5.12) act as protectors as well. Furthermore, yardangs covered by salt crusts (e.g. the W group, Fig. 5.12) usually display much more streamlined forms than others. Some yardangs carved in harder rocks usually exhibit irregular weathered tops (Laity, 2011), which are found in the L-R type yardangs of the Qaidam Basin that developed in sandstone interbedded with calcareous mudstones (Fig. 5.10(c)).

The MFF is a light-toned friable layered deposit (e.g. Morgan *et al.*, 2015; Tanaka, 2000; Weitz *et al.*, 2010). Radar data suggest the MFF materials could be dry, low-density deposits (e.g. volcanic ash) or ice-rich deposits in terms of their dielectric properties (Carter *et al.*, 2009; Harrison *et al.*, 2010; Head and Kreslavsky, 2001; Kerber *et al.*, 2008; Kerber and Head, 2010, 2012; Keszthelyi and Jaeger, 2008; Watters *et al.*, 2007). The W-H type yardangs distributed in the AZP area show evidence of cracks, blocks (Fig. 5.6(d)), or infrequent craters with ejected rubbles (Fig. 5.6(c)), suggesting an indurated homogeneous material. These cracks and blocks may be caused by the difference in temperature, varying greatly from day to night. However, the surface of floors exhibits greater roughness than the W-B type yardangs in the MFF area (Fig. 5.6(a)), which are similar to the fluted ventifacts, implying an indurated floor layer with lateral heterogeneity. Finely layered deposits are visible on the yardangs (Fig. 5.6(a')), indicating that the MFF materials have a multilayered structure with different competent layers, and yardangs are developed in the soft layers, which are consistent with the observations of terraced craters in MFF. Previous studies suggest that these terraced craters may be related to a subsurface ice-rich layer (e.g. Bramson *et al.*, 2015; Kadish *et al.*, 2009; Ormö *et al.*, 2013).

Except for the MFF exposure area, yardangs are also found developing in the upper unit with layered sediments in the Gale Crater (Le Deit *et al.*, 2011). Thomson *et al.* (2011) suggest that the upper mount material may be related to the MFF material, and were supported by Zimbelman and Scheidt (2012) and Dapremont *et al.* (2014). The S-T type yardangs are developed in the finely layered deposits, while their surface layers look much rougher, indicating a more resistant layer (Fig. 5.14(b)), which has also been found in the Qaidam Basin (Fig. 5.14(a)).

5.2.4.2 *Wind*

As reported in previous studies (Halimov and Fezer, 1989; Li *et al.*, 2016; Rohrmann *et al.*, 2013), the wind is the dominant driver of yardangs' erosion in the Qaidam Basin. Both the abrasion and deflation are included (Breed *et al.*, 1989; Whitney, 1983), where the former plays an important

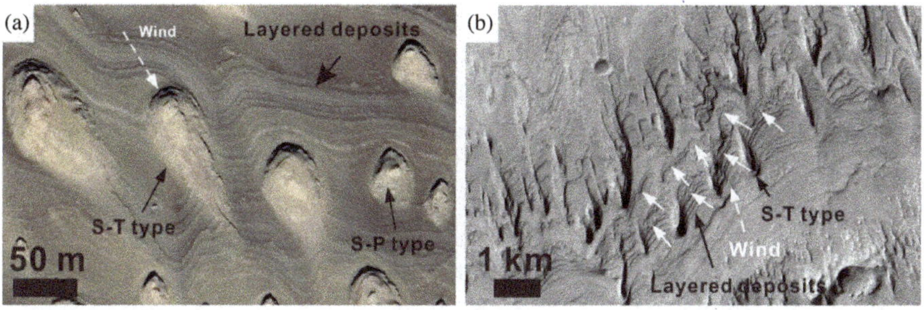

Fig. 5.14. (a) The S-T and S-P type yardangs shown in Fig. 7(a) in the Qaidam Basin. (b) The S-T type yardangs in Gale Crater (white dashed arrow indicates the wind directions, and white solid arrows mark the boundary between thin layer and antiresistant layers). CTX image, f20_043750_1747_xn_05s221w.

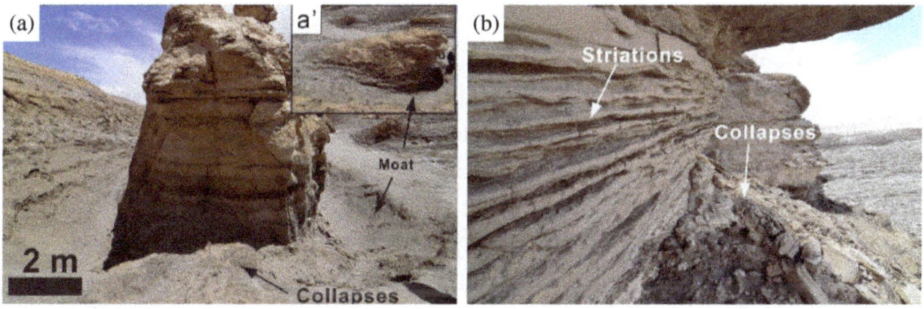

Fig. 5.15. (a) Moat on the upwind face and flanks (38.49°N, 91.16°E). Collapsed material occurs in the upwind direction. (b) Striations on the flank (38.50°N, 91.16°E).

role. Wind abrasion aids in the deepening and widening of topographic lows into troughs (Ward, 1979). It is most active within the lower 1~3 m above ground, resulting in the development of steep, often undercut, upwind faces (Fig. 5.15(a')) (Bosworth *et al*., 1922; Dong *et al*., 2004; Donner and Embabi, 2000; Grolier *et al*., 1980; Hobbs, 1917; Laity, 2011).

In the Qaidam Basin, moats and flank striations are obvious on many steep upwind faces (Fig. 5.7(b), 5.15(a), and 5.15(b)), indicating the dominant wind direction and turbulent erosion (McCauley *et al*., 1977; Whitney, 1983). Unlike abrasion, the deflation functions across the entire surface of the yardangs all the time. In fact, these two processes always co-operate in yardang fields. However, many yardangs in the Qaidam Basin, such as the W-B type (Fig. 5.12(a)), the W-L type (Fig. 5.12(f)), and the L-B type yardangs (Fig. 5.9), usually show a thick (~50 cm), hard, salt crust, which appears to have protected the underlying horizontally layered lacustrine sediments from further wind erosion, including abrasion and deflation (Fig. 5.12(b)). Dark streaks are visible in the W-H type yardangs of AZP area on Mars indicating a thick dust covering on the surface, which reveals that both the abrasion and deflation perhaps are not active in this area at present.

Most studies support that yardangs are formed by a unidirectional strong wind (Hobbs, 1917; McCauley *et al.*, 1977; Niu *et al.*, 2011), while opposite bidirectional winds, with one stronger, can also account for their formation (Laity, 2011; Ritley and Oontuya, 2004). Usually, the upwind side broad tops are eroded into small ridges, while their downwind areas are still unaltered, this can be used as an indicator of prevailing wind (Fig. 5.9(b) and 5.11(b)). We found the L-B type yardangs in Lucus Planum on Mars have been eroded to small ridges on both sides (Fig. 5.2(a) and 5.2(d)), indicating that they have been eroded by two opposite winds. The amount of small ridges in the NW part outnumber those in the SE part, implying that the NW wind may be stronger than the SE wind, or the NW wind works a longer time. However, the reason why the wind changed directions during the yardangs' forming is still not clear.

In the Qaidam Basin, the M-E type yardangs develop in the downwind direction of the L group yardangs that are eroded by bidirectional winds. Their dominant long axis orientations range from NNW to N (Wang *et al.*, 2018), which coincides with the upwind direction of the L group yardangs (Fig. 5.8(a) and 5.8(e)), and they were segmented by NW winds (Fig. 5.8(e)). The included angles between these two directions vary from 40° to nearly vertical. The M-E type yardangs developed in Zephyria Planum on Mars also formed under bidirectional winds with an angle of about 60° (Fig. 5.1(d)). When the angle between these two winds ranged from 30° to 45°, then usually faceted yardangs with curvilinear crestlines formed (Fig. 5.3), which are widely distributed on the Martian surface (Bradley, 2002).

In the north Zephyria Planum, yardangs developed in different layers by multidirectional winds are observed (Fig. 5.16). Some L-B type paleo-yardangs are excavated to the surface with over layered L-N type yardangs (Fig. 5.16(e)). Their included angles range from ~40° to ~60°, indicating the dominant wind direction changing dramatically over different periods. The variation of yardang orientations also has been found in the same layer in the eastern region of this area (Fig. 5.16(b), 5.16(c), and 5.16(d)).

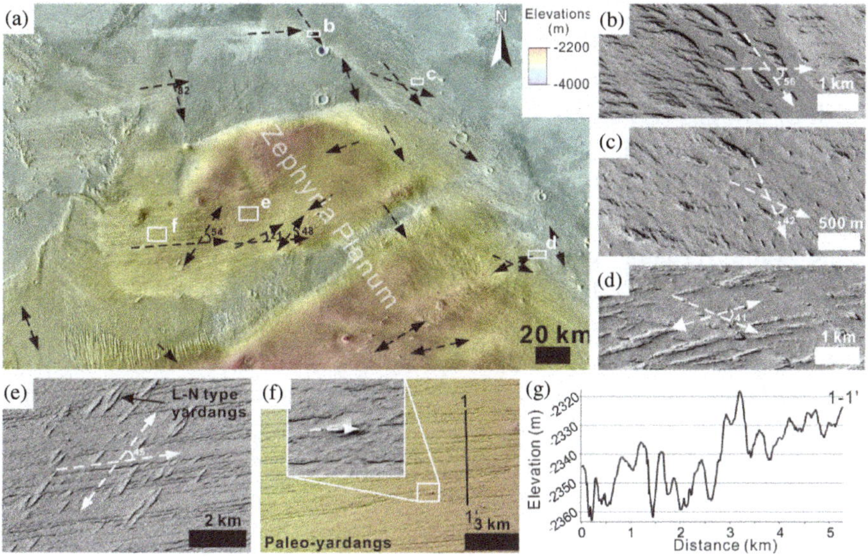

Fig. 5.16. (a) Yardangs and paleo-yardangs in the North Zephyria Planum, THEMIS Day IR image (Edwards *et al.*, 2011) superposed on HRSC DTMs (h2110_0000_dt4, h2121_0000_dt4, h2143_0000_dt4, h2154_0000_dt4, h2165_0001_dt4, h2176_0000_dt4), black dashed lines mark the orientation of yardangs. (b), (c), and (d) The variation of yardang orientations at the eastern area in (a) (white dashed lines mark the orientations of yardangs, CTX images, b09_013302_1844_xn_04n208w, b06_011957_1865_xn_05n207w, f23_044844_1818_xn_01n205w). (e) and (f) L-B type paleo-yardangs shown in (a). (g) Elevation profile along 1-1' in figure (f).

Their included angles various from ~40° to ~60°, and present wind directions revealed from the sand ripples are NWW-SEE. The variation of yardang orientations suggests that the dominant wind directions have changed and last for enough time to form yardangs.

5.2.4.3 *Water*

Though yardangs are wind erosion features, water erosion definitely exerts an influence during their evolution (Dong *et al.*, 2012; Laity, 2011; Li *et al.*, 2016; Wang and Ha, 2009; Wang *et al.*, 2016; Xia, 1985; Xiao *et al.*, 2017). For example, temporary runoff from surrounding mountains may cut the surface into pieces that provide passages to the wind, while rainfall modifies the surface of the yardangs by gullying (Fig. 5.7(b) and

5.7(d)), both of which may accelerate the evolution of yardangs during the embryonic stage (Halimov and Fezer, 1989; Li *et al.*, 2016). The water table is also an important factor. Their fluctuations can determine the depth of troughs (Laity, 2011).

The climate of Mars became cold and dry during the Hesperian epoch (Carr,1999; Clifford, 1993; Ehlmann *et al.*, 2011; Fassett and Head, 2008; Head, 2002; Golombek *et al.*, 2006; Treiman *et al.*, 1995; Xiao, 2013). We found some yardangs coexisting with fluvial landforms. The M-I type yardangs that developed in Aeolis Dorsa show dendritic shape (Fig. 5.1(a)). In the Gunjur Crater, eastern Aeolis Planum, paleo MFF yardangs have been excavated from delta deposits (Fig. 5.17(b) and 5.17(d)), and remnant yardangs have also been found on the delta surface (Fig. 5.17(c)), suggesting that the fluvial activity was still active during the formation or reformation of MFF. But there is no evidence showing whether the fluvial activity has influenced the formation of yardangs, and the preservation of the water is unknown.

5.2.4.4 *Other factors*

In addition to lithology, wind, and water erosion, several other factors may have controlled the formation and evolution of yardangs in the Qaidam Basin. The presence of salts in the Qaidam Basin implies that salt solution and weathering may have played a significant role during yardang formation and modification. The solution features are common on yardangs that are developed within playas (Fig. 5.7(d)). They usually occur on the surface of leeward sides, and away from the windward abrasion areas, implying that a wetter past existed when the solution features were formed and the present drier climate has ablated the salt solution products from the windward sides (Vincent and Kattan, 2006). Mass wasting processes, very common both in the Qaidam Basin on Earth and in MFF and Gale Crater on Mars, contribute a lot to the modification and erosion of the yardangs, which are manifested as detached boulders around the yardangs (Fig. 5.7(d) and 5.6(c)). On the other hand, impact activities, in most cases, accelerate the erosion of yardangs (Fig. 5.6(c)) on Mars, but the overlapping relationship between yardangs and impact structures are not common. Furthermore, the formation of some S-R type yardangs on Mars might also be related to the impact (Fig. 5.18). It has been well demonstrated that impact cratering on competent rocks normally cause fracturing and excavation, thus decreasing the porosity of

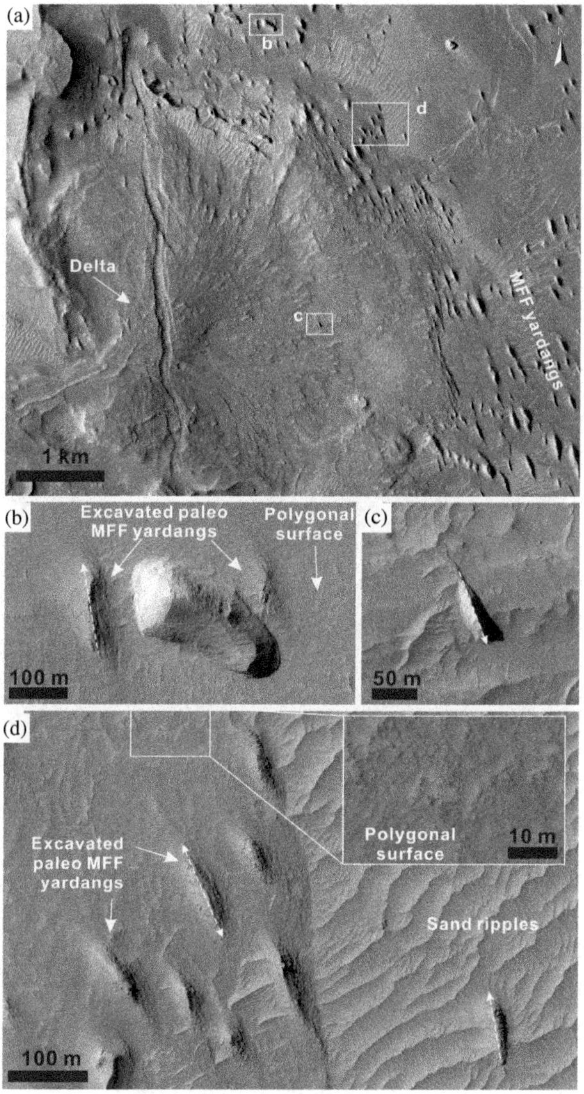

Fig. 5.17. (a) Delta and yardangs in the Gunjur Crater (CTX image, d15_033108_1798_xn_00s213w). (b) The excavated paleo MFF yardangs (white dashed line marks the orientation of yardang, HiRISE image, psp_007883_1800_red). (c) The remnant yardang on the surface of delta deposits (white dashed line marks the orientation of yardang, HiRISE image, psp_007883_1800_red). (d) The excavated yardangs (white dashed line marks the orientation of yardang, HiRISE image, psp_007883_1800_red). Polygonal features are obvious on the floor layer.

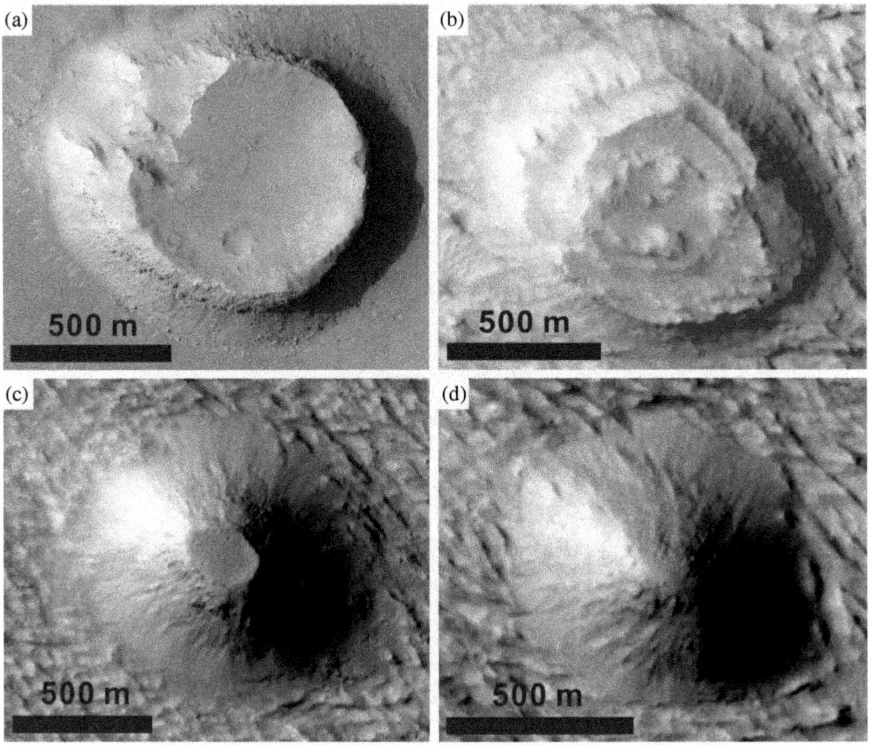

Fig. 5.18. The formation of S-R type yardangs that are related to craters (CTX image, f07_038607_1821_xn_02n215w, f02_036497_1830_xn_03n216w).

target materials (Melsoh, 1989). However, it has been recently noticed that when target features high porosity, impact cratering actually causes the substantial collapse of pores and melting, which decrease the porosity and enhance the strength (Scott and Wilson, 2005; Wünnemann *et al.*, 2006). Such impact hardening occurs beneath the final crater floor, which has been used to interpret the observed larger-than-predicted gravity anomaly beneath some lunar highland impact craters (e.g. Milbury *et al.*, 2015). For the S-R yardangs on Mars, it might be possible that previous impact craters that have formed within the MFF, which might feature high porosity (e.g. Carter *et al.*, 2009; Ojha and Lewis, 2018; Watters *et al.*, 2007), have caused a decrease of porosity beneath the crater floor. In this scenario, materials beneath the crater floor will be slightly stronger than the surrounding materials, thus continual weathering could cause such cone-like yardangs in their upper layers.

5.3 Dunes

5.3.1 *Dunes on Mars*

5.3.1.1 *Distribution of dunes on Mars*

Multiple remote sensing data indicate that there are many loose sediments on the Martian surface. Under the action of the Martian wind, these loose sediments are transported to different locations, forming a large number of sand dune fields. On Earth, the dune fields are mainly distributed in the middle latitudes along the zonal direction. While sand dune fields on Mars are mainly distributed surrounding the Arctic ice sheet, and scattered in low-lying areas, such as craters and troughs (Fig. 5.19). Hayward *et al.* (2014) mapped the dune field on Mars based on the THEMIS VIS, MOC NA, MRO CTX, and MRO HiRISE images. The total area of dune fields is ~975,000 km^2 and their areal extent is summarized in Table 5.1.

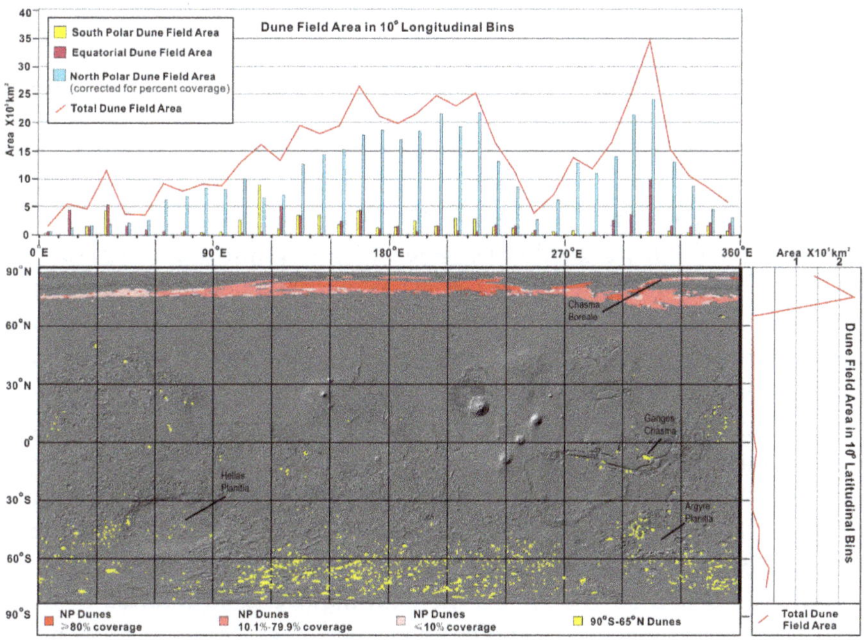

Fig. 5.19. Global distribution of dark dune fields (>1 km^2) on Mars (Background MOLA Hillshade, Simple Cylindrical Projection, Mars 2000). Bar charts emphasize polar concentrations and overall uneven distribution of dune fields (modified from Hayward *et al.*, 2014).

Table 5.1. Summary of the area of sand dunes in EQ, NP, and SP regions on Mars. A small number of dune fields were omitted because satellite imagery was incomplete when our initial mapping was done (Hayward *et al.*, 2014).

Region	Latitude(°)	Area (km^2)	Area × percent coverage (km^2)
North pole (NP)	+65~+90	845,000	380,000
Equatorial (EQ)	+65~−65	70,000	70,000
South pole (SP)	−65~−90	60,000[a]	60,000
Total	+90~−90	975,000	510,000

Notes: [a]This does not include the ~15,000 km² of sand sheets and smaller dune fields that would not have been mapped using criteria followed in the earlier (EQ and NP) versions. Similar exclusions are likely in the equatorial and north polar regions but have not been quantified.

For the NP region, the mapped dune fields include both significant areas and sparsely covered areas. The coverage densities of the dunes in red areas are 80–100%, areas in dark pink with less complete coverage are 10–80%, and areas of widely scattered barchan dunes are in light pink. Within the ~845,000 km² of dune fields in the NP region, only ~10% or less (~90,000 km²) areas are covered by dunes. Considering the percent dune coverage of each area, the actual area covered by dunes in the NP will be reduced to ~380,000 km² and the global total is reduced to ~510,000 km².

The EQ region accounts for ~70,000 km² of the dune fields. The SP contains ~60,000 km² of dune fields >1 km² (shown on map) and an additional ~15,000 km² of sand deposits and smaller dune fields (not shown in Fig. 5.19). For the EQ and SP regions, the individual dune field areas are much smaller and coverage by dunes is more uniform, so the percent dune coverage is not an issue.

The spatial distribution of dune fields across Mars is uneven both across latitudes and longitudes (Fig. 5.19). The latitudinal disparity is quite dramatic, with 28% of mapped dune fields (by area) located between 80°N and 90°N, and another 47% between 70°N and 80°N. Together they account for 75% of the dune field area on Mars. The latitudinal zone between 60°S and 70°S contains the next highest concentration of dune fields, 8%. The adjacent latitudinal band between 70°S and 80°S holds 7%. Together they account for 15% of dune fields. Combined, these NP and SP regions have 90% of the dune field area.

From the longitudes, there is a broader peak in dune field abundance from ~160°E to 230°E and a narrow peak between 310°E and 320°E.

There is a broader low between 0°E and 60°E and a fairly abrupt low in abundance between 250°E and 270°E. From the latitudes, a distinct peak is obvious between ~65°N and ~90°N. These distribution patterns are due to the dominant north polar dunes.

5.3.1.2 *Morphology of dunes on Mars*

The main types of dunes on Mars are transverse dunes, including barchan dunes, barchan dune chains, and transverse ridges. Other types are relatively few, such as linear dunes, star dunes, dome dunes, etc. Parabolic dunes have not yet been found.

Transverse dunes: The transverse dunes are formed in a single wind environment. Their typical morphological characteristics are that the windward slope is convex and gentle, while the concave leeward slope is steep, and the two slopes intersect into a curved sand ridge. The plan view is a crescent shape (Fig. 5.20(a)), with only one leeward surface (McKee, 1979). The main wind direction can be judged from the gentle slope to the steep slope according to the shape of the dune. The barchan dune chains formed by connecting of the barchan dunes (Fig. 5.20(b)). Compared to barchan dunes, barchan dune chains are often formed in areas with abundant sand sources.

Crescent sand dunes: When the crescent lines of the barchan dune chains are relatively straight, they appear as transverse ridges (Fig. 5.20(c)).

Linear dunes: The linear dunes are usually formed in a bidirectional winds environment, exhibiting as lines, with two leeward surfaces (McKee, 1979). Compared with the transverse dunes, linear dunes have symmetrical ridges. Usually the ridge lines are like sharp peaks, and the leeward surfaces on both sides of the base are convex. Some linear dunes have "Y" shaped nodes that open out towards the wind. There are also some linear dunes whose ridges meander or undulate.

Star dunes: They are usually formed in a multidirectional wind environment. This type of sand dune has a central high point, and the apex of the center is connected to three or more ridges, showing as a star-shape on the plane view (McKee, 1979) (Fig. 5.21(b)). Therefore, there are three or more leeward surfaces, which can be explained by the combination of

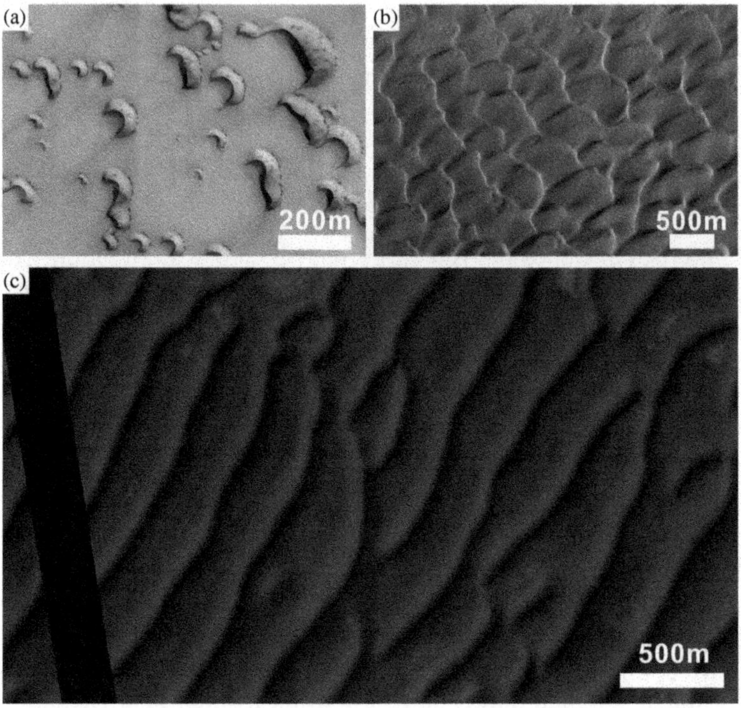

Fig. 5.20. (a) Barchan dunes (HiRISE image, esp_033228_2560_red); (b) Barchan dune chains (Transverse dunes) (CTX image, b21_017762_1891_xn_09n292w); (c) Transverse ridges (HiRISE image, esp_032185_1165_red).

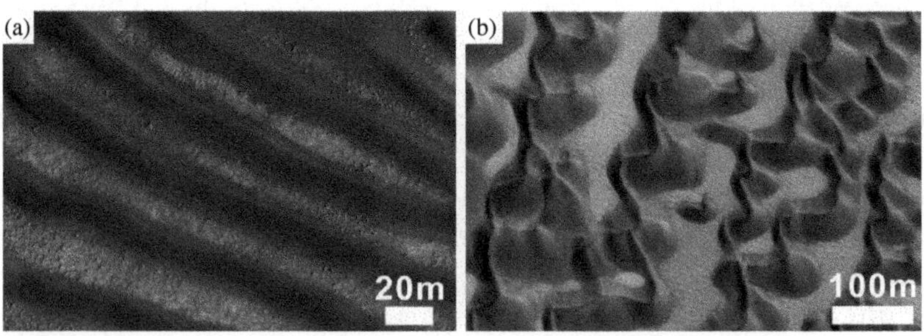

Fig. 5.21. (a) Linear dunes (HiRISE image, esp_0016036_1370_red); (b) Star dunes (HiRISE image, esp_018742_2565_red).

multiple wind directions. Star dunes are less distributed on the surface of Mars. When multiple star dunes are connected to each other, grid dunes are formed.

Dome dunes: They are usually formed in a single wind direction environment, the geometric shape is similar to a single pie-shaped dune, generally without external slip surfaces (McKee, 1979) (Fig. 5.20(e)). They are usually small in size, and distributed at the edge of the dunes or connected to crescent and shorter linear dunes.

As mentioned above, the surface of Mars is extremely dry and cold, and various dune landforms are formed under the combined effect of the Martian atmosphere and surface materials. In addition to common types of dunes, some special types of dunes are found on the surface of Mars, and some types of dunes are unique, without counterparts on earth.

Frosted dunes: When the winter leaves and the early spring comes on Mars, the dunes in the southern hemisphere of Mars just get enough sunlight to start defrosting their seasonal cover of carbon dioxide. Spots form where pressurized carbon dioxide gas escapes to the surface (Fig. 5.23). Dunes cover much of this terrain, and some large boulders lie on flat areas between the dunes.

"Morse code" sand dunes: A circular depression (probably an old and infilled impact crater) has limited the amount of sand available for dune formation and influenced local winds. As a result, the dunes here form distinct dots and dashes. The "dashes" are linear dunes formed by bidirectional winds, which are not traveling parallel to the dune. Instead, the

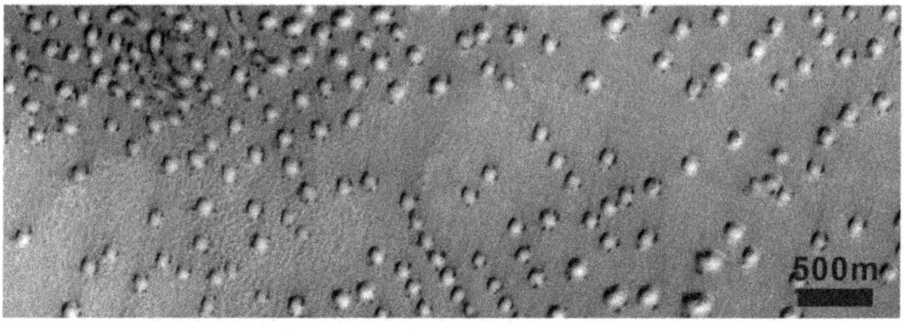

Fig. 5.22. Dome dunes on Mars (HiRISE image, esp_050886_2565_red).

Fig. 5.23. Frosted dunes on Mars. This image was captured by HiRISE at 15:31 Mars time on March 27, 2016. Image credit: NASA/JPL/University of Arizona (https://www.nasa.gov/image-feature/frosted-dunes-on-mars).

combined effect of winds from two directions at right angles to the dunes funnels material into a linear shape. The smaller "dots" (called "barchanoid dunes") occur where there is some interruption to the process forming those linear dunes.

5.3.2 *Dunes in the Qaidam Basin*

This section will introduce the main dune types in the Qaidam Basin. According to the morphological classification, there are mainly barchan dunes, barchan dune chains, parabolic dunes, linear dunes, and star-shaped dunes.

5.3.2.1 *Transverse dunes*

Barchan dunes are independently distributed, with widths and heights ranging from 50–300 m and 2–5 m, respectively, while their convex windward slopes are gently dipping, and their concave leeward slopes are steep (Fig. 5.24(a)), similar in shape and size to their counterparts on Mars (Fig. 5.20(a)).

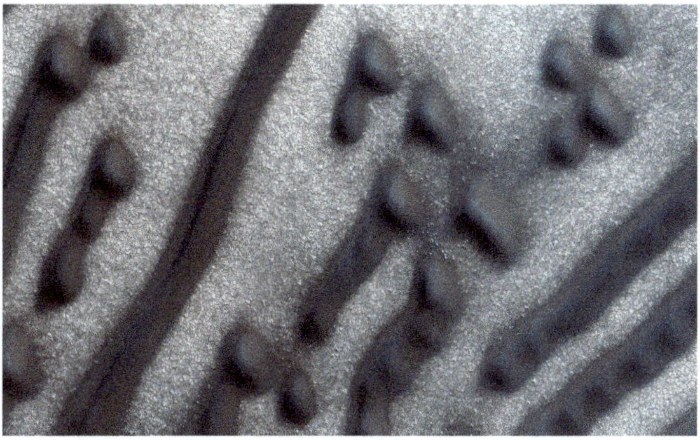

Fig. 5.24. The Morse code dunes on Mars. This image was captured by HiRISE on February 6, 2016. Image Credit: NASA/JPL/University of Arizona (https://www.nasa.gov/image-feature/martian-morse-code).

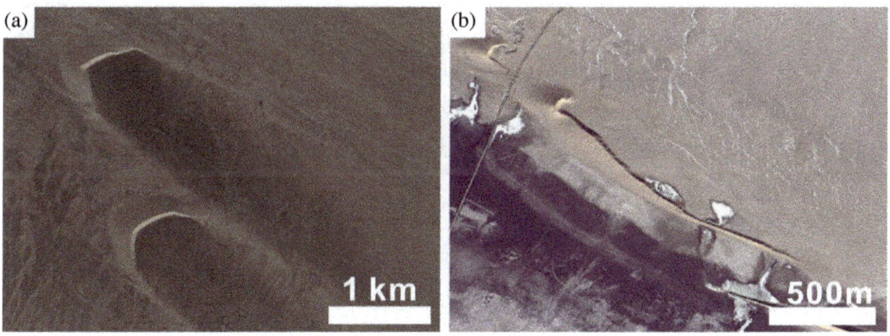

Fig. 5.25. Barchan dunes in the Qaidam Basin. (a) The two arms of the dune have better symmetry, the windward side is gentle, the leeward side is steep, and the crescent is open to the leeward side. The composition of the sand grains is mainly minerals such as quartz and feldspar. The extension of the arms of the barchan dunes indicates that the wind direction is from the upper left to lower right (37.81°N, 90.98°E). (b) Asymmetric barchan dunes (37.07°N, 95.45°E), linear with one-arm extension, possibly related to bottom brine water (Xiao *et al.*, 2017).

5.3.2.2 *Barchan dune chains*

Barchan dune chains are formed in areas rich in sand sources (Fryberger and Dean, 1979) and are usually several hundred meters to several kilometers long (Fig. 5.27(a)).

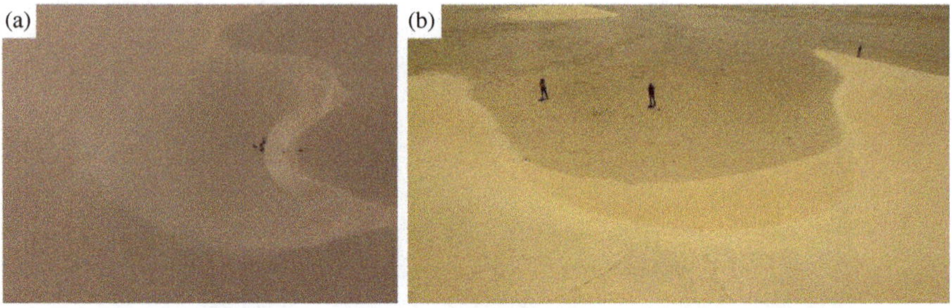

Fig. 5.26. (a) Overview of a single barchan dune in the Qaidam Basin. The arms of the crescent are asymmetric, about 40 m in length, about 40 m in width, and about 10 m in height. The wind direction is from left to right (the black spots are the team members). (b) A leeward surface of a barchan dune. The small piece of sand sheet at the bottom of the leeward slope is a sand skirt.

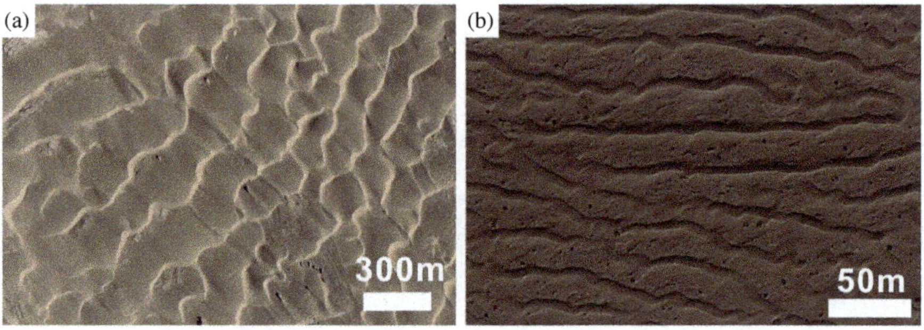

Fig. 5.27. (a) The barchan dune chains in the Qaidam Basin (38.22°N, 91.36°E). (b) Transverse sand ridges with relatively straight crestlines (37.79°N, 92.69°E).

5.3.2.3 *Parabolic dunes*

The parabolic dunes are usually developed in the east of the Qaidam Basin. They are typically characterized by a recessed windward slope, a leeward slope protruding, two wing angles pointing in the direction of the windward, and the plane view is parabolic. Parabolic dunes belong to horizontal dunes or semi-fixed dunes. The arms of the dunes are fixed by plants, and the sediment in the middle continues to form an anti-crescent shape in the wind (Fig. 2.28).

5.3.2.4 *Linear dunes*

Linear dunes mostly occur in the north part of the Qarhan Salt Lake and Sanhu areas (three lakes, namely east Taijinaier Lake, west Taijinaier

Fig. 5.28. The parabolic dunes in the east of the Qaidam Basin (36.56°N, 97.48°E), the wind direction is from left to right.

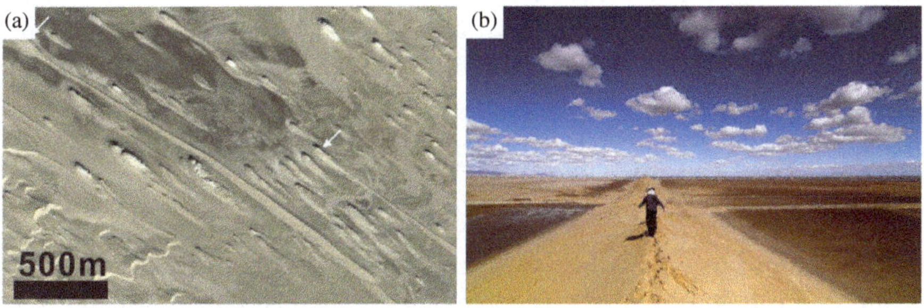

Fig. 5.29. The linear dunes in the Qaidam Basin. (a) A group of linear dunes in the Qaidam Basin, where linear dunes are arranged in parallel. The white arrow points to the yardangs (37.15°N, 95.23°E). (b) The two wings of a linear dune present symmetry (37.08°N, 95.42°E), which is the main feature that distinguishes it from the transverse ridge above.

Lake, and Senie Lake). In some cases, the linear dunes co-exist with barchan dune chains, barchan dunes, or yardangs (Fig. 5.29(a)). Linear dunes at the northern Qarhan Basin are northwest-trending (296°), ~12 km long, 30–200 m wide, 5–30 m high, and 100–600 m apart (Fig. 5.29). The linear dunes at the Sanhu area (Fig. 3.22(c)) are also northwest-trending (311°) but relatively smaller in size (1–5 km long, 40–100 m wide, 3– 20 m high, and 50–1400 m apart).

Fig. 5.30. The star dunes (a) and the grid dunes (b) in the Qidam Basin (36.80°N, 92.70°E).

5.3.2.5 *Star dunes*

Star dunes are rare with respect to other types in the Qaidam Basin (Fig. 5.30(a)). They are generally distributed among large barchan dune chains, forming due to the local multidirection wind environment. When the star dunes appear in groups, they form the grid/honeycomb dune (Fig. 5.30(b)).

5.4 Summary

In this chapter, we described the classification, distribution, evolution of aeolian landforms, including yardangs and dunes, both in the Qaidam Basin on Earth and on Mars. The yardangs in the Qaidam Basin are classified into 11 types out of 4 main groups based on their morphologies and distributions. They are mainly distributed in the central-eastern and northwestern regions of the basin. We have found similar yardang forms on Mars. The wind is the dominant driver of yardangs' erosion in these two study areas. Independent of topographic effects, the variation of yardangs' dominant orientations in different layers of these two study areas probably indicate a polytropic and complex wind regime when they were formed. The evolution processes of yardangs in the Qaidam Basin are divided into four stages (embryonic, adolescent, mature, and receding). The yardangs on Mars appear to have experienced a similar multi-stage evolution, during which the evolution processes may have been interrupted at some stages due to differences in the competence of rocks and/or change of environment. Dunes in the Qaidam Basin can be divided into 7 types, of which 6 types have been found on Mars. Some dune types are unique on Mars and we haven't found their terrestrial analogs on Earth, which may be related to their formation environment. In summary, both morphological and geometrical characteristics show that they have more similarities than differences, indicating the Qaidam Basin can be listed as an analog site for Mars.

Acknowledgments

We are grateful for funding from the Science and Technology Development Fund (FDCT) of Macau (No. 121/2017/A3), National Natural Science Foundation of China (No. 41830214 and 41773061), MOST Special Fund from the State Key Laboratory of Geological Processes and Mineral Resources, China University of Geosciences (No. MSFGPMR05), and the Fundamental Research Funds for the Central Universities, China University of Geosciences (Wuhan) (Nos. CUGYCJH18-01 and CUG2017G02). We would like to thank Dr. Harald Hiesinger, Dr. Dennis Reiss, Dr. Zhiyong Xiao, Dr. Jiannan Zhao, and Dr. Goro Komatsu for suggestions and constructive discussions. Dr. Le Qiao, Dr. Ziye Cheng, Dr. Ting Huang, Dr. Yanan Dang, Ms. Zhuqing Xue, Mr. Limin Zhao, Mr. Yu Wei, Mr. Ping Zhu, Mr. Bo Wang, Mr. Hansheng Liu, Mr. Jiawei Zhao, Mr. Siyuan Zhao, and Mr. Ning Zhang are thanked for fieldwork assistance.

References

Bagnold, R. (1941). *The Physics of Blown Sand and Desert Dunes*. Chapman and Hall Ltd, London.

Bosworth, T. O., Woods, H. and Vaughan, T. W. (1922). *Geology of the Tertiary and Quaternary Periods in the North-west Part of Peru*. Macmillan and Company, limited.

Bourke, M. C., Balme, M., Lewis, S., Lorenz, R. D. and Parteli, E. (2019). *Planetary Aeolian Geomorphology*, 261–286.

Bradley, B. A. (2002). Medusae Fossae Formation: New perspectives from Mars global surveyor. *J. Geophys. Res.*, **107**, 2–1–2–17.

Bramson, A. M., Byrne, S., Putzig, N. E., Sutton, S., Plaut, J. J., Brothers, T. C. and Holt, J. W. (2015). Widespread excess ice in Arcadia Planitia, Mars. *Geophys. Res. Lett.*, **42**, 6566–6574.

Breed, C. S., McCauley, J. F. and Whitney, M. I. (1989). Wind erosion forms. *Arid Zone Geomorphology*, 284–307.

Brookes, I. A. (2001). Aeolian erosional lineations in the Libyan Desert, Dakhla Region, Egypt. *Geomorphology*, **39**, 189–209.

Carr, M. H. (1999). Retention of an atmosphere on early Mars. *J. Geophys. Res. Planet*, **104**, 21897–21909.

Carr, M. H. (2007). *The Surface of Mars*. Cambridge University Press.

Carter, L. M., Campbell, B. A., Watters, T. R., Phillips, R. J., Putzig, N. E., Safaeinili, A., Plaut, J. J., Okubo, C. H., Egan, A. F., Seu, R., Biccari, D. and Orosei, R. (2009). Shallow radar (SHARAD) sounding observations of the Medusae Fossae Formation, Mars. *Icarus*, **199**, 295–302.

Chamberlain, T. E., Cole, H. L., Dutton, R. G., Greene, G. C. and Tillman, J. E. (1976). Atmopspheric Measurements on Mars: the Viking Meteorology Experiment. *Bull. Am. Meteorol. Soc.*, **57**, 1094–1104.

Clifford, S. M. (1993). A Model for the Hydrologic and Climatic Behavior of Water on Mars. *J. Geophys. Res. Planet*, **98**, 10973–11016.

Cooke, R. U., Warren, A. and Goudie, A. S. (1993). *Desert Geomorphology*. CRC Press.

Dapremont, A., Allen, C., and Runyon, C. (2014). The Gale Crater Mound in a Regional Geologic Setting: Comparison Study of Wind Erosion in Gale Crater and Within a 1000 KM Radius, AGU Fall Meeting Abstracts, p. 1855.

Dong, Z., Lv, P., Lu, J., Qian, G., Zhang, Z. and Luo, W. (2012). Geomorphology and origin of Yardangs in the Kumtagh Desert, Northwest China. *Geomorphology*, 139–140, 145–154.

Dong, Z., Wang, H., Liu, X. and Wang, X. (2004). A wind tunnel investigation of the influences of fetch length on the flux profile of a sand cloud blowing over a gravel surface. *Earth Surf. Proces. Land.*, **29**, 1613–1626.

Donner, J. and Embabi, N. S. (2000). The significance of yardangs and ventifacted rock outcrops in the reconstruction of changes in the late Quaternary wind regime in the western desert of Egypt [L'importance des yardangs et des affleurements de roches modelées par le vent dans la reconstitution des changements du régime des vents au Quaternaire supérieur dans le désert libyque, Egypte]. *Quaternaire*, **11**, 179–185.

Ehlmann, B. L., Mustard, J. F., Murchie, S. L., Bibring, J.-P., Meunier, A., Fraeman, A. A. and Langevin, Y. (2011). Subsurface water and clay mineral formation during the early history of Mars. *Nature*, **479**, 53–60.

Embabi, N. (1999). Playas of the western desert, Egypt. *Studies of Playas in the Western Desert of Egypt*, 5–47.

Fassett, C. I. and Head, J. W. (2008). Valley network-fed, open-basin lakes on Mars: Distribution and implications for Noachian surface and subsurface hydrology. *Icarus*, **198**, 37–56.

Golombek, M. P., Grant, J. A., Crumpler, L. S., Greeley, R., Arvidson, R. E., Bell, J. F., Weitz, C. M., Sullivan, R., Christensen, P. R., Soderblom, L. A. and Squyres, S. W. (2006). Erosion rates at the Mars exploration rover landing sites and long-term climate change on Mars. *J. Geophys. Res. Planet*, **111**, E12S10, doi: 10.1029/2006JE002754.

Grotzinger, J., Hayes, A., Lamb, M. and McLennan, S. (2013). Sedimentary processes on earth, Mars, titan, and Venus. *Comparative Climatology of Terrestrial Planets*, **1**, 439–472.

Halimov, M. and Fezer, F. (1989). Eight yardang types in Central-Asia. *Zeitschrift fur Geomorphologie*, **33**, 205–217.

Harrison, S. K., Balme, M. R., Hagermann, A., Murray, J. B. and Muller, J. P. (2010). Mapping Medusae Fossae Formation materials in the southern highlands of Mars. *Icarus*, **209**, 405–415.

Hayward, R. K., Fenton, L. K. and Titus, T. N. (2014). Mars Global Digital Dune Database (MGD3): Global dune distribution and wind pattern observations. *Icarus*, **230**, 38–46.

Head, J. and Kreslavsky, M. A. (2001). Medusae Fossae Formation as volatile-rich sediments deposited during high obliquity: An hypothesis and tests, Conference on the Geophysical Detection of Subsurface Water on Mars.

Head, J. W., Kreslavsky, M. A. and Pratt, S. (2002). Northern lowlands of Mars: Evidence for widespread volcanic flooding and tectonic deformation in the Hesperian Period. *J. Geophys. Res. Planets*, 107.

Hobbs, W. H. (1917). The Erosional and degradational processes of deserts, with especial reference to the origin of desert depressions. *Ann. Am. Assoc. Geogr.*, 7, 25–60.

Kadish, S. J., Barlow, N. G. and Head, J. W. (2009). Latitude dependence of Martian pedestal craters: Evidence for a sublimation-driven formation mechanism. *J. Geophys. Res. Planets*, **114**, E10001, doi: 10.1029/2008JE003318.

Kerber, L. (2016). Controls on the Morphology of Yardangs on the Earth and Mars, Lunar and Planetary Science Conference, p. 2708.

Kerber, L., Head, J. and Forget, F. (2012). Aeolian Features in the Medusae Fossae Formation: A HiRISE Survey, Third International Planetary Dunes Workshop: Remote Sensing and Data Analysis of Planetary Dunes, pp. 58–59.

Kerber, L., Head, J., Madeleine, J., Wilson, L. and Forget, F. (2008). Modeling ash dispersal from Apollinaris Patera: Implications for the Medusae Fossae formation. Lunar and Planetary Science Conference, p. 1881.

Kerber, L. and Head, J. W. (2010). The age of the Medusae Fossae Formation: Evidence of Hesperian emplacement from crater morphology, stratigraphy, and ancient lava contacts. *Icarus*, **206**, 669–684.

Keszthelyi, L. and Jaeger, W. (2008). HiRISE observations of the Medusae Fossae formation, Lunar and Planetary Science Conference, p. 2420.

Kieffer, H. H., Martin, T. Z., Peterfreund, A. R., Jakosky, B. M., Miner, E. D. and Palluconi, F. D. (1977). Thermal and albedo mapping of Mars during the Viking primary mission. *J. Geophys. Res.*, **82**, 4249–4291.

Laity, J. E. (2011). Wind erosion in drylands. *Arid Zone Geomorphology: Process, Form and Change in Drylands*, Third Edition, 539–568.

Le Deit, L., Hauber, E., Fueten, F., Pondrelli, M., Rossi, A., Mangold, N., van Gasselet, S., Massé, M. and Jaumann, R. (2011). Geological analysis of Gale Crater on Mars, EPSC-DPS Joint Meeting 2011.

Li, J., Dong, Z., Qian, G., Zhang, Z., Luo, W., Lu, J. and Wang, M. (2016). Yardangs in the Qaidam Basin, northwestern China: Distribution and morphology. *Aeolian Res.*, **20**, 89–99.

Li, M., Fang, X., Yi, C., Gao, S., Zhang, W. and Galy, A. (2010). Evaporite minerals and geochemistry of the upper 400m sediments in a core from the Western Qaidam Basin, Tibet. *Quatern. Int.*, **218**, 176–189.

McCauley, J., Grolier, M. and Breed, C. (1977). Yardangs. *Geomorphology in Arid Regions*, 233–269.

McKee, E. D. (1979). *A Study of Global Sand Seas* (Vol. 1052), US Government Printing Office.

Melosh, H. J. (1989). *Impact Cratering: A Geologic Process*. Research supported by NASA. New York, Oxford University Press (Oxford Monographs on Geology and Geophysics, No. 11), 1989, **253**, p. 11.

Milbury, C., Johnson, B. C., Melosh, H. J., Collins, G. S., Blair, D. M., Soderblom, J. M., Nimmo, F., Bierson, C. J., Phillips, R. J. and Zuber, M. T. (2015). Preimpact porosity controls the gravity signature of lunar craters. *Geophys. Res. Lett.*, **42**, 9711–9716.

Morgan, G. A., Campbell, B. A., Carter, L. M. and Plaut, J. J. (2015). Evidence for the episodic erosion of the Medusae Fossae Formation preserved within the youngest volcanic province on Mars. *Geophys. Res. Lett.*, **42**, 7336–7342.

Niu, H., Qu, J., Li, X., Dong, G., Zhang, K. and Han, Q. (2011). Review and prospect of Yardangs landforms research. *Adv. Earth Sci.*, **26**, 516–527.

Ojha, L. and Lewis, K. (2018). The Density of the Medusae Fossae Formation: Implications for its Composition, Origin, and Importance in Martian History. *J. Geophys. Res. Planet*, 123, 1368–1379.

Ormö, J., Rossi, A. P. and Housen, K. R. (2013). A new method to determine the direction of impact: Asymmetry of concentric impact craters as observed in the field (Lockne), on Mars, in experiments, and simulations. *Meteorit. Planet. Sci.*, **48**, 403–419.

Read, P. L. and Lewis, S. R. (2004). *The Martian Climate Revisited: Atmosphere and Environment of a Desert Planet*. Springer Science & Business Media.

Ritley, M. K. and Oontuya, E. (2004). Yardangs and dome dunes northeast of Tavan Har, Gobi, Mongolia, GSA Abstracts with Programs, p. 33.

Rohrmann, A., Heermance, R., Kapp, P. and Cai, F. (2013). Wind as the primary driver of erosion in the Qaidam Basin, China. *Earth Planet. Sci. Lett.*, **374**, 1–10.

Scott, E. R. D. and Wilson, L. (2005). Meteoritic and other constraints on the internal structure and impact history of small asteroids. *Icarus*, **174**, 46–53.

Tanaka, K. (2000). Dust and ice deposition in the martian geologic record. *Icarus*, **144**, 254–266.

Thomson, B. J., Bridges, N. T., Milliken, R., Baldridge, A., Hook, S. J., Crowley, J. K., Marion, G. M., de Souza Filho, C. R., Brown, A. J. and Weitz, C. M. (2011). Constraints on the origin and evolution of the layered mound in Gale Crater, Mars using Mars Reconnaissance Orbiter data. *Icarus*, **214**, 413–432.

Treiman, A. H., Fuks, K. H. and Murchie, S. (1995). Diagenetic layers in the upper walls of Valles Marineris, Mars: Evidence for drastic climate change since the mid-Hesperian. *J. Geophys. Res. Planet*, **100**, 26339–26344.

Viúdez-Moreiras, D., Gómez-Elvira, J., Newman, C., Navarro, S. and Marin, M. (2017). Gale Wind Speed Weibull Distribution Based on the First Two Years of REMS Wind Data, The Sixth International Workshop on the Mars Atmosphere: Modelling and observation, Granada, Spain, p. 1108.

Vincent, P. and Kattan, F. (2006). Yardangs on the Cambro-Ordovician Saq Sandstones, North-West Saudi Arabia. Zeitschrift für Geomorphologie, NF, 305–320.

Wang, J., Fang, X., Appel, E. and Song, C. (2012). Pliocene-pleistocene climate change at the NE Tibetan Plateau deduced from Lithofacies variation in the drill core SG-1, Western Qaidam Basin, China. *J. Sedimen. Res.*, **82**, 933–952.

Wang, J., Xiao, L., Reiss, D., Hiesinger, H., Huang, J., Xu, Y., Zhao, J., Xiao, Z. and Komatsu, G. (2018). Geological features and evolution of Yardangs in the Qaidam Basin, Tibetan Plateau (NW China): A Terrestrial Analogue for Mars. *J. Geophys. Res. Planets*, **123**, 2336–2364.

Wang, S. and Ha, S. (2009). Advances in the study of geomorphology and process of wind erosion landforms. *J. Earth Sci. Environ.*, 100–105.

Wang, Y., Wu, F., Zhang, X., Zeng, P., Ma, P., Song, Y. and Chu, H. A. O. (2016). Formation and evolution of yardangs activated by Late Pleistocene tectonic movement in Dunhuang, Gansu Province of China. *J. Earth Sys. Sci.*, **125**, 1603–1614.

Ward, A. W. (1979). Yardangs on Mars: Evidence of recent wind erosion. *J. Geophys. Res.*, **84**, 8147.

Watters, T. R., Campbell, B., Carter, L., Leuschen, C. J., Plaut, J. J., Picardi, G., Orosei, R., Safaeinili, A., Clifford, S. M. and Farrell, W. M. (2007). Radar sounding of the Medusae Fossae Formation Mars: Equatorial ice or dry, low-density deposits? *Science*, **318**, 1125–1128.

Weitz, C. M., Milliken, R. E., Grant, J. A., McEwen, A. S., Williams, R. M. E., Bishop, J. L. and Thomson, B. J. (2010). Mars Reconnaissance Orbiter observations of light-toned layered deposits and associated fluvial landforms on the plateaus adjacent to Valles Marineris. *Icarus*, **205**, 73–102.

Whitney, M. I. (1983). Eolian features shaped by aerodynamic and vorticity processes. *Developments in Sedimentology*, **38**, 223–245.

Wunnemann, K., Collins, G. S. and Melosh, H. J. (2006). A strain-based porosity model for use in hydrocode simulations of impacts and implications for transient crater growth in porous targets. *Icarus*, **180**, 514–527.

Xia, X. (1985). Review to the Scientific Investigation in Lop Nor, Xinjiang. *Arid Zone Res.*, **1**, 1–5.

Xiao, L. (2013). *Planetary Geology*. Geological Publishing House, Beijing.

Xiao, L., Wang, J., Dang, Y., Cheng, Z., Huang, T., Zhao, J., Xu, Y., Huang, J., Xiao, Z. and Komatsu, G. (2017). A new terrestrial analogue site for Mars research: The Qaidam Basin, Tibetan Plateau (NW China). *Earth-Science Rev.*, **164**, 84–101.

Zimbelman, J. R. and Scheidt, S. P. (2012). Hesperian age for western medusae fossae formation, Mars. *Science*, **336**, 1683–1683.

https://doi.org/10.1142/9789811224027_0006

Chapter 6

Polygons

Yanan Dang[*,†], Long Xiao[*,†,‡], and Yi Xu[*,†]

[*]*State Key Laboratory of Lunar and Planetary Sciences, Macau University of Science and Technology, Macau, PR China*

[†]*CNSA Macau Center for Space Exploration and Science, Macau, PR China*

[‡]*State Key Laboratory of Geological Processes and Mineral Resources, Planetary Science Institute, School of Earth Sciences, China University of Geosciences, Wuhan 430074, PR China*

6.1 Introduction

Polygonal patterned ground, or polygonal terrain, is the distinct and asymmetrical or symmetrical natural pattern of geometric shapes (Mangold, 2005). The formation of polygonal patterned grounds often involve cracking in response to stress in sediments in which they are located, and are quite common on Earth (Fig. 6.1). They appear in periglacial regions, dried lakes, dried farmlands, and solidified lava surface and interior. Many studies have been conducted to study the morphology and formation of these features (e.g. Lachenbruch, 1962; Neal *et al.*, 1968; Weinberger, 1999).

Polygonal patterned grounds often provide much information about the composition of sedimentary deposits and associated processes responsible for their formation, which have a close relationship to (global or regional) climatic and environmental changes. Because of their different formation mechanisms, polygonal features are accordingly often given different names.

Fig. 6.1. Polygonal patterned grounds formed under Earth's different conditions. (a) Ice-wedge polygons of Alaska. (b) Desiccation cracks in the Qaidam Basin (Photo by Y. Dang, July, 2018). (c) Hexagonal columnar cooling joints exposed in the Devil's Postpile National Monument of California. (d) Salt polygons in Uyuni salt lakes, in Bolivia.

Sources: (a) https://www.usgs.gov/media/images/eroding-shoreline-along-alaskas-coast-0; (c) https://epod.usra.edu/blog/2011/06/devils-postpile-tile-pattern.html; (d) https://www.aracari.com/wp-content/uploads/2015/10/Uyuni-Horizon-Long1.jpg.

6.1.1 *Thermal contraction polygon*

Thermal contraction-induced polygons, often seen in periglacial regions on Earth (Fig. 6.1(a)) and across a large fraction of Mars's surface (Carr, 2007; Levy *et al.*, 2010), are complex landforms involving solid–liquid phase change of water due to seasonal temperature variation. Seasonal contraction of an ice-cemented soil will result in high tensile stresses. When the tensile stresses are large enough to overcome the tensile strength of the soil, then a series of fractures begin to form in the soil to release the stress. These fractures intersect with each other forming thermal contraction polygons. Repeated seasonal melting of the ice in these fractures may cause these cracks to be filled or mixed with such materials as dust and/or loose sand, or ice, resulting in the formation of sand wedges or ice wedges (El Maarry *et al.*, 2010; Marchant and Head, 2007; Mellon *et al.*, 2008).

The thermal contraction cracking mechanism determines the final appearance (e.g. size, morphology, and orientation) of polygons. However, the formations of thermal contraction polygons are always subject to various environmental factors (Marchant and Head, 2007; Ulrich *et al.*, 2011). Generally, thermal contraction polygons on Earth have an average diameter ranging between a few meters and a few decameters. According to the topographic variation from the center to their rim, thermal contraction polygons can be divided into two types defined as low- and high-centered polygons (Fig. 6.2). Topographically, low-centered polygons are clearly distinguishable from high-centered polygons. Low-centered polygons display outlining furrows expressed by raised lineation (ridge-like rims) (Fig. 6.2(a)). Upturned rims are created by thermal expansion, accompanied by lateral growth and shifting of the crustal material. This active layer overlying a growing ice/sand wedge will melt seasonally (Mackay, 1980). However, as the ice melts, the ice-wedge degrades, and the troughs outlining polygons become larger and deeper, forming high-centered polygons (Fig. 6.2(b)).

6.1.2 *Desiccation (Shrinkage) crack polygon*

Desiccation cracks (Fig. 6.1(b)) are commonly formed as shrinkage cracks due to the evaporation of water from the surface of clay-rich sediments (Goehring *et al.*, 2010). They are commonly called mud cracks as a consequence of the slow drying-out of muddy sediments that have been exposed to the action of sun and wind for a period of time. Naturally, the mud cracks begin to form when a wet, muddy sediment starts to dry and shrink. The shrinkage of the top layer will cause the development of a strain, given that the physical status of the material below undergoes almost no change in size (Goehring, 2013). As the strain becomes larger with time, cracks form in the upper dried surface, through which the strain is relieved. The growth of individual cracks leads to their connection and intersection with each other, ultimately forming a polygonal patterned ground.

6.1.3 *Lava polygon*

This type of polygonal pattern is mainly caused by cracking due to thermal contraction involved in volcanic processes such as is often observed

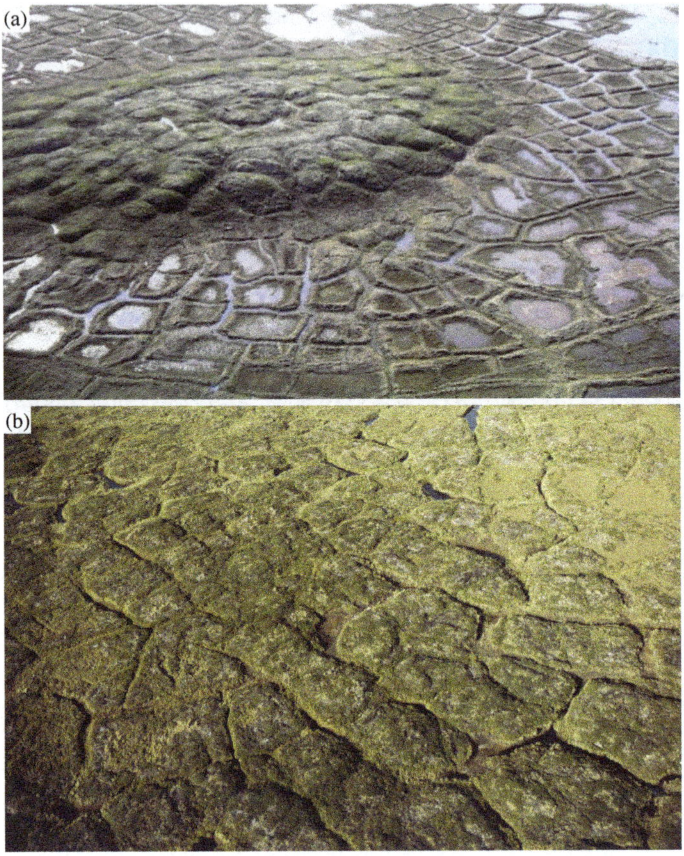

Fig. 6.2. (a) Low-centered polygons near Tuktoyaktuk, Northwest Territories, Canada. (b) Aerial view of the tundra showing high-centered polygons caused by freezing and thawing (credit: Dan Guravich/Science Source).

Sources: (a) https://commons.wikimedia.org/wiki/File:Melting_pingo_wedge_ice.jpg; (b) https://www.sciencesource.com/archive/Tundra-Polygons-SS2358258.html#.

in basaltic columns (Fig. 6.1(c)). In ideal situations, basaltic lava will be uniform in terms of composition and density, causing the lava to cool evenly (Nishiwaki, 2009). The cooling takes place both from the bottom up and the center outward (Budkewitsch and Robin, 1994). Thus, this contraction would cause regular fractures taking the form of regular hexagons at the top. The whole process is called columnar jointing.

6.1.4 *Salt polygon*

Salt polygons (Fig. 6.1(d)) are present in dried salt lakes, which are usually called playas. Under the condition of a playa undergoing drying, there is often a salt crust formed through convection due to water evaporation and mineral precipitation. These polygons are characterized by their uplifted rims (or pressure ridges) that formed from a subsurface process of lateral salt displacement (e.g. Artieda *et al.*, 2015, and references therein; Christiansen, 1963; Lowenstein and Hardie, 1985) due to crystallization pressures (Handford, 1991), or to thermal contraction and expansion (Dellwig, 1968; Handford, 1991; Tucker, 1981). The durability of salt polygons is in most cases limited by seasonal rainfall or flooding events, which often destroy the surface morphology and then start a new cycle of the formation and evolution of salt polygons (Artieda *et al.*, 2015).

The formation of salt polygons cannot happen without the involvement of water, which is a significantly important factor for the existence, evolution, and preservation of biological systems. The widespread occurrence of salt polygons in terrestrial playas represents strong evidence of climatic changes from wet to arid conditions at global or regional scales, thus making the Martian terrains with lacustrine/playa deposits important candidate sites to search for evidence for the habitable environments in past or current history of Mars (Dang *et al.*, 2018).

6.2 Polygonal Terrains on Mars

Polygonal terrains are some of the most widely distributed landforms on Mars, and have been observed and studied for many years (e.g. El Maarry *et al.*, 2015b; Helfenstein and Mouginis-Mark, 1980; Kerber *et al.*, 2017; Levy *et al.*, 2009; Mutch *et al.*, 1977; Seibert and Kargel, 2001; Stein *et al.*, 2018). Images from both landers and orbiting spacecraft show that the Martian polygons (Fig. 6.3) vary in both size and morphology (Levy *et al.*, 2009; Mangold, 2005). The average diameters of these polygons range from a few centimeters to a few kilometers, and they occur at almost all latitudes, suggesting multiple formative processes and environments (El Maarry *et al.*, 2010; Hiesinger and Head, 2000; Levy *et al.*, 2009; McGill and Hills, 1992; Pechmann, 1980).

Polygons on Mars would have formed by the isotropic contraction of the substrate, whether under conditions of freeze–thaw cycles, wetting

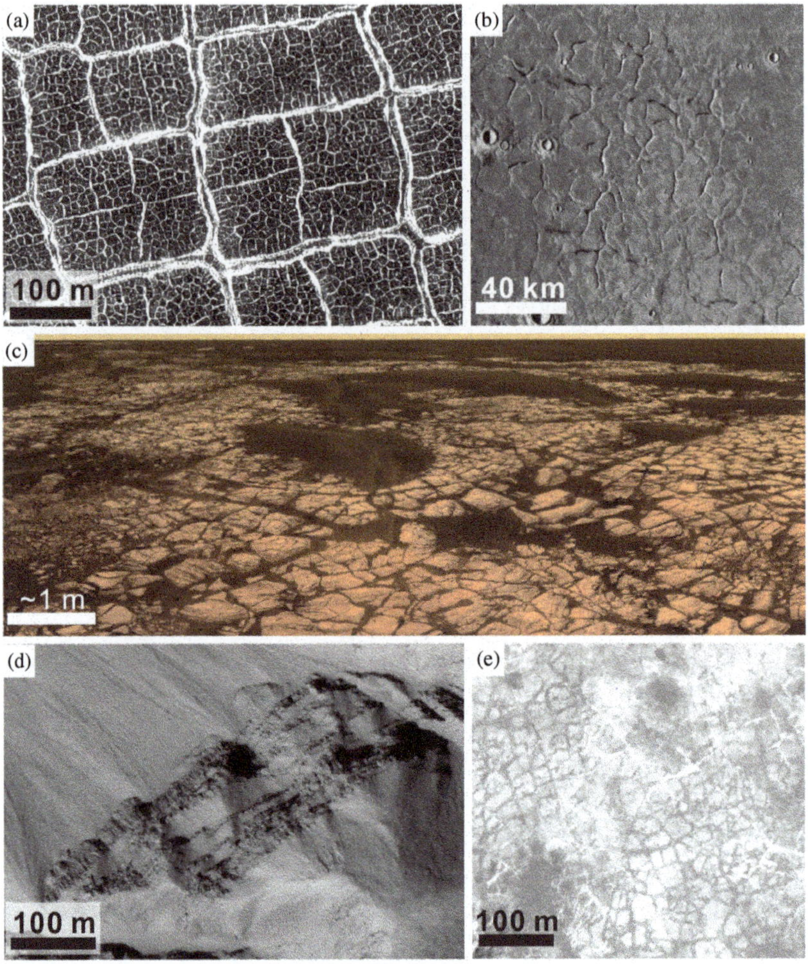

Fig. 6.3. Polygonal terrains observed on Mars. (a) Ice-related polygons. The polygon edges are bounded by troughs or ridges covered by bright frost relative to their darker, frost-free interiors. (HiRISE image IDs: ESP_016641_2500). (b) Tectonic-caused giant polygons in Utopia Planitia. Viking Orbiter image 538A08. (c) Desiccation polygons at Meridiani Planum captured by NASA's Opportunity Rover. (d) Columnar jointing exposed on an unnamed crater wall in the Marte Vallis region on Mars (HiRISE image IDs: PSP_006774_2020). (e) Polygons observed in chloride-bearing terrain on Mars (HiRISE image IDs: ESP_042008_1515).

and drying cycles, or simply cooling. They can vary in size from a few centimeters up to tens of kilometers. In general, the small-sized (<~25 m) polygons are thought to be related to icy permafrost (Fig. 6.3(a)), and they are actually latitude dependent on Mars (Levy *et al*., 2009; Pina, 2014). V-shaped ice wedges trigger the creation of cracks, which are arranged in polygonal networks, by volume change due to the transition of water from a liquid to a solid. The cracks may be filled with ice or sand/dust or their mixture, which varies according to local geological settings (Marchant and Head, 2007). Martian volatile activity is a very important process in response to the erosional activity of water and ice. Sand and dust are trapped and released as volatiles undergo their sublimation and melting/freezing processes.

Giant polygons (Fig. 6.3(b)) larger than 1 km in size are present on Martian northern plains in Utopia and Acidalia, and are much larger than polygons found on Earth. Their formation is associated with extensional tectonic activities that are restricted to specific geological settings (such as subsidence and uplift). In this scenario, polygons are bounded by polygonal oriented extensional faults, as seen within 27 marine basin structures on Earth (Cartwright and Dewhurst, 1998). Formation of polygonal faults in response to compaction caused by drying up of wet sediments is often used to explain the formation of terrestrial polygons of tectonic origin (Cartwright and Dewhurst, 1998; Cartwright and Lonergan, 1996; Lonergan and Cartwright, 1999).

A great variety and a number of desiccation polygons were observed by Mars Exploration Rover (MER) Opportunity at Meridiani (Fig. 6.3(c)). These polygons are usually less than 1 m in size. They are flat, lacking raised rims or centers, and mostly occur in light-toned aeolian sandstones relative to the surrounding terrain (El Maarry *et al*., 2014; Mutch *et al*., 1977).

Milazzo *et al*. (2009) reported the discovery of columnar jointing (Fig. 6.3(d)) on the surface of Mars, which is a distinctive feature of volcanism. They found many outcrops of columnar jointing within a well-preserved, unnamed crater in Marte Vallis. These columns are 30–40 m tall and about 2 m wide. Columnar jointing is a pattern of cracking in rocks that forms slender columns, typically six-sided. On Earth, the jointing occurs when lava or magma comes into contact with a cool, flat surface. The lava contracts and cracks extend perpendicular to the cooling front, usually due to chilled water infiltration.

Polygonal cracks (Fig. 6.3(e)) in chloride-bearing terrains (e.g. Osterloo *et al.*, 2008, 2010), indicative of a lacustrine origin and associated aqueous processes (El Maarry *et al.*, 2013; Goodall *et al.*, 2000), are located in the ancient southern highlands on Mars. They usually range in size from a few meters to tens of meters.

These polygonal landforms are characteristically different in origin depending on their morphology and location. In this chapter, we mainly focus on the salt polygons, referring to a cracking process in response to stress in playa sediments in which they are located. Playas are dried ancient salt lakes filled with lacustrine deposits, which record rich information about water related geological processes and aqueous environments that are closely related to the origin of life. Thus, the study of salt polygons on Mars has been the key to understanding the planets' climate evolution, origin of life, climatic change, and habitability.

6.2.1 *Distribution of salt polygons on Mars*

Martian salt polygons are mainly concentrated in chloride-bearing terrains. Up to now, more than 600 locations (Fig. 6.4) containing

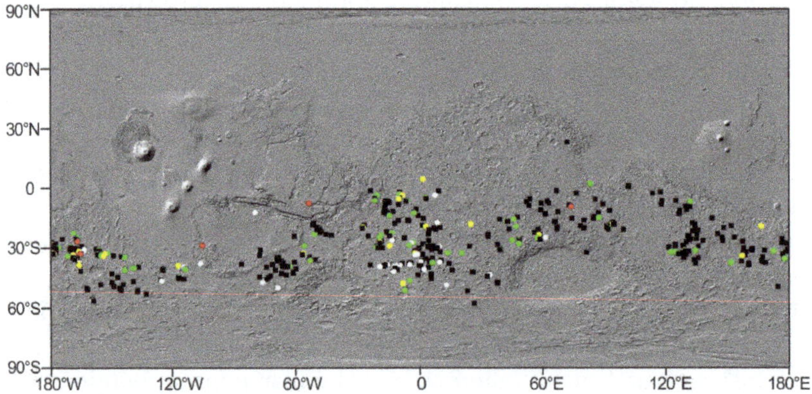

Fig. 6.4. Global distribution of 642 chloride-bearing terrains on Mars from database built by Osterloo *et al.* (2010). All dots represent the localities of chloride-bearing depositions. Green dots show the locations of crack polygons (83 regions). Red dots show the locations of raised rim polygons (9 regions). Yellow dots represent the locations of both fracture polygons and raised rim polygons (37 regions). White dots exhibit the region where no polygons have been observed (41 regions). Black dots indicate the areas where there is no HiRISE image coverage (472 regions). The context map is Mars Orbiter Laser Altimeter (MOLA) instrument shaded relief data.

chloride-bearing deposits have been identified in the Martian southern highlands (Osterloo *et al.*, 2008, 2010). The terrain units at these sites are dated to be Middle Noachian to the Early Hesperian in age, and most of them appear as small, spatially isolated patches in local shallow closed basins. One of the prominent geomorphological characteristics of these chloride-bearing terrains is the polygonal features within them; we call them salt polygons. The salt polygons are formed by the beginning process of cracking followed by later inflation process resulting from salt crystal growth and doming effect due to intensive, repeated evaporation (Dang *et al.*, 2018 and references therein).

Based on 447 frames of the High Resolution Imaging Science Experiment (HiRISE) images covering 170 sites of these 642 locations identified by Osterloo *et al.* (2010), we divide the salt polygons that occurred within chloride-bearing deposits into two types geomorphologically characterized by cracked and raised rims, respectively.

6.2.2 *Morphology of salt polygons on Mars*

Around 120 sites of chloride-bearing terrains are found to contain fractured polygons (Fig. 6.5). Most of them are located between 0°S to 30°S across the Martian southern highlands. Nearly all polygonal features in Noachian Terra fall in this category. The polygonal fractures usually take the form of three- to six-sided polygons with generally straight sides (Figs. 6.5(a)–(c)) with some displaying arc-shaped borders (Fig. 6.5(d)), forming polygon systems in regular random orthogonal pattern. Small polygons are sometimes contained in the interior of larger ones (Fig. 6.5(b)). Both the low- and high-relief centers are observed in the same terrain (Fig. 6.5(c)). The length of these polygonal fractures has a range from ~2 m to ~30 m, while their widths vary from ~0.2 m to ~2 m. The crack regions look relatively dark compared to the adjacent terrain, indicating that dark materials (e.g. sand and dust) were filled inside the cracks.

Generally, about 46 sites (Fig. 6.4, red and yellow dots) are found to have polygons with raised rims in the southern highlands of Mars. They are characterized by remarkable raised rims (Fig. 6.6). Most of them have a size range from ~ 10 to 30 m. These raised rims intersect with each other forming quadrangles and hexagons usually in a right angle manner. Raised rims are generally brighter than the interior. Their width is generally estimated to be between 0.5 and 2 m. Due to the limitation in coverage of HiRISE DTM data, the height of the rim cannot be measured

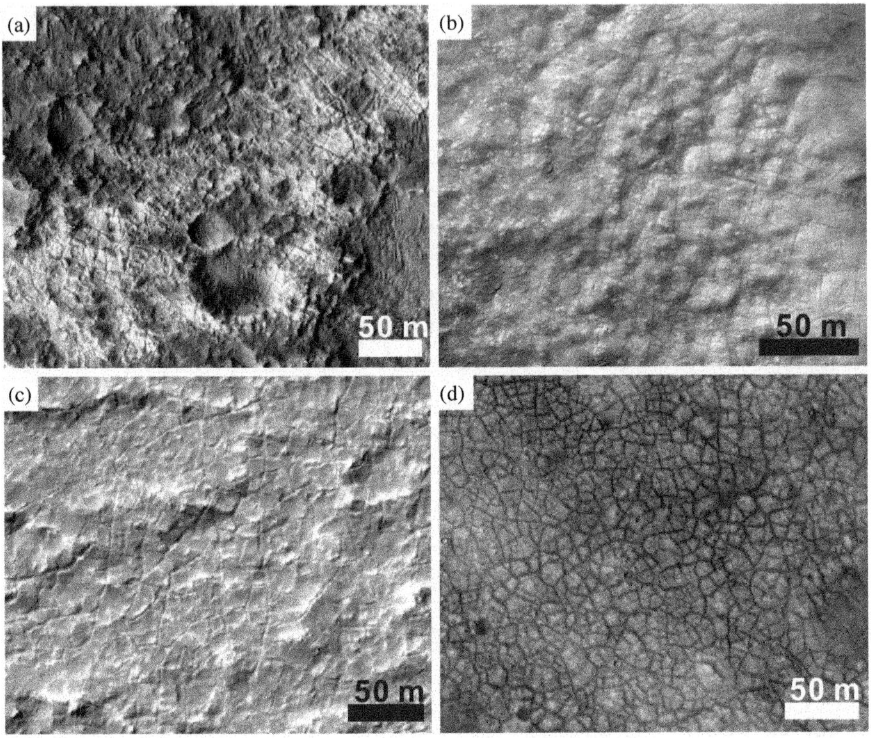

Fig. 6.5. Polygonal cracked patterns in chloride-bearing depositions on Mars. HiRISE Image IDs: (a) ESP_012012_1625; (b) ESP_016569_1405; (c) ESP_034732_1600; (d) PSP_007095_2020.

directly. A simple measurement technique using incidence angle of the HiRISE data and the shadow information indicates that most of the raised rims are less than 1 m in height (Ye *et al.*, 2019).

The chloride-bearing terrains containing polygons are always found to be located at low-topography areas such as the floor of craters degraded to varying degrees (Figs. 6.7 and 6.8). Figure 6.7 shows an example of salt polygons occurring in the floor of a ~130-km-diameter, degraded crater (Location: 191.15°E, 31.28°S; Fig. 6.7(a)), whose southwest crater rim is breached (Fig. 6.7(b)). A distributary channel system appears to originate from the breached rim region, suggesting that the regional liquid water activity at or near the surface is most likely. Commonly, old impact craters are ideal places for storage of a static body of water (Goudge *et al.*, 2015). Topographic advantage, chloride-rich deposits, salt polygons, and the

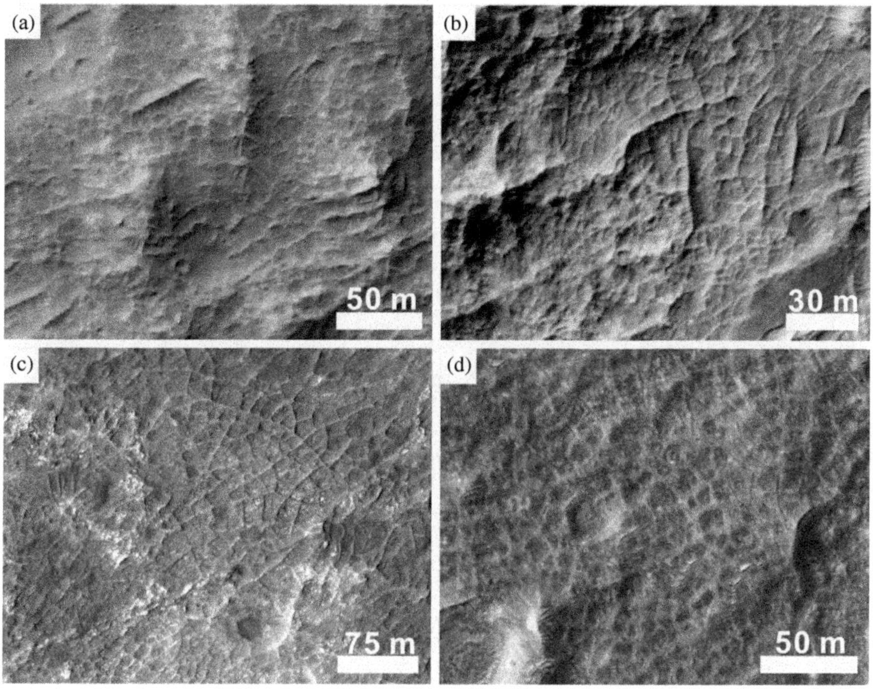

Fig. 6.6. Polygons defined by raised rims in chloride-bearing depositions on Mars. HiRISE image IDs: (a) ESP_043434_1495; (b) ESP_045252_1480; (c) ESP_047343_1610; (d) ESP_023716_1420.

traceable signs of water activity provide lines of evidence for potential lacustrine/playa environments on an early wet Mars.

The salt cracked polygons and the ones with raised rims are sometimes observed to coexist in some areas (Fig. 6.8). A number of cracked polygons (Fig. 6.8(c)) with high-relief centers are concentrated in a chloride-bearing terrain on the upper wall slope of a valley (Fig. 6.8(b)) located in the western arcuate mountain region of the Thaumasia Highlands at the southeast margin of Tharsis. Honeycomb patterned polygons also occur on the middle section of this elongated terrain (Fig. 6.8(d)). Honeycomb patterned polygons are raised rim salt polygons with a lower interior. Both types of polygons are <50 m in length, but the latter is relatively smaller than the former. Many locations on such chloride-bearing terrains display honeycomb-like polygons, but the mechanism for their formation is still unclear.

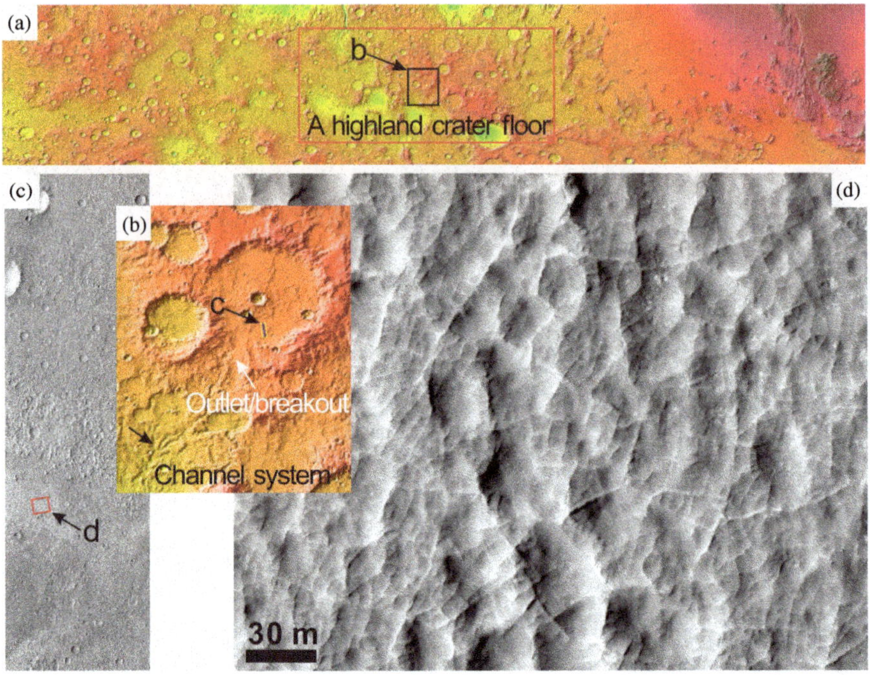

Fig. 6.7. Crater floor salt polygons in the chloride-bearing terrains on Mars. The location of the crater is indicated from the colored MOLA shaded relief map, as seen in (a) and (b). A frame of high resolution HiRISE image (c), HiRISE image ID: ESP_0115471475) and the enlarged view (d) show the location and the detail of the raised rim salt polygons.

Several factors are thought to account for the main morphological differences between negative cracked and raised-rim salt polygons. The first factor could be related to the thickness of the deposit. The desiccated chloride-bearing deposits (Osterloo *et al.*, 2008, 2010) are scattered in small isolated patches across the southern highlands. They may be the result of a short-lived water body, making the water-saturated deposits significantly thin. Raised-rim salt polygons are usually located in large depressions such as impact basins, illustrated in the examples in Figs. 6.7 and 6.8. They would have formed in a thick deposit, which has undergone long-term evaporation and precipitation processes. The composition of deposits is another important factor determining the final appearances of salt polygons. For example, clay-rich deposits tend to produce large cracks due to a significant volume shrinkage (El Maarry *et al.*, 2014), while NaCl-rich deposits favor formation of small polygons with raised

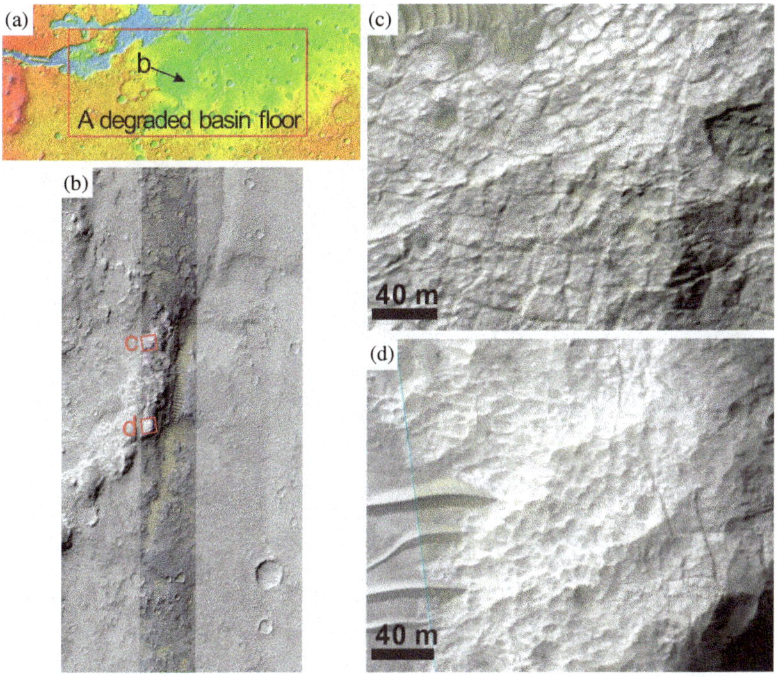

Fig. 6.8. Coexistence of crater-floor salt cracked polygons (c) and honeycomb-patterned polygons (d) located in chloride-bearing highland terrain (a and b). HiRISE image ID: ESP_044165_1610.

rims under the condition of continuous evaporation and supply of underground water (Dang *et al.*, 2018). Other factors such as the specific soil properties of the deposits, hydrological history, and ground permeability will also affect the diversity of polygonal landforms. They should be treated individually or combined for the cases formed in different geological settings.

A potential relationship between the spatial distribution of salt polygons and topography is observed in some geological settings with chloride-bearing sediments. For example, in the irregular depressions of the northern margin of the Terra Sirenum region, polygons on the relatively higher terraces of the depression are clearly much larger than those located in the lower floor (Fig. 6.9). In Earth playa environments, such polygon size–topography relationships are also observed (Dang *et al.*, 2018). This is because the topography within the terrestrial playa environment controls different mineralogical precipitations primarily due to

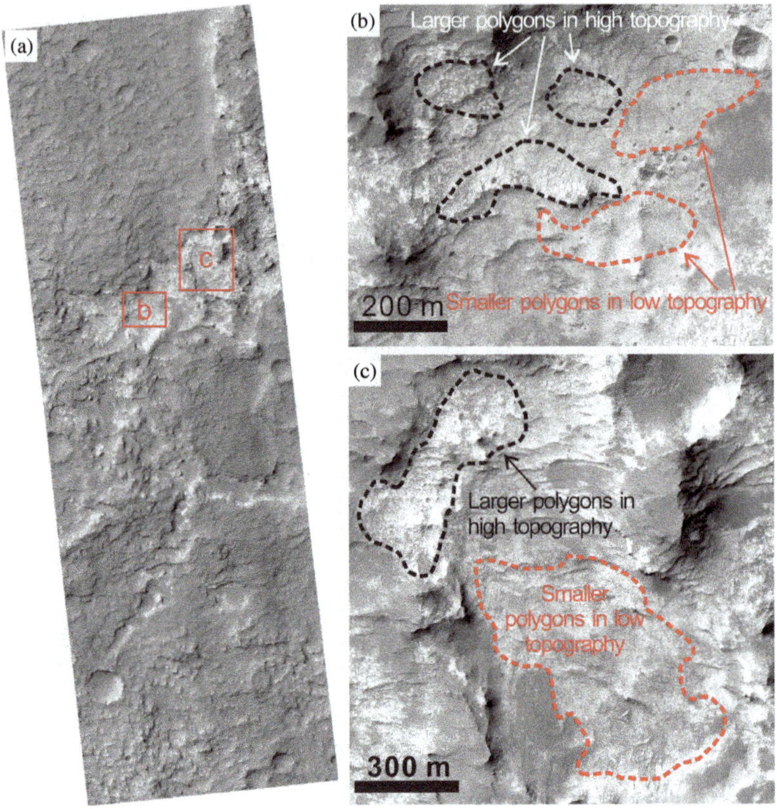

Fig. 6.9. Polygons size–topography relationship in irregular depression with infilling of bright chloride-bearing sediments. (a) The geological setting and locations of these irregular depressions (HiRISE ID: ESP_042364_1510). (b) Distribution of varying-size polygons in the depression indicated by the lower black box in (a). (c) Distribution of varying-size polygons in the depression indicated by the upper black box in (a). Larger-sized polygons occurred in the high chloride-bearing regions indicated by black arrows, while red arrows point to smaller-sized polygons in low topography.

evaporation processes (Komatsu *et al.*, 2007), which may also apply to the Martian sedimentary environments like the example indicated in Fig. 6.9.

6.3 Polygonal Ground in the Qaidam Basin

The Qaidam Basin (90°16'~ 99°16'E, 35°00'~ 39°20'N) is one of the highest elevation hyperarid inland basins on Earth, and positioned on the northeast corner of the Qinghai–Tibet Plateau. It is considered to be a unique Mars analog site on Earth (Dang *et al.*, 2018; Anglés and Li, 2017; Cheng *et al.*, 2017; Xiao *et al.*, 2017; Kong *et al.*, 2014b; Mayer *et al.*, 2009; Wang and Zheng, 2009), considering that both share a diversity of geomorphologically similar landforms and a comparable environment condition illustrated as dry, cold, and with high ultraviolet (UV) radiation. The polygonal ground in the Qaidam Basin represents a typical morphological surface feature that developed in salt sedimentary deposits. They are mainly distributed within dried salt lakes, such as Gasikule, Dalangtan, Chahansilatu, Kunteyi, Mahai, Yiliping, and Qarhan, with a variety of morphologies and sizes. Salt polygons can form in different stages of lake evolution under arid environments. The analog study of salt polygons on Earth and their counterparts on Mars can help in understanding the past polygon-formation-related hydrological processes occurring on Mars.

The present climate in the Qaidam Basin is hyperarid with evaporation/precipitation ratios exceeding 100:1 (Chen and Bowler, 1986), cold and with high UV radiation, which is similar to the surface condition of present-day Mars (Kminek and Bada, 2006; Kong *et al.*, 2018). The lithology of upper sediment layers in the basin (above 400 m to the top) is characterized by cycles of thick gray-white salt and massive-laminated mudstone-siltstone mostly containing gypsum crystals (Li *et al.*, 2010; Wang *et al.*, 2012). The occurrence of NaCl-dominated thick salt layers in the Qaidam Basin indicates a more arid climate (Li *et al.*, 2010). A long-term, general drying climatic trend in the Qaidam Basin since 1.2 Ma and 0.6 Ma (Han *et al.*, 2014) has caused the basin lake system to shrink and turn into salt lakes, finally ending up in salt-encrusted playas.

A wide variety of Martian-like polygons have been found in the Qaidam Basin. Since 25,000 yr B.P, the Qaidam Basin has been experiencing a climatic change from wet to dry (Chen and Bowler, 1986). The crater floor polygons found on Mars are hypothesized to be indicators of a climate change from wet (Pollack *et al.*, 1987) to dry (playas/lacustrine) (El Maarry *et al.*, 2010), similar to the conditions in the Qaidam Basin (Chen and Bowler, 1986; Huang *et al.*, 1993). The relationship between distinct polygon morphologies and their sediment composition could tell us about paleolake water properties and further shed light on environmental and climatic changes that have occurred on Earth and Mars.

6.3.1 *Distribution of polygons in the Qaidam Basin*

We selected the Dalangtan Playa in the Qaidam Basin at the foot of the Altyn-Tagh Mountain as the focus of our investigation based on the following considerations: (1) the western part of the basin is much drier than the eastern part (annual mean precipitation ranging from 100 mm in the basin east to less than 20 mm in its west, Han *et al.*, 2014), an environment closer to the current Mars environment; (2) the Dalangtan Playa is one of the paleo-depocenters of the western Qaidam Basin dating from the Oligocene (Kong *et al.*, 2014(a)); (3) the western part of the basin has large expanses of almost barren land, salt marshes, and salt domes (Zhu, 1985); (4) the Dalangtan Playa has been highly recommended as one of the most promising Mars analog sites because of its climatic and environmental conditions similar to that of the equatorial region of Mars, suggesting an astrobiological potential and/or the possibility of past life on Mars (Huang *et al.*, 2018; Kong *et al.*, 2018; Sobron *et al.*, 2018; Wang *et al.*, 2018); (5) crucially, a wide range of polygonal terrain types can be found scattered across most areas of the playa. In addition, the field investigation of polygons offers easier access by land and potentially with fewer hazards than other comparable sites.

The Dalangtan salt polygons occur across a wide range of scales. In order to simplify the analysis of the complex polygonal system, these polygons can be divided into three distinct types according to their sizes (Fig. 6.10). The small-sized polygons (<10 m) are mainly distributed in the western Dalangtan playa with lower flat topography or within the interior of larger ones. The medium-sized polygons (10–100 m) are usually irregular in shape, with some having discontinuous sinuous rims, which we ascribe to the inhomogeneities in the salt crust (Christiansen, 1963) or the degradation caused by wind or water activity. The large-sized polygons (>100 m) appear to be concentrated in patches with a discrete distribution. Larger polygons are mainly located on higher topographies; in particular the largest polygons are found at the outer edge of the playa (Fig. 6.10).

6.3.2 *Morphology of salt polygons in the Qaidam Basin*

Generally, salt polygons in the Qaidam Basin playas have a size range from centimeters to hundreds of meters in diameter, and most of them have raised rims. They commonly take the form of triangle, quadrangle,

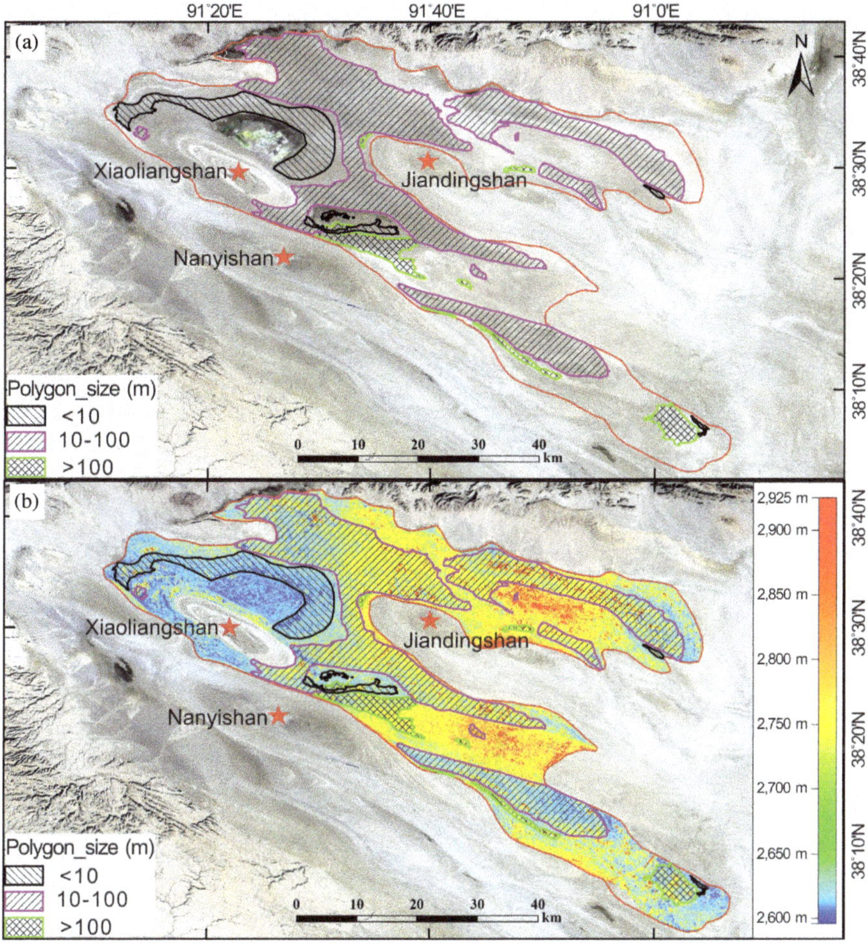

Fig. 6.10. Distribution of salt polygons in the Dalangtan Playa. (a) The distribution of the three types of the salt polygons (according to their size). (b) The distribution of the three types of salt polygons with elevation map. Satellite image credit: Google Earth. The elevation data is generated from ASTER GDEM (http://www.gscloud.cn). North is up in all figures.

and complex-shaped polygons, depending on their edge/side number and arrangement pattern. They tend to topographically concentrate in relatively lower areas and be spatially organized into a variety of patterns (Fig. 6.11). Regular (Figs. 6.11(a)–(c)) and irregular (Figs. 6.11(d)–(f)) random orthogonal polygons (Neal *et al.*, 1968) are predominant in the

Qaidam Basin. Compound polygons are a common pattern that we define as a set of different size polygons assembled in a containment relationship (large polygons contain small ones in their interior, Fig. 6.11(c)). Though salt polygons in terrestrial playa conditions look similar

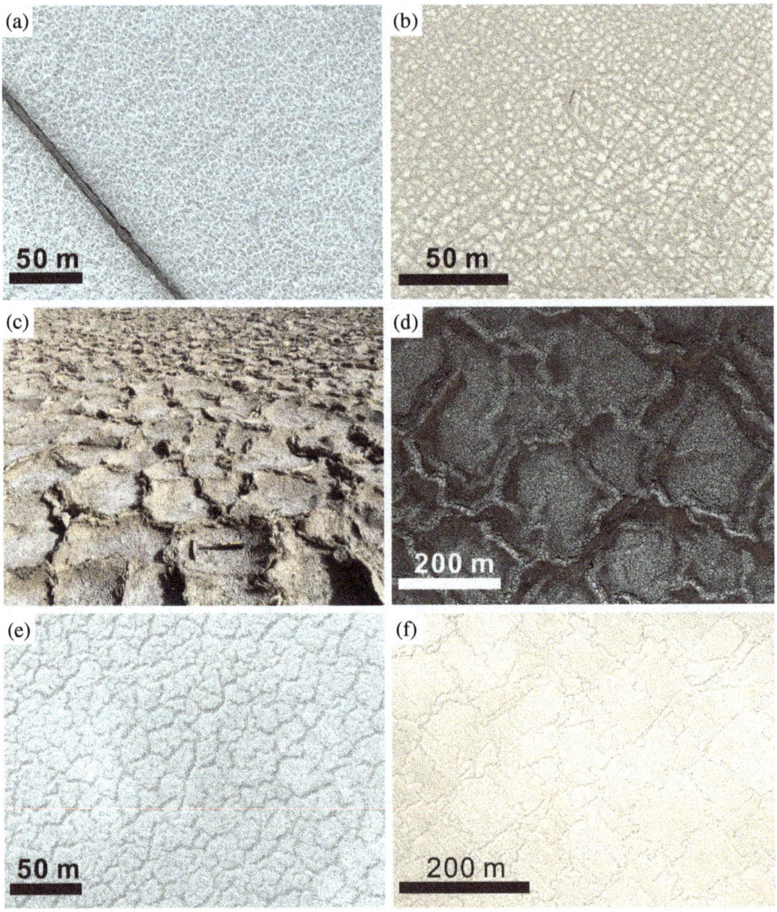

Fig. 6.11. Spatial patterns of different sized and shaped polygons in the Qaidam Basin: (a) regular polygons (91.475°E, 38.459°N); (b) spiderweb polygons (92.073°E, 38.110°N); (c) honeycomb polygons (91.376°E, 38.488°N); (d) compound polygons (91.512°E, 38.411°N); (e) and (f) irregular polygons (91.620°E, 38.434°N; 92.017°E, 38.126°N). Image credit: Google Earth (a, b, d-f) and photograph (c) taken in July 2019 during the field trip. All subfigures show a sun direction from southeast to northwest (lower-right to top-left).

(e.g. Christiansen, 1963; Neal *et al*., 1968; Lowenstein and Hardie, 1985), other interesting and distinct spatial patterns are also observed in the Qaidam Basin. For example, a series of polygons are connected by linear structures, which intersect each other forming giant networks (Fig. 6.11(b)), while numerous typical polygons showing relatively lower interiors and higher rims are assembled closely forming a honeycomb structure (Fig. 6.11(c)). The clustered occurrences of regular (Fig. 6.11(a) and irregular shape polygons (Figs. 6.11(e) and (f)) together with their associated spatial patterns make the formation settings of the Qaidam salt polygons rather complex.

The final appearances of polygons appear to be dominated by a series of contributing factors including regional and local topography, water level change, temperature variation, periodic evaporation, strength and thickness of the tabular-shaped salt or sedimentary units, brine activity (such as seasonal capillary upflow) and injection of foreign materials transported from other areas around the basin. Moreover, these factors acting upon each other add to the complexity of the diversity of polygonal terrains. However, they share polygonal fracture patterns; this is the result of contraction of a layer within a media that undergoes volume change. Although the formative process of irregular polygons is still unclear, they appear to be arranged in a more random manner (Figs. 6.11(e) and (f)).

Salt polygons in the Qaidam Basin are characterized by raised rims, which are measured to have widths between a few centimeters to tens of meters and heights between 10 centimeters to 2 meters. Figure 6.12(a) shows an example of a ~120-m-diameter polygon rim in the Dalangtan Playa. The rim appears as a broad arch, with a width of ~4 m, and a height of ~30 cm. The top surface of the rim is rough and uneven, with fractures and pores in the center. When we turned over the top salt crust, we saw stalactite-like halite crystals (Fig. 6.12(b)) at the bottom of the masses. Fracture system at or around the top center of the rim suggests a dynamic convection place for matter and energy exchange between interior and exterior, driven by evaporation.

6.3.3 *Formation and evolution of salt polygons in the Qaidam Basin*

A complete saline pan cycle (lifespan from a salt lake to playa) can be divided into three stages (Lowenstein and Hardie, 1985). At the flooding stage, a low, flat zone is filled with a standing body of water collected

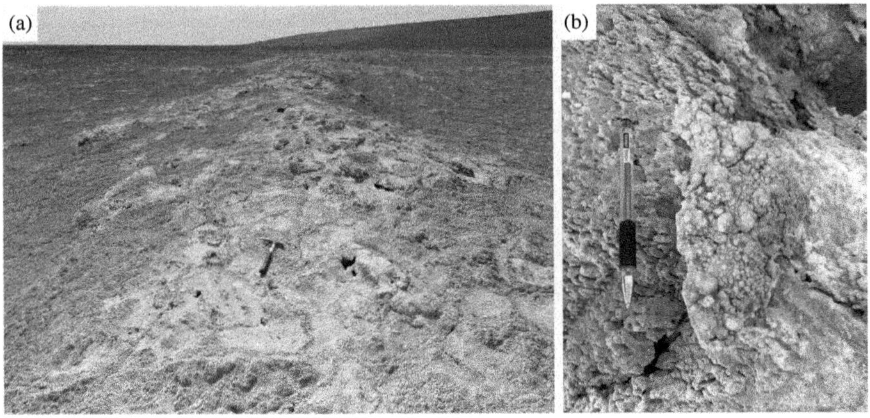

Fig. 6.12. (a) A rim of a large-sized salt polygon in Dalangtan Playa, the hammer (~30 cm long) is placed for scale; (b) Stalactite-like halite crystals beneath the rim of (a), a pen with a length of ~14 cm is shown for scale. Photos were taken in July of 2015 in Dalangtan Playa.

from surrounding rivers/streams or ground water flows or rainfall activity. Original floor surface features and structures are deformed or modified, even damaged. Resetting of the lake bottom floor marks the beginning of deposition process in a new cycle. Partial saline crust and other sedimentary materials dissolve in the water, followed by mud deposition and salt mineral crystallization due to gravity and saturation as water evaporates. This duration may be lengthy, but this quiet time is always fed by new injections of later flows from different sources.

The continuous evaporation and salt crystallization lead to the shrinkage of the saline lake, even drying out. And at this point, the desiccation stage is reached. Ultimately, a residual brine is left, which is dense, saturated with halite, and sustained by perennial surface flow or groundwater inflow or both. Its bottom is covered by the salt–mud mixture sediment layers. However, the evaporation never stops, and the brine continues to evolve with time. The evaporation-driven force upward leads to the growth of void-filling and displacive halite within the vadose and upper phreatic zones. This dynamic process can probably be explained by the evaporative pumping mechanism (Goehring and Fourrière, 2014; Hsü and Siegenthaler, 1969). Finally, pervasive salt crystallization and its growth in the subsurface causes varying levels of fracturing and buckling of the pan's surface because of its volume expansion, causing the development of a network of meter-scale polygons with ridge-like raised rims (Lowenstein and Hardie, 1985). The polygon cracks become the passageway for dynamic convection between systems (driven by evaporation), leading to the development of a spongy efflorescence of halite. This halite crystallization phase lasts throughout the entire course of evaporation until disturbed by the next flooding event. The repeated flooding event will renew the cycle of stages (Lowenstein and Hardie, 1985).

The desiccation stage is most important for the formation and evolution of salt polygons on the salt-encrusted surface. A steady upward flow of salty water caused by surface evaporation can result in precipitation near the surface. This results in a situation where the less salty (source) water is underlain by heavy salt-saturated water. A vertical salt gradient in the porous soil then builds up, which is gravitationally unstable and causes vertical convection. The dynamic process of the porous medium convection causes surface deformation, forming salt polygons. Salt crystallization occurs along the places where the convective downwellings actively take place below, and this is clearly observed in previous laboratory experiments (Fig. 6.13) (Goehring and Fourrière, 2014).

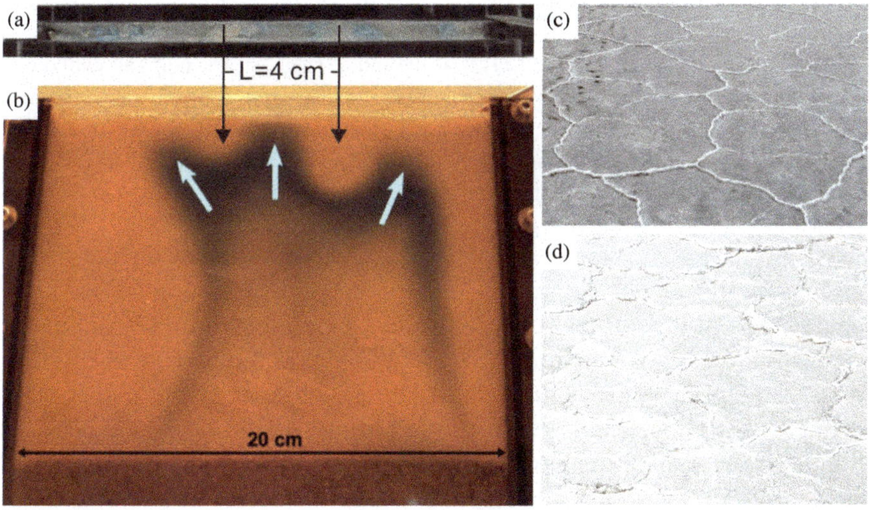

Fig. 6.13. The observable convective downwellings occurring in a porous medium saturated with salt solution (modified from Goehring and Fourrière, 2014). Downwelling convective plumes in subsurface zone below the salt ridge are observed. (a) Top view; (b) Side view; (c) Salt polygons on a beach in Namibia; (d) Salt polygons in the Dalangtan Playa, image was taken in July of 2018.

Fig. 6.14. The deformation of salt crust at the initial stage of salt polygon evolution: (a) the domes formed in the weak zone of the crust, and their growth into fin- or island-like structures; (b) an enlarged example of domical structures on which a long fracture is clearly seen; (c) a doming point and associated two anticlinal folds grow radially at about 120 degrees, as can be seen in detail (d). Photos were taken in July of 2016 in Yiliping Playa.

During this stage, the salt crust continues to evolve until it becomes separated from the soil below. Eventually, it develops into a fin-like salt dome (Fig. 6.14(a)). The doming effect will cause the formation of crust fractures (Fig. 6.14(b)) (Christiansen, 1963). At these dome places, two or three anticlinal folds will grow radially at about 120 degrees (Fig. 6.14(c)). With extension of cracks in which the growth of salt crystals continues, the domes are extended along the linear fractures to form fin- or island-like features. When they further grow and connect with each other, the incipient elevated rims of salt polygons begin to form (Fig. 6.14(d)).

As evaporation advances, precipitation, growth, and displacement of salt crystals in the salt crust can cause planar isotropic positive stresses to develop, which will lead to polygonal fracturing of the upper layer of perennial salt. Eventually, reticulate or polygonal strain systems form and develop with time. Continuous and violent evaporation will cause the steady upward flow of saturated brine, causing the salt crust to continue to grow and develop. This will result in the annual expansion of the salt crust, and finally, produce polygonal patterns of small folds and fractures (thrusts).

The inhomogeneities in the salt crust or weak places with higher porosity level are thought to be the reason for sinuous or arcuate raised rims to develop. The lateral displacement of salt crust will produce pressure ridge-like features. Maximum thrust displacement of the feature often takes place at about right angles. Minor fractures in the thrust zone tend to appear at the places with maximum curvature (Fig. 6.15), as illustrated in the study of Christiansen (1963). The pattern of this sinuous fold system may become rounded or curved because of later weathering effects.

The fractured polygon rims always intersect with out-growing fractures at an angle of about 90 degrees. When an out-growing fracture develops to approach a rim of another polygon at angles smaller than a

Fig. 6.15. The growth and angular or curved pattern of salt polygon rims. Photos were taken in July of 2016 in Yiliping Playa.

Fig. 6.16. The right angle or orthogonal pattern of polygon rims. Photos were taken in July of 2016 in Yiliping Playa.

right angle, it will change its direction and meet with the rim at about right angles, forming the T-shaped orthogonal system, as observed in Fig. 6.16.

The thickness and the strength of the salt crust are the two main factors that control the size of the salt polygons. The variation of polygonal rim numbers is probably associated with inhomogeneities of the sediment composition and governed by the extent of strain within the salt crust. Therefore, polygons with fewer than six sides tend to concentrate in places where the sediment is more homogeneous in composition, but sometimes coexist with hexagonal polygons.

Polygonal fold systems are frequently found to concentrate in the thinner salt crust (Fig. 6.14). Their formation starts as raised dome features, along which two or three anticlinal folds form and extend to grow radially at about 120 degrees. This fold system is often accompanied by the development of the fracture systems on their tops, as well as other polygonal fold systems arranged in biradial, triradial, orthogonal, and sinuous patterns.

The formation of polygonal fold systems can be simulated in the laboratory using a sheet of wet, tabular-shaped clay under desiccation conditions at room temperature. Laboratory simulation and field observations show that the development of negative (tensile) stress in the clay layers causes the formation of polygonal fracture systems. Thus, a similar situation for the thrust and fold systems formed in the salt crust is expected. There is a broad range for the ratio between layer thickness and polygon width. A thickness to width ratio of 1:12 for polygons developed in 3 mm thick clay layers is observed and measured at room temperature (Christiansen, 1963).

Polygon rim growth and development: The surface evaporation fed by groundwater permits a continuous growth of polygonal rims, forming patterned ground characterized by polygons of variable sizes ranging from meters to hundred meters in size. Their raised rims are up to ~2 meters (Fig. 6.17) with abundant gas holes and voids below the surface. Stalactite-like salt flakes can be observed at the bottom side of the crust, providing evidence for brine upward migration in porous soils and subsequent salt crystallization through the rim fissure.

Lowering and degradation of polygon rims: If the source brine at depth is massively decreased in volume with time, this will cause a significant

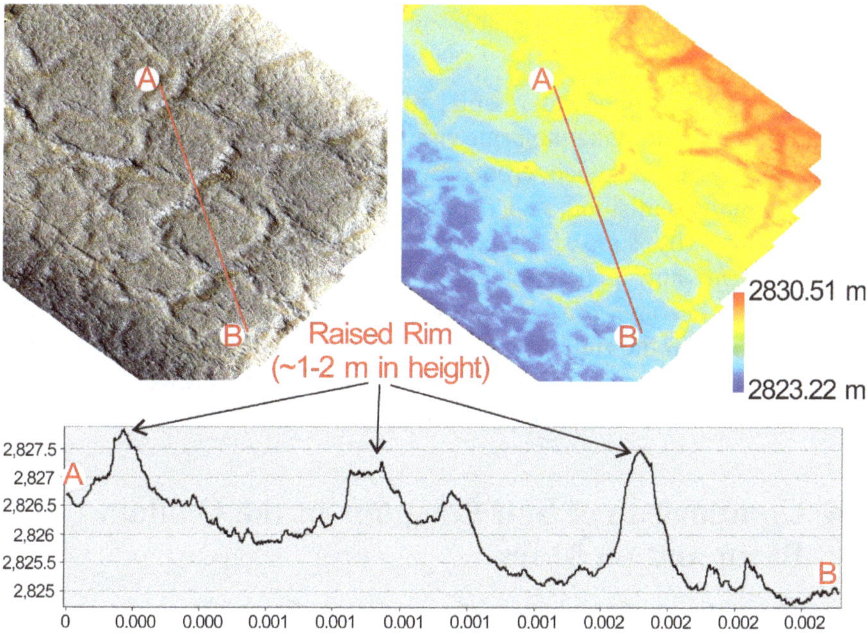

Fig. 6.17. The topographic profile of the large polygons in Dalangtan Playa using the UAV-DEM.

Fig. 6.18. Fractured and collapsed polygon rims with degradation at the late stage of polygon evolution. Photos were taken in July of 2016 in Dalangtan Playa.

lowering of the groundwater table, ultimately reaching a point where evaporation cannot make the brine rise to the surface. The growth of rims will gradually be weakened, eroded by exposure to wind and rain activity, even eventually vanishing. As a dynamic place, the rim zone produces salt solution tubes by subsurface convection. Subsequent collapses occur because of no new generation of vertical tabular-shaped salt crystals. Raised rims become lower and even partially disappear (Fig. 6.18). Digging a hole in the rim, we can see the interior void and a thick weathered layer overlying the salt crust, indicating that the growth of the salt crystals in the fracture has stopped for a long time.

6.4 Comparison of Salt Polygons in the Qaidam Basin and on Mars

6.4.1 *Morphological observations*

In order to equally compare with salt polygons in the Qaidam Basin, we also divide Martian salt polygons into three classes on the basis of their size, similar to the cases in the Dalangtan Playa of the Qaidam Basin. The subsequent sections describe them in detail.

Remote sensing (and confirmation by field survey) shows that the small-sized salt polygons in the Qaidam Basin mainly take the form of regular shapes such as triangles and quadrangles, similar to their counterparts on Mars (Fig. 6.19). For instance, Fig. 6.19(c) shows numerous meter-size polygonal cracks in the northern circum-Hellas region. They are interpreted to have formed in Fe/Mg-smectite/chlorite mixed-layer sedimentary deposits of potentially lacustrine origin (El Maarry *et al.*, 2014). Some typical Dalangtan polygons are spatially arranged to align in a line (red arrows in Fig. 6.19(b)), appearing to be controlled by localized

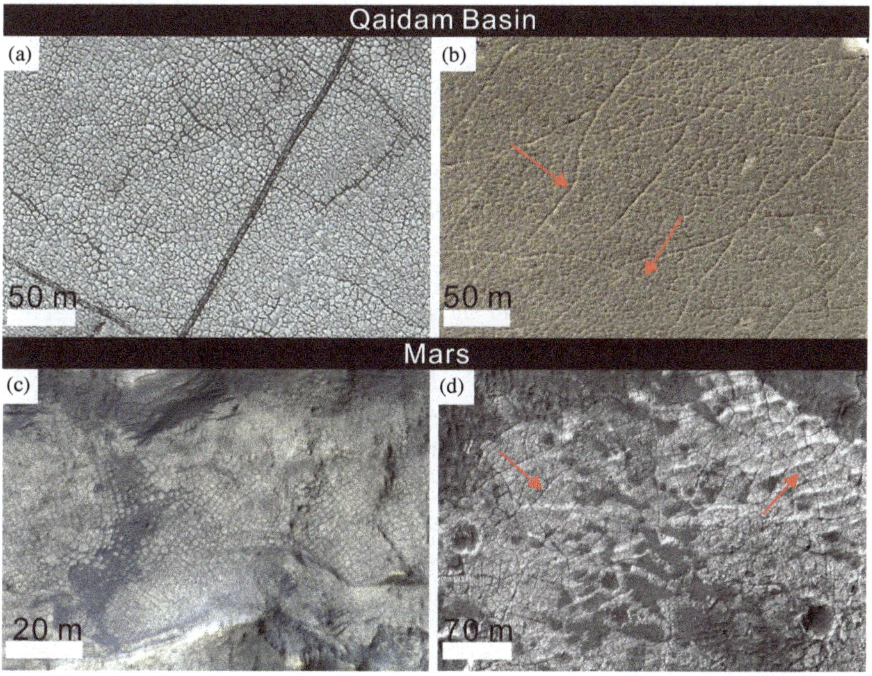

Fig. 6.19. Small-sized salt polygons in the Dalangtan Playa ((a) and (b)) and their counterparts on Mars ((c) and (d)). Red arrows in figures b and d point to the linear tectonic structures. Image sources and locations: (a) Google Earth, 91.532°E, 38.405°N; and (b) Google Earth, 91.414°E, 38.584°N. (c) Chloride-bearing terrain in the northern circum-Hellas region, HiRISE image ID: ESP_016259_1535; 67.656°E, 26.471°S. (d) Possible chloride-rich terrain, HiRISE image ID: ESP_028197_1580; 309.135°E, 21.961°S. North is up in all figures.

linear tectonic structures. Such structurally controlled patterns can also be seen in the polygonal terrain that occurred within chloride-bearing deposits on Mars (El Maarry *et al.*, 2013; Osterloo *et al.*, 2008) (Fig. 6.19(d)). They share some similarities, including: (1) a polygon system of regular random orthogonal patterns (Neal *et al.*, 1968); (2) a fracture system that tends to consist of fracture joints at about 120° and 90° (Neal *et al.*, 1968; Christiansen, 1963); (3) the intersection of linear structures developing a three- or four-fracture system. Therefore, we propose that these polygons on both Earth and Mars share somewhat similar processes for their formation. Red arrows in Fig. 6.19(b) and d point to the longer and larger linear fractures in both locations, which are thought to have formed first and later subdivided and intersected by the smaller fractures. For the Dalangtan salt polygons, these linear tectonic structures are raised ridges resulting from the combination of the variation of the field stresses of the sediments (composition, bulk strength, and thickness) during dehydration and subsurface salt crystal growth and resulting doming effect as intensive evaporation continues. For the ones on Mars indicated by Fig. 6.19(d), they are most likely formed by the dehydration of highly hydratable salts since they are located within diverse chloride-rich terrains (e.g. El Maarry *et al.*, 2013; Osterloo *et al.*, 2008). In addition, we propose that the dehydration duration and the evaporation or sublimation process are also important factors for controlling their final appearance and pattern. This is because the Dalangtan polygon-formation is the consequence of a slow, continuous, and repeated process; however, for their counterparts on Mars, it is still unclear whether Mars has experienced a gradual transition or a fast, catastrophic process of the environmental change from early wet and warm to today's dry and cold state. However, the Martian water cycle might probably never have been identical to that on Earth (e.g. Carter *et al.*, 2015; Kargel, 2004), although some processes and characteristics might have been comparable.

A number of Dalangtan polygons with lower interior floors and relatively higher rims are concentrated within a confined area, forming a honeycomb-like pattern (Figs. 6.20(a) and (b)). Fig. 6.20(d) shows similar honeycomb pattern polygons on the upper wall slope of a valley located in the western arcuate mountain region of the Thaumasia Highlands at the southeast margin of Tharsis. Some other locations in chloride-bearing terrains also display honeycomb-like polygons (Fig. 6.20(c)), but the mechanism for their formation is still unclear.

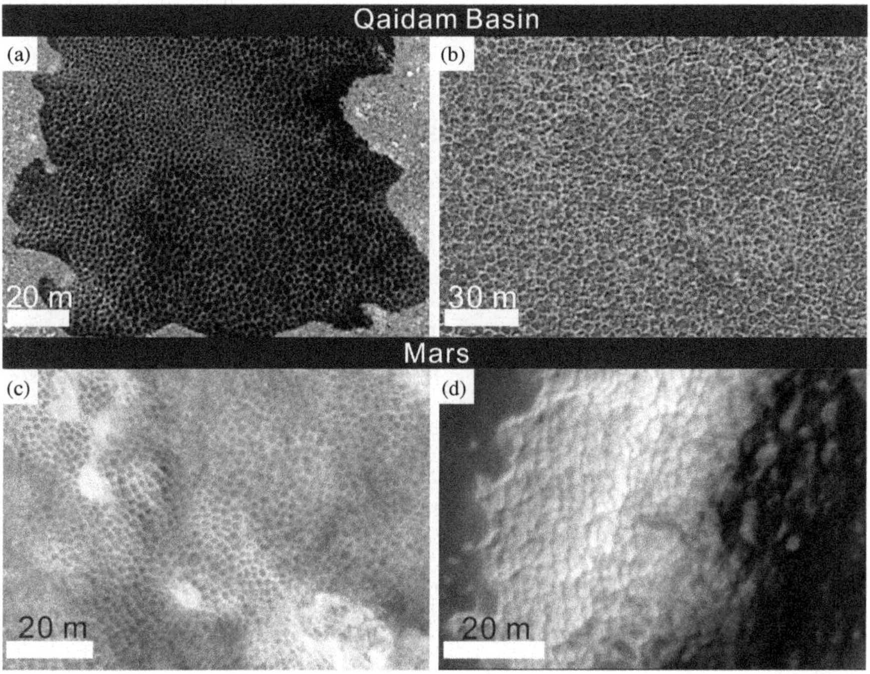

Fig. 6.20. Honeycomb-like salt polygons in the Dalangtan Playa and their counterparts on Mars. Image sources and locations: (a) Google Earth, 91.551°E, 38.425°N; and (b) Google Earth, 91.431°E, 38.452°N;. (c) Possible chloride-bearing terrain in Mawrth Vallis region, HiRISE image ID: ESP_011884_2045; 341.048°E, 24.138°N. (d) White-colored chloride-bearing deposit on the upper wall slope of a valley, HiRISE image ID: ESP_017729_1510; 254.930°E, 28.669°S. North is up in all figures.

The medium-sized polygons in the Qaidam Basin exhibit a significant variation in appearance, and most show some characteristic aspects resembling those in Martian chloride-bearing terrains. Some salt polygons consist of linear rims that intersect with each other in a quasi-orthogonal manner (Figs. 6.21(a) and (c)). This typical formation may result from the secondary fracture development at right angles to the initial primary fracture, leading to a plan-view of orthogonal intersections of the patterned ground (Weinberger, 2001; Loope and Haverland, 1988; Christiansen, 1963; Neal *et al.*, 1968). In some cases, small-sized polygons can be seen in the interior of medium-sized polygons (Figs. 6.21(b) and (d)). However, the key difference is that these

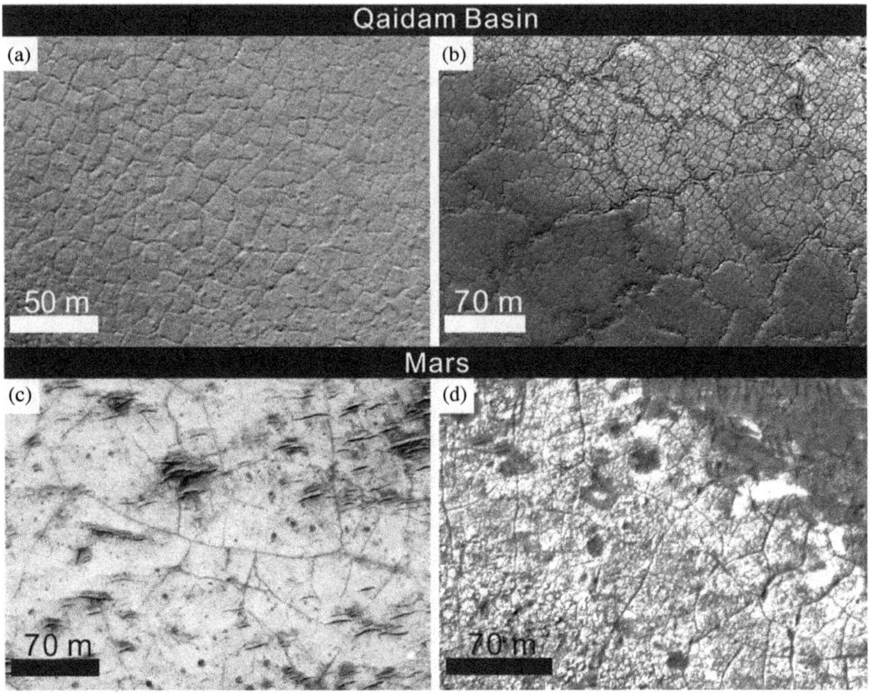

Fig. 6.21. Medium-sized polygons in the Dalangtan Playa and the similar polygonal patterns in chloride-bearing terrains in the southern highland on Mars. Image sources and locations: (a) Google Earth, 91.933°E, 38.200°N; and (b) Google Earth, 91.862°E, 38.355°N. (c) HiRISE image ID: ESP_024922_1570; 75.477°E, 22.812°S. (d) HiRISE image ID: ESP_028197_1580; 309.142°E, 21.964°S. North is up in all figures.

terrestrial salt polygons are more irregular in shape than those on Mars. The medium-sized polygons, the predominant polygon style in the Dalangtan playa, have arcuate or curved rims in a random manner exhibiting no preferred orientation (irregular random orthogonal pattern; Neal *et al.* (1968)) (Fig. 6.21(b)), while the Martian ones are mainly controlled by linear primary fractures forming a regular random orthogonal polygon configuration (Fig. 6.21(d)). Nevertheless, small-sized polygons in their interiors (Fig. 6.21(b)) appear to be much more regular, forming regular random orthogonal patterns.

Compressional structures are observed in some localities in the Dalangtan Playa and in some chloride-bearing regions on Mars. These

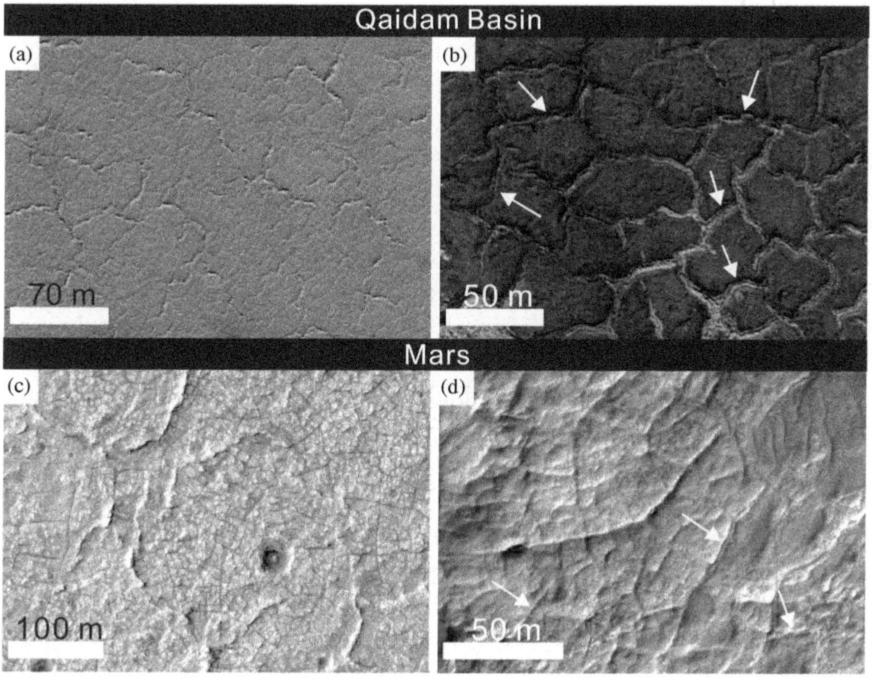

Fig. 6.22. Compressional structures observed in the Qaidam Basin and on Mars. Left column (a) and (c) display sinuous ridges (compressional structures) across the polygonal terrains in both Dalangtan Playa on Earth and chloride-bearing terrains on Mars; Right column (b) and (d) show raised polygon rims with summit fractures along their strikes, the white arrows in the figures point to the linear summit fractures or collapses following the strikes of the rims. Image sources and locations: (a) Google Earth, 91.848°E, 38.189°N; and (b) Google Earth, 91.855°E, 38.223°N. (c) HiRISE image ID: PSP_007770_1770; 351.505°E, 3.136°S; (d) HiRISE image ID: PSP_008720_1610; 332.214°E, 18.805°S. North is up in all figures.

large and long sinuous ridges that occur on both Earth and Mars (Figs. 6.22(a) and (c)) are interpreted to have undergone intra-mineral contraction resulting from great volume changes produced during dehydration, in particular the deposition of salts in the subsurface by evaporation of water contained in rock or sediment pores and ascent of brine into the drying layer (e.g. Komatsu *et al.*, 2007). The presence of summit fractures on these terrestrial and Martian raised polygonal ridges (white arrows, Figs. 6.22(b) and (d)) indicates that they may have once been dynamic and linear conduits/fissures for material and energy exchange between internal and external systems.

Chlorides commonly precipitate during the evaporation of surface water or groundwater. The patterned ground with small to medium-sized polygons in the chloride-bearing terrains on Mars makes a stronger case for evaporation of a standing body of water. These light-toned, polygonally fractured chloride deposits lie within Early Noachian to Early Hesperian geological units (Osterloo *et al.*, 2008, 2010) in the southern highlands of Mars. Polygons have a size range from several to more than 100 m, with some showing subdivision into smaller polygons (Figs. 6.21(d) and 6.22(d)). The large size range of the desiccation crack-like polygons suggests a varying thickness of the stressed zone, revealing a probably different local water history and conditions (such as material composition and permeability) at the time of deposition (El Maarry *et al.*, 2010).

The large-sized salt polygons tend to concentrate at the outer edge of the Dalangtan playa (Fig. 6.10). Some of them cluster fairly close together to form polygon-concentrated patches, showing discrete distributions. By the end of the Pliocene, the salt-bearing layers in the Qaidam Basin were up to 1000 m thick, with the total thickness of the halite sequence being more than 200 m (e.g. Chen and Bowler, 1986; Xia *et al.*, 2001). Generally, large-sized salt polygons are mainly located at relatively higher topographic areas, where the stressed layer subject to fracturing is sufficiently thick to create large polygons (El Maarry *et al.*, 2014; Neal *et al.*, 1968). Figures 6.23 and 6.24 present examples of large-sized polygons that exist in the Dalangtan playa and on Mars. These large Dalangtan polygons (Figs. 6.23(a) and (b)) appear to have the same structure as those of similar-sized Martian polygons (Figs. 6.23(c) and (d)).

The large-sized Dalangtan polygons commonly have a rim varying in size from 4 to 6 m wide (Figs. 6.24(a) and (b)), and some display secondary small-sized salt polygons in their interiors. Varying lengths of their raised rims are commonly observed. For example, the large polygons at the foot of Nanyishan have raised rims with side lengths that vary ranging from 20 to 150 m (Fig. 6.24(b)). We also found raised rim polygons in chloride-bearing terrains on Mars as shown in Figs. 6.24(c) and (d). The white salt ridges are generally associated with saturated brines (e.g. Neal *et al.*, 1968), whereas their summit fractures (or cracks) are usually dark because of shadows or infilling of weathered products (such as sand and dust) transported from other places.

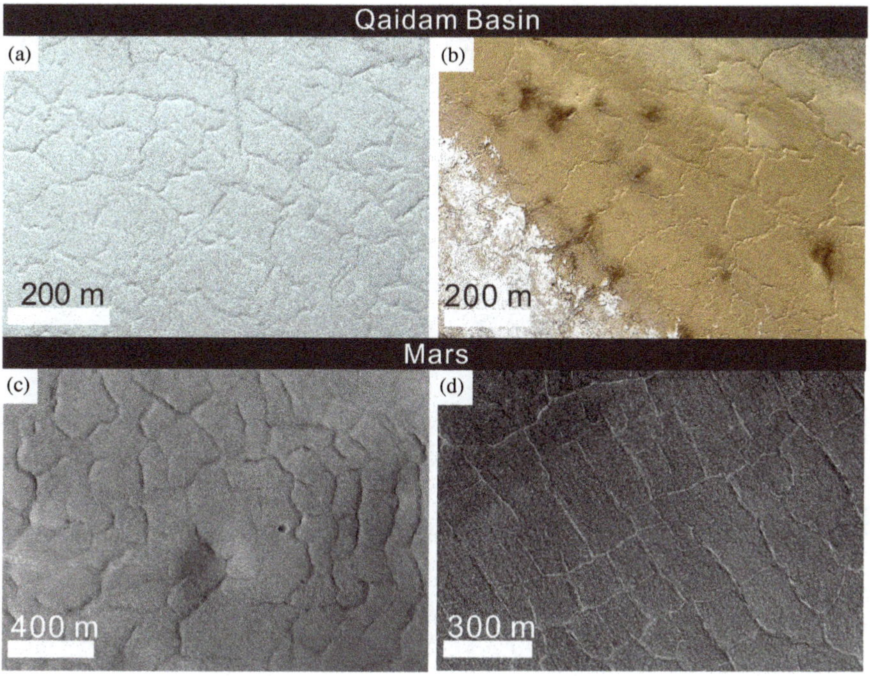

Fig. 6.23. Large-sized salt polygons in the Dalangtan Playa and their counterparts occurred in crater floors on Mars. Image sources and locations: (a) Google Earth, 91.529°E, 38.394°N; and (b) Google Earth, 91.802°E, 38.214°N. (c) HiRISE image ID: PSP_001942_2310; 341.644°E, 50.655°N. (d) HiRISE image ID: PSP_007539_2420; 170.943°E, 61.824°N. North is up in all figures.

About 46 locations of raised ridge polygons were identified in chloride-bearing terrains across the southern highlands of Mars (black circles in Fig. 6.4). Most appear to cluster in and around Terra Sirenum (Glotch *et al.*, 2010). This kind of polygonal feature is characterized by prominent raised ridges (Figs. 6.22(d) and .24(d)). Most of these polygons have sizes of tens of meters, with some exceptions having lengths up to hundreds of meters (Fig. 6.23). The regions of the raised ridges are usually relatively brighter than their surroundings. Their rim widths are commonly between ~0.5 and 2 m, even up to about 10 meters for the large-sized polygons shown in Fig. 6.24(a). We were unable to measure the height of the rims directly, because of a lack of HiRISE Digital Terrain Model and stereo pair coverage. Instead, we used incidence angle of the scene and number of the pixels of the shadow region of the rims to estimate their heights.

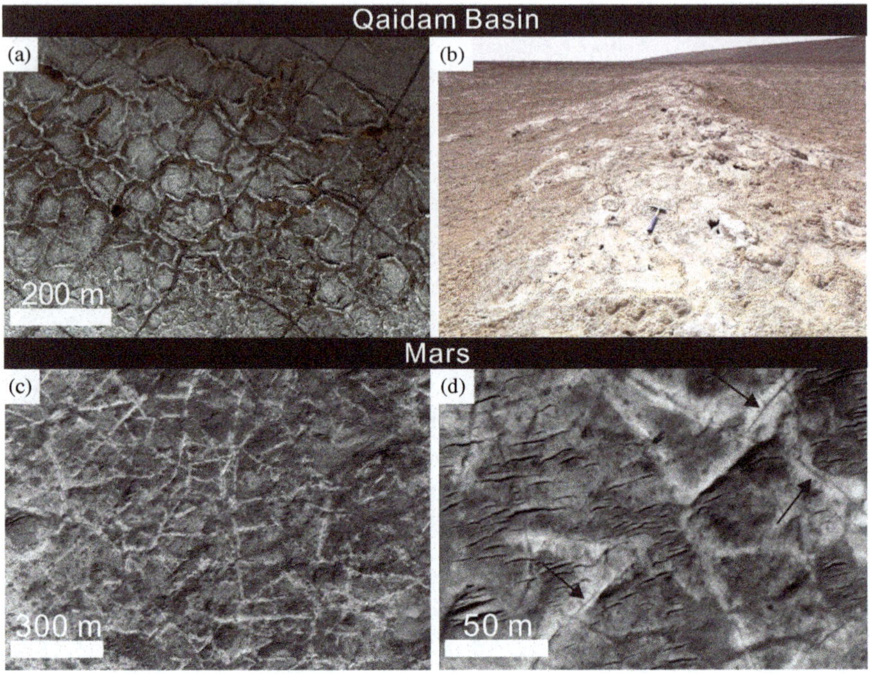

Fig. 6.24. Left column (a) and (c) show other examples of large-sized polygons in the Qaidam Basin and on Mars, respectively. Right column (b) and (d) display the enlarged views of the raised rims in the Qaidam Basin and on Mars, respectively. Black arrows in figure (d) point to the linear dark top fractures relative to their surrounding bright raised rim zones. Image sources and locations: (a) Google Earth, 91.512°E, 38.410°N; (b) Image taken in July of 2015 by the author; (c) and (d) HiRISE image ID: ESP_024922_1570; 75.493°E, 22.781°S. North is up in all figures.

The heights are generally less than ~1 m, consistent with a range similar to measured cases for Dalangtan salt polygons.

A significantly morphological similarity that should be noted is that the linear negative fractures or collapses along the strike coexist for some small to large-sized polygons in the Dalangtan playa and chloride-bearing terrains on Mars. A series of different-sized polygons with raised rims and summit fractures or cracks are contained in the Middle Noachian to Hesperian (2.9 billion years ago) aged Martian highland units (Osterloo *et al.*, 2008, 2010). Their distribution within chloride-containing salty sediments in low areas and their bright colors distinct from surroundings are consistent with the general morphological and compositional characteristics of the

Dalangtan salt polygons. On Earth, chlorides are known to form through aqueous processes (Goodall *et al.*, 2000). Thus, similar processes are expected to be responsible for the formation of chloride deposits on Mars. This provides strong evidence of the presence of surface or subsurface water on ancient Mars (e.g. Osterloo *et al.*, 2010). The Earth-like evaporation-triggered playa sediment cracks, and different levels of salt mineral upwelling and convection along zones of weakness of deposits, could have once been involved in the formation and development of these polygons on Mars.

6.4.2 *Clues for depositional environments on Mars*

The study of the salt polygons in the Qaidam Basin reveals the important role of mineral assemblages in their formation, and this may also apply to Martian salt polygons. The physical properties of soil or sediments and the presence/absence of salts greatly influence the desiccation rate of the medium (El Maarry *et al.*, 2014; Plummer and Gostin, 1981). The clay minerals are the main contributors to the development of large desiccation cracks on playa surfaces (El Maarry *et al.*, 2014). If the soil or sediment is richer in clay minerals, it will produce enhanced shrinking with desiccation (El Maarry *et al.*, 2014). Therefore, in addition to the thickness of the sediment deposits, the formations of numerous salt polygons of different sizes and patterns are significantly controlled by mineral assemblages, indicating a diversity of sedimentary environments during the evolutionary periods of salt lakes in the Qaidam Basin.

With the cycled lowering of lake water-table levels, more salt becomes concentrated in isolated flat floors of salt ponds (low topography). The salt deposits dominated by chloride-bearing materials are thicker in such low topography than anywhere else outside these negative relief areas owing to their higher solubility (e.g. Komatsu *et al.*, 2007). Therefore, the small-sized salt polygons are expected to accumulate on these flat floors due to the high salt content of a thick solid crust, limiting the emergence of large polygonal networks (El Maarry *et al.*, 2015(a)). Furthermore, lower topography has a higher frequency of inundation by rain or groundwater movement than the higher topography, which would lead to the frequent destruction of polygons, resetting the regrowth of polygons. It is assumed that the above-mentioned conditions would have also existed in the Martian geological context.

We have detected such a polygon size–topography relationship in a chloride-bearing highland unit on Mars. Figure 6.9 shows several spatially connected irregular depressions with exposures of bright-colored chloride-bearing sediments, at the northern margin of the Terra Sirenum region. The salt polygons on the relatively higher terraces of the depression are clearly much larger than those located in the lower floor. Within and around the Terra Sirenum region, six sites of chloride deposits (10–50 km^2) in the lowest topographic levels of an inter-crater basin (300–400 km) have been interpreted as individual salt flats (Davila *et al.*, 2011). The connecting channels between the salt flats provide evidence of a common origin, such as evaporation of water (Davila *et al.*, 2011). This may prove that both the Earth and Mars would have experienced similar lacustrine environments and associated processes for salt polygon formation at some time in their geological histories. We suggest that this type of correlation (polygon size–topography–composition) should be further investigated as important geological diagnostic tools.

6.5 Implications for Martian Environmental and Climatic Change

6.5.1 *Implications for climate change*

Mars is considered to have had a globally active hydrological cycle (Baker *et al.*, 1991; Carr, 1996 and references therein) during the billion years after the formation of its solid crust. The presence of paleolakes (Salese *et al.*, 2016; Zhao and Xiao, 2016; Goudge *et al.*, 2015; Komatsu *et al.*, 2009; Fassett and Head, 2008b), valley networks, and associated deltas (Baker *et al.*, 1991; Carr, 1986; Fassett and Head, 2008a; Gulick, 2001; Hynek *et al.*, 2010), explosive volcanism (Blasius and Cutts, 1981; Halevy and Head, 2014; Huang and Xiao, 2014; Hynek *et al.*, 2003; Kerber *et al.*, 2012; Wilson and Head, 2007), and the discovery of the surface mineralogical diversity, including hydrous phyllosilicates (Ehlmann *et al.*, 2009; Murchie *et al.*, 2009; Mustard *et al.*, 2008; Wray *et al.*, 2009) and salts (Ehlmann *et al.*, 2008; Langevin *et al.*, 2005; Osterloo *et al.*, 2008, 2010; Wray *et al.*, 2010, 2011), all reveal a potential warmer and wetter climate on early Mars (>3 Ga; Carr and Head, 2010; Carter *et al.*, 2015) and this in stark contrast to the present-day dry-and-cold Mars.

Likewise, the Qaidam Basin has experienced a climatic change (more recently however) from wet to dry conditions since 25,000 yr B.P., and approximately 14,000 yr B.P. when the lakes finally dried up (Chen and Bowler, 1986; Cheng *et al.*, 2017). The wetter and warmer climatic conditions that prevailed in the past are preserved in a series of paleohydrological features such as fluvial valleys and lacustrine sediments, which are common in the basin today (Xiao *et al.*, 2017). The climatic change from the wetter to drier conditions also resulted in the extensive formation of the polygons in the basin. Based on the striking resemblance between the salt polygons in the Qaidam Basin and those found within crater floors and chloride-bearing terrains on Mars, we hypothesize that the widespread presence of the chloride-bearing materials that occur on the terrains from the Early Noachian to the Early Hesperian ages in the southern highlands (Osterloo *et al.*, 2008, 2010) is a strong indication of the general climatic change from wetter to drier condition in the Noachian–Hesperian transition. This is consistent with a very dry present-day environment on Mars and a great number of preserved surface expressions and mineralogical diversity as evidence of an ancient hydrological circle and regimes (Masson *et al.*, 2001; Wray *et al.*, 2009, 2010, 2011). However, many questions remain as to the exact environments (such as atmospheric condition and availability of liquid water on the surface and subsurface) involved in such climate change and the timing of the polygon formation on Mars.

Some craters hosting polygons are of Amazonian age, and most craters at high latitude are to some degree superposed by dust–ice mantle deposits (Head *et al.*, 2003). According to the work of El Maarry *et al.* (2010), desiccation cracks form when a volatile-rich material dries up creating stresses through changing surface tension between soil grains. Many Martian landforms are linked to ice deposits (e.g. Kargel, 2004). Impact may have modified localized subsurface structure and thermal gradient (supplied heat from impact generated melt-bearing deposits) resulting in an active zone where a phase change from subsurface ice to volatiles could take place. In addition, numerous craters at low (warmer) latitudes show simple, only partially filled troughs, which is more indicative of "one-time" desiccation processes (El Maarry *et al.*, 2010). Therefore, for Amazonian aged craters, crater-floor polygons may be just a result of localized geological setting or intermittent climate change rather than an indication of global climatic change.

6.5.2 *Hydrology of lacustrine/playa environments on Mars and its implications for polygons formation*

The general high abundance of halite content within the Dalangtan playa sediments indicates that salt polygons are commonly associated with chloride-bearing terrains that have a lacustrine/playa origin. A similar conclusion has been reached by numerous recent studies of the chloride-bearing materials on Mars (El Maarry *et al*., 2013; Glotch *et al*., 2010; Osterloo *et al*., 2008, 2010; Ruesch *et al*., 2012). As noted in our study, the size and morphology of salt polygons in the Qaidam Basin are similar to those polygons in impact crater floors and depressions associated with chlorides on Mars (El Maarry *et al*., 2010, 2013). It has been suggested that there were many ancient salt lakes, which had later turned to drier playas based on the occurrence of polygons and evaporate minerals (El Maarry *et al*., 2013).

Resemblance of some peculiar morphologies of salt polygons in the Qaidam Basin and on Mars is also apparent. Most polygons in the Qaidam Basin have raised rims; in particular those large-sized polygons with considerably raised rims. Interestingly, we also find similar raised-rim salt polygons in many Martian locations of chloride-bearing terrains across a large fraction of the southern highlands. Furthermore, polygonal ridge networks with different origins are found in many locations and geological contexts across Mars (Kerber *et al*., 2017). Linear ridges intersect with each other to form polygons with uplifted rims, and some (particularly those at the same scale as that in the Dalangtan Playa) are at present believed to be the result of chemical accretion or sedimentary infilling along pre-existing fractures by near-surface water circulation (Kerber *et al*., 2017). A significant difference is that sedimentary infilling of pre-existing fractures on Mars is interpreted to be a liquid process, while the raised, intersecting ridges in the Dalangtan playa are a consequence of endogenous processes by evaporation. The identification of chloride salts (Glotch *et al*., 2010, 2016; Murchie *et al*., 2009; Osterloo *et al*., 2008, 2010; Ruesch *et al*., 2012; Wray *et al*., 2009) together with the detection of desiccation cracks (El Maarry *et al*., 2013, 2014) across a variety of geological contexts such as possible paleolakes, inverted channels, distributary fans, and crater fills provide strong evidence of the desiccated origin of the deposits on Mars. However, since sodium chloride is devoid of diagnostic spectral absorption features in the VNIR or MIR spectral ranges, its abundance is difficult to derive from spectral unmixing techniques of remotely-sensed data or other analytical approaches based on current methodologies (Glotch *et al*., 2010; Murchie *et al*., 2009). Sample analyses in this work may provide clues to solve this problem. Some types of polygons with raised rims in many locations, mostly distributed in the Martian southern highlands, are thought to be formed as a result of fractures filled by sedimentary deposits due to groundwater upwellings (Kerber *et al.*, 2017). They may also have similar mineral abundances with the polygons in the Qaidam Basin. We propose that the study of these Martian polygon terrains correlated with chloride salt deposits should be focused on to provide key evidence of paleolakes evolution that might have occurred on Mars.

Chloride-bearing deposits are identified to have formed on different aged terrains with diverse geological settings. Mars surface polygonally fractured deposits with chloride-bearing materials in the southern highlands are light-toned relative to their surroundings (Figs. 6.19, 6.20, 6.21, 6.22, 6.23 and 6.24). They are commonly topographically lower than the surrounding terrain and are distributed globally in small ($<\sim 25$ km^2) patches, occurring in the Middle Noachian to Early Hesperian terrains (Osterloo *et al.*, 2008). This suggests that near-surface water was globally widespread in the early Martian history. The precipitation of chlorides usually happens under the conditions of intensive evaporation of surface water or groundwater and volcanic outgassing (Osterloo *et al.*, 2008). The polygonally fractured deposits, occurring in many Hesperian to Amazonian craters (Tanaka *et al.*, 2014) across low to middle latitudes of Mars surface (El Maarry *et al.*, 2010) are interpreted to be formed in desiccated environments of ancient lakes on Mars (El Maarry *et al.*, 2010). The impact events of these floor-fractured craters might have re-melted the near-surface salt-ice mixture materials and resulted in their accumulation in the craters forming ancient lakes and/or water-rich sediments in localized areas. The geomorphology of younger craters with fluidized ejecta blanket (Carr *et al.*, 1977; Mouginis-Mark, 1987) and more recently recurring slope lineae (RSL) (Chevrier and Rivera-Valentin, 2012; Levy, 2012; McEwen *et al.*, 2011, 2014; Ojha *et al.*, 2014, 2015) all indicate that the ice-rich and/or salt–ice mixture materials (producing enough brine to account for the recently observed flow features) still exist in today's Mars's surface/subsurface. Though RSL were recently interpreted as dry mass wasting (granular flows) (Dundas *et al.*, 2017), the strong seasonality associated with warm slopes and the detection of hydrated salts are still consistent with some role for water in their initiation (Dundas *et al.*, 2017). Therefore, the remelting of the salt–ice mixture materials and subsequent evaporation could have contributed to the formation of polygonal landforms in many locations. The different age units on which salt polygons formed suggest that the intense evaporation environment was globally widespread in the past, but the lifetime of water storage in different areas (such as in craters) is believed to have been different (e.g. Cabrol and Grin, 1999; El Maarry *et al.*, 2010; Forsythe and Blackwelder, 1998; Scott *et al.*, 1995; Wharton *et al.*, 1995).

6.6 Summary

This chapter is focused on the observations of the diverse polygonal patterned terrains in the dried salt lakes in the Qaidam Basin, western China, and the comparison with their counterparts in potential ancient lake-like environments on Mars. The dried salt lakes on Earth feature many polygon characteristics (including their size, spatial organization pattern, morphology, composition of sediments, and relation to topography) that are similar to those in salt-bearing terrains on Mars. In view of some similarities of polygonal structures on the Earth and Mars, similar processes are predicted to be responsible for the formation of these polygonal terrains on both planets. On Earth, salt deposits are known to form through the aqueous processes of efflorescence (i.e. when a salt forms a coating over porous material due to evaporation) and precipitation. The discovery of the terrestrial patterned terrains analogous to those on Mars provides further evidence of the presence of a liquid-water-related process early in the history of Mars.

The climatic change from the wetter to drier conditions resulted in the extensive formation of the polygonal structures in playas. Based on the striking resemblance between the polygons in the Qaidam Basin and those found in the chloride-bearing terrains on Mars, we hypothesize that the widespread presence of the chloride-bearing materials that mainly occur in the terrains from the Early Noachian to the Early Hesperian ages in the southern highlands is a strong indication of the general climatic change from wetter to drier condition in the Noachian–Hesperian transition. This is consistent with a very dry present-day environment on Mars and a great number of preserved surface expressions and mineralogical diversity as evidence of an ancient hydrological circle and regimes. It is not certain, however, that liquid water is needed to form the polygonal ground. We believe some endogenous processes such as ice sublimation (volatile-rich material) from depth would also have contributed to the initiation of the formation of the polygonal ground, even in sediments superposed by younger Amazonian deposits.

Acknowledgments

This work was supported by the Science and Technology Development Fund (FDCT), Macau SAR (Grant No. 121/2017/A3), the pre-research Project on Civil Aerospace Technology (D020101) of CNSA, and the National Natural Science Foundation of China, China (41830214). The imagery data used for Mars are available on the Planetary Data System Geosciences Node. We thank Dr. Feng Zhang for his great help during the writing of this chapter.

References

Anglés, A. and Li, Y. (2017). The western Qaidam Basin as a potential Martian environmental analogue: an overview. *J. Geophys. Res. Planets*, **122**, 856–888.

Artieda, O., Davila, A., Wierzchos, J., Buhler, P., Rodríguez-Ochoa, R., Pueyo, J. and Ascaso, C. (2015). Surface evolution of salt-encrusted playas under extreme and continued dryness, *Earth Surf. Proces. Land.*, **40**, 1939–1950.

Baker, V. R., Strom, R. G., Gulick, V. C., Kargel, J. S., Komatsu, G., Kale, V. S. (1991). Ancient oceans, ice sheets and the hydrological cycle on Mars. *Nature*, **352**(6336), 589–594.

Blasius, K. R. and Cutts, J. A. (1981). Topography of Martian central volcanoes. *Icarus*, **45**, 87–112.

Budkewitsch, P. and Robin, P.-Y., (1994). Modelling the evolution of columnar joints. *J. Volcanol. Geothermal Res.*, **59**, 219–239.

Cabrol, N. A. and Grin, E. A., (1999). Distribution, classification, and ages of Martian impact crater lakes. *Icarus*, **142**, 160–172.

Carr, M., Crumpler, L., Cutts, J., Greeley, R., Guest, J. and Masursky, H. (1977). Martian impact craters and emplacement of ejecta by surface flow. *J. Geophys. Res.*, **82**, 4055–4065.

Carr, M. H. (1986). Mars: A water-rich planet? *Icarus*, **68**, 187–216.

Carr, M. H. (1996). *Water on Mars*, Oxford University Press, New York, p. 229.

Carr, M. H. (2007). *The Surface of Mars*, Cambridge University Press.

Carr, M. H. and Head, J. W. (2010). Geologic history of Mars. *Earth and Planet. Sci. Lett.*, **294**, 185–203.

Carter, J., Loizeau, D., Mangold, N., Poulet, F. and Bibring, J.-P. (2015). Widespread surface weathering on early Mars: A case for a warmer and wetter climate. *Icarus*, **248**, 373–382.

Cartwright, J. and Dewhurst, D. (1998). Layer-bound compaction faults in fine-grained sediments. *Geol. Soc. Am. Bull.*, **110**, 1242–1257.

Cartwright, J. and Lonergan, L. (1996). Volumetric contraction during the compaction of mudrocks: A mechanism for the development of regional-scale polygonal fault systems. *Basin Res.*, **8**, 183–193.

Chen, K. and Bowler, J. M. (1986). Late Pleistocene evolution of salt lakes in the Qaidam Basin, Qinghai Province, China. *Palaeogeography, Palaeoclimatology, Palaeoecology*, **54**, 87–104.

Cheng, Z., Xiao, L., Wang, H., Yang, H., Li, J., Huang, T., Xu, Y. and Ma, N. (2017). Bacterial and archaeal lipids recovered from subsurface evaporites of Dalangtan playa on the Tibetan Plateau and their astrobiological implications. *Astrobiology*, **17**, 1112–1122.

Chevrier, V. F. and Rivera-Valentin, E. G. (2012). Formation of recurring slope lineae by liquid brines on present-day Mars. *Geophys. Res. Lett.*, **39**, L21202, doi: 10.1029/2012GL054119.

Christiansen, F. W. (1963). Polygonal fracture and fold systems in the salt crust, Great Salt Lake Desert, Utah. *Science,* **139**, 607–609.

Dang, Y., Xiao, L., Xu, Y., Zhang, F., Huang, J., Wang, J., Zhao, J., Komatsu, G. and Yue, Z. (2018). The Polygonal Surface Structures in the Dalangtan Playa, Qaidam Basin, NW China: Controlling Factors for Their Formation and Implications for Analogous Martian Landforms. *J. Geophys. Res. Planets*, **123**, 1910–1933.

Davila, A. F., Gross, C., Marzo, G. A., Fairén, A. G., Kneissl, T., McKay, C. P., and Dohm, J. M. (2011). A large sedimentary basin in the Terra Sirenum region of the southern highlands of Mars. *Icarus*, 212, 579–589.

Dellwig, L. F. (1968). Significant features of deposition in the Hutchinson Salt, Kansas, and their interpretation. *Geol. Soc. Am. Spec. Pap.*, **88**, 421–428.

Dundas, C. M., McEwen, A. S., Chojnacki, M., Milazzo, M. P., Byrne, S., McElwaine, J. N. and Urso, A. (2017). Granular flows at recurring slope lineae on Mars indicate a limited role for liquid water. *Nat. Geosci.,* **10**, 903.

Ehlmann, B. L., Mustard, J. F., Murchie, S. L., Poulet, F., Bishop, J. L., Brown, A. J., Calvin, W. M., Clark, R. N., Des Marais, D. J. and Milliken, R. E. (2008). Orbital identification of carbonate-bearing rocks on Mars. *Science,* **322**, 1828–1832.

Ehlmann, B. L., Mustard, J. F., Swayze, G. A., Clark, R. N., Bishop, J. L., Poulet, F., Des Marais, D. J., Roach, L. H., Milliken, R. E. and Wray, J. J. (2009). Identification of hydrated silicate minerals on Mars using MRO-CRISM: Geologic context near Nili Fossae and implications for aqueous alteration. *J. Geophys. Res. Planets*, **114**, E00D08, doi: 10.1029/2009JE003339.

El Maarry, M. R., Markiewicz, W. J., Mellon, M. T., Goetz, W., Dohm, J. M. and Pack, A. (2010). Crater floor polygons: Desiccation patterns of ancient lakes on Mars? *J. Geophys. Res. Planets*, **115**, E10006, doi: 10.1029/2010JE003609.

El Maarry, M. R., Pommerol, A. and Thomas, N. (2013). Analysis of polygonal cracking patterns in chloride-bearing terrains on Mars: Indicators of ancient playa settings. *J. Geophys. Res. Planets*, **118**, 2263–2278.

El Maarry, M. R., Pommerol, A. and Thomas, N. (2015a). Desiccation of phyllosilicate-bearing samples as analog for desiccation cracks on Mars: Experimental setup and initial results. *Planet. Space Sci.*, **111**, 134–143.

El Maarry, M. R., Watters, W., McKeown, N. K., Carter, J., Dobrea, E. N., Bishop, J. L., Pommerol, A. and Thomas, N. (2014). Potential desiccation cracks on Mars: A synthesis from modeling, analogue-field studies, and global observations. *Icarus*, **241**, 248–268.

El Maarry, M. R., Watters, W. A., Yoldi, Z., Pommerol, A., Fischer, D., Eggenberger, U. and Thomas, N. (2015b). Field investigation of dried lakes in western United States as an analogue to desiccation fractures on Mars. *J. Geophys. Res. Planets*, **120**, 2241–2257.

Fassett, C. I. and Head, J. W. (2008a). The timing of martian valley network activity: Constraints from buffered crater counting. *Icarus*, **195**, 61–89.

Fassett, C. I. and Head, J. W. (2008b). Valley network-fed, open-basin lakes on Mars: Distribution and implications for Noachian surface and subsurface hydrology. *Icarus*, **198**, 37–56.

Forsythe, R. D. and Blackwelder, C. R. (1998). Closed drainage crater basins of the Martian highlands: Constraints on the early Martian hydrologic cycle. *J. Geophys. Res. Planets*, **103**, 31421–31431.

Glotch, T. D., Bandfield, J. L., Tornabene, L. L., Jensen, H. B. and Seelos, F. P. (2010). Distribution and formation of chlorides and phyllosilicates in Terra Sirenum, Mars. *Geophys. Res. Lett.*, **37**.

Glotch, T. D., Bandfield, J. L., Wolff, M. J., Arnold, J. A. and Che, C. (2016). Constraints on the composition and particle size of chloride salt-bearing deposits on Mars. *J. Geophys. Res. Planets*, **121**, 454–471.

Goehring, L. (2013). Evolving fracture patterns: Columnar joints, mud cracks and polygonal terrain. *Philosophical Transactions of the Royal Society A*, **371**, 20120353.

Goehring, L., Conroy, R., Akhter, A., Clegg, W. J. and Routh, A. F. (2010). Evolution of mud-crack patterns during repeated drying cycles. *Soft Matter*, **6**, 3562–3567.

Goehring, L. and Fourrière, A. (2014). Dynamics of salt playa polygons. *AGU Fall Meeting Abstracts*, p. 3770.

Goodall, T. M., North, C. P. and Glennie, K. W. (2000). Surface and subsurface sedimentary structures produced by salt crusts. *Sedimentology*, **47**, 99–118.

Goudge, T. A., Aureli, K. L., Head, J. W., Fassett, C. I. and Mustard, J. F. (2015). Classification and analysis of candidate impact crater-hosted closed-basin lakes on Mars. *Icarus*, **260**, 346–367.

Gulick, V. C. (2001). Origin of the valley networks on Mars: A hydrological perspective. *Geomorphology*, **37**, 241–268.

Halevy, I. and Head, J. W. (2014). Episodic warming of early Mars by punctuated volcanism. *Nat. Geosci.*, **7**, 865.

Han, W., Fang, X., Ye, C., Teng, X. and Zhang, T., (2014). Tibet forcing Quaternary stepwise enhancement of westerly jet and central Asian aridification: Carbonate isotope records from deep drilling in the Qaidam salt playa, NE Tibet. *Glob. Planet. Change*, **116**, 68–75.

Handford, C. R. (1991). Chaper 1 Marginal Marine Halite: Sabkhas and Salinas, *Developments in Sedimentology*, Elsevier, pp. 1–66.

Head, J. W., Mustard, J. F., Kreslavsky, M. A., Milliken, R. E. and Marchant, D. R. (2003). Recent ice ages on Mars. *Nature*, **426**, 797–802.

Helfenstein, P. and Mouginis-Mark, P. (1980). Morphology and distribution of fractured terrain on Mars, Lunar and Planetary Science Conference, pp. 429–431.

Hiesinger, H. and Head, J. W. (2000). Characteristics and origin of polygonal terrain in southern Utopia Planitia, Mars: results from Mars Orbiter Laser Altimeter and Mars Orbiter Camera data. *J. Geophys. Res. Planets*, **105**, 11999–12022.

Hsü, K. J. and Siegenthaler, C. (1969). Preliminary experiments on hydrodynamic movement induced by evaporation and their bearing on the dolomite problem. *Sedimentology*, **12**, 11–25.

Huang, J., Salvatore, M. R., Edwards, C. S., Harris, R. L. and Christensen, P. R. (2018). A complex fluviolacustrine environment on early Mars and its astrobiological potentials. *Astrobiology*, **18**, 1081–1091.

Huang, J. and Xiao, L. (2014). Knobby terrain on ancient volcanoes as an indication of dominant early explosive volcanism on Mars. *Geophys. Res. Lett.*, **41**, 7019–7024.

Huang, Q., Ku, T. L. and Phillips, F. M. (1993). Evolutionary characteristics of lakes and paleoclimatic undulation in the Qaidam Basin, China. *Chin. J. Oceanol. Limnol.*, 34–45.

Hynek, B. M., Beach, M. and Hoke, M. R. (2010). Updated global map of Martian valley networks and implications for climate and hydrologic processes. *J. Geophys. Res. Planets*, **115**, E09008, doi: 10.1029/2009JE003548.

Hynek, B. M., Phillips, R. J. and Arvidson, R. E. (2003). Explosive volcanism in the Tharsis region: Global evidence in the Martian geologic record. *J. Geophys. Res. Planets*, **108**, E9, 5111, doi: 10.1029/2003JE002062.

Kargel, J. S. (2004). Mars-A warmer, Wetter Planet, Springer Science & Business Media.

Kerber, L., Dickson, J. L., Head, J. W. and Grosfils, E. B. (2017). Polygonal ridge networks on Mars: Diversity of morphologies and the special case of the eastern Medusae Fossae Formation. *Icarus*, **281**, 200–219.

Kerber, L., Head, J. W., Madeleine, J.-B., Forget, F. and Wilson, L. (2012). The dispersal of pyroclasts from ancient explosive volcanoes on Mars: Implications for the friable layered deposits. *Icarus*, **219**, 358–381.

Kminek, G. and Bada, J. L. (2006). The effect of ionizing radiation on the preservation of amino acids on Mars. *Earth Planet. Sci. Lett.*, **245**, 1–5.

Komatsu, G., Di Achille, G., Popa, C., Di Lorenzo, S., Rossi, A. P. and Rodriguez, J. A. P. (2009). Paleolakes, paleofloods, and depressions in Aurorae and Ophir Plana, Mars: Connectivity of surface and subsurface hydrological systems. *Icarus*, **201**, 474–491.

Komatsu, G., Ori, G. G., Marinangeli, L. and Moersch, J. E. (2007). Playa environments on Earth: possible analogs for Mars, In: Chapman, M.G. (ed.), *The Geology of Mars: Evidence from Earth-Based Analogs, The Geology of Mars: Evidence from Earth-Based Analogs*. Cambridge University Press, pp. 322–348.

Kong, F., Zheng, M., Hu, B., Wang, A., Ma, N. and Sobron, P. (2018). Dalangtan Saline Playa in a Hyperarid Region on Tibet Plateau-I: Evolution and Environments. *Astrobiology*, **18**(10), 1243–1253.

Kong, W. G., Zheng, M. P. and Kong, F. J. (2014a). Brine Evolution in Qaidam Basin, Northern Tibetan Plateau, and the Formation of Playas as Mars Analogue Site. Lunar and Planetary Science Conference, p. 1228.

Kong, W. G., Zheng, M. P., Kong, F. J. and Chen, W. X. (2014b). Sulfate-bearing deposits at Dalangtan Playa and their implication for the formation and preservation of Martian salts. *Am. Mineral.*, **99**, 283–290.

Lachenbruch, A.H., 1962. Mechanics of thermal contraction cracks and ice-wedge polygons in permafrost. *Geolog. Soc. Am. Spec. Pap.*, **70**, 1–66.

Langevin, Y., Poulet, F., Bibring, J.-P., Gondet, B., 2005. Sulfates in the north polar region of Mars detected by OMEGA/Mars Express. *Science*, **307**, 1584–1586.

Levy, J. (2012). Hydrological characteristics of recurrent slope lineae on Mars: Evidence for liquid flow through regolith and comparisons with Antarctic terrestrial analogs. *Icarus*, **219**, 1–4.

Levy, J., Head, J. and Marchant, D. (2009). Thermal contraction crack polygons on Mars: Classification, distribution, and climate implications from HiRISE observations. *J. Geophys. Res. Planets*, **114**, E01007, doi: 10. 1029/2008JE003273.

Levy, J. S., Marchant, D. R. and Head, J. W. (2010). Thermal contraction crack polygons on Mars: A synthesis from HiRISE, Phoenix, and terrestrial analog studies. *Icarus*, **206**, 229–252.

Li, M., Fang, X., Yi, C., Gao, S., Zhang, W. and Galy, A. (2010). Evaporite minerals and geochemistry of the upper 400m sediments in a core from the Western Qaidam Basin, Tibet. *Quat. Int.*, **218**, 176–189.

Lonergan, L. and Cartwright, J. A. (1999). Polygonal faults and their influence on deep-water sandstone reservoir geometries, Alba Field, United Kingdom central North Sea. *AAPG Bull.*, **83**, 410–432.

Loope, D. B. and Haverland, Z. E. (1988). Giant desiccation fissures filled with calcareous eolian sand, Hermosa Formation (Pennsylvanian), southeastern Utah. *Sedimentary Geology*, **56**, 403–413.

Lowenstein, T. K. and Hardie, L. A. (1985). Criteria for the recognition of salt-pan evaporites. *Sedimentology*, **32**, 627–644.

Mackay, J. (1980). Deformation of ice-wedge polygons, Garry Island, Northwest Territories. Current research, part A, *Geological Survey of Canada*, Paper **80**, 287–291.

Mangold, N. (2005). High latitude patterned grounds on Mars: Classification, distribution and climatic control. *Icarus*, **174**, 336–359.

Marchant, D. R. and Head, J. W. (2007). Antarctic dry valleys: Microclimate zonation, variable geomorphic processes, and implications for assessing climate change on Mars. *Icarus*, **192**, 187–222.

Masson, P., Carr, M. H., Costard, F., Greeley, R., Hauber, E. and Jaumann, R. (2001). Geomorphologic evidence for liquid water. *Space Sci. Rev.*, **96**, 333–364.

Mayer, D. P., Arvidson, R. E., Wang, A., Sobron, P. and Zheng, M. P. (2009). Mapping minerals at a potential Mars analog site on the Tibetan Plateau, Lunar and Planetary Science Conference, p. 1877.

McEwen, A. S., Dundas, C. M., Mattson, S. S., Toigo, A. D., Ojha, L., Wray, J. J., Chojnacki, M., Byrne, S., Murchie, S. L. and Thomas, N. (2014). Recurring slope lineae in equatorial regions of Mars. *Nat. Geosci.*, **7**, 53–58.

McEwen, A. S., Ojha, L., Dundas, C. M., Mattson, S. S., Byrne, S., Wray, J. J., Cull, S. C., Murchie, S. L., Thomas, N., Gulick, V. C. (2011). Seasonal flows on warm Martian slopes. *Science*, **333**, 740–743.

McGill, G. E. and Hills, L. S. (1992). Origin of giant Martian polygons. *J. Geophys. Res. Planets*, **97**, 2633–2647.

Mellon, M. T., Arvidson, R. E., Marlow, J. J., Phillips, R. J. and Asphaug, E. (2008). Periglacial landforms at the Phoenix landing site and the northern plains of Mars. *J. Geophys. Res. Planets*, (1991–2012) **113**, E00A23, doi: 10.1029/2007JE003039.

Milazzo, M. P., Keszthelyi, L. P., Jaeger, W. L., Rosiek, M., Mattson, S., Verba, C., Beyer, R. A., Geissler, P. E., McEwen, A. S. and Team, H. (2009). Discovery of columnar jointing on Mars. *Geology*, **37**, 171–174.

Mouginis-Mark, P. J. (1987). Water or ice in the Martian regolith?: Clues from rampart craters seen at very high resolution. *Icarus*, **71**, 268–286.

Murchie, S. L., Mustard, J. F., Ehlmann, B. L., Milliken, R. E., Bishop, J. L., McKeown, N. K., Noe Dobrea, E. Z., Seelos, F. P., Buczkowski, D. L. and Wiseman, S. M. (2009). A synthesis of Martian aqueous mineralogy after 1 Mars year of observations from the Mars Reconnaissance Orbiter. *J. Geophys. Res. Planets*, **114**, E00D06, doi: 10.1029/2009je003342.

Mustard, J. F., Murchie, S., Pelkey, S., Ehlmann, B., Milliken, R., Grant, J., Bibring, J.-P., Poulet, F., Bishop, J. and Dobrea, E. N. (2008). Hydrated silicate minerals on Mars observed by the Mars Reconnaissance Orbiter CRISM instrument. *Nature*, **454**, 305–309.

Mutch, T. A., Arvidson, R. E., Binder, A. B., Guinness, E. A. and Morris, E. C. (1977). The geology of the Viking Lander 2 site. *J. Geophys. Res.*, **82**, 4452–4467.

Neal, J. T., Langer, A. M. and Kerr, P. F. (1968). Giant desiccation polygons of Great Basin playas. *Geol. Soc. Am. Bull.*, **79**, 69–90.

Nishiwaki, H. (2009). Columnar Joints and Landscapes, *Introduction to the Landforms and Geology of Japan*.

Ojha, L., McEwen, A., Dundas, C., Byrne, S., Mattson, S., Wray, J., Masse, M. and Schaefer, E. (2014). HiRISE observations of recurring slope lineae (RSL) during southern summer on Mars. *Icarus*, **231**, 365–376.

Ojha, L., Wilhelm, M. B., Murchie, S. L., McEwen, A. S., Wray, J. J., Hanley, J., Massé, M. and Chojnacki, M., (2015). Spectral evidence for hydrated salts in recurring slope lineae on Mars. *Nat. Geosci.*, **8**, 829–832.

Osterloo, M. M., Anderson, F. S., Hamilton, V. E. and Hynek, B. M. (2010). Geologic context of proposed chloride-bearing materials on Mars. *J. Geophys. Res. Planets*, **115**, E10012, doi: 10.1029/2010JE003613.

Osterloo, M. M., Hamilton, V. E., Bandfield, J. L., Glotch, T. D., Baldridge, A. M., Christensen, P. R., Tornabene, L. L. and Anderson, F. S. (2008). Chloride-bearing materials in the southern highlands of Mars. *Science*, **319**, 1651–1654.

Pechmann, J. C. (1980). The origin of polygonal troughs on the northern plains of Mars. *Icarus*, **42**, 185–210.

Pina, P. (2014). Polygonal pattern analysis on Mars based on Svalbard analogues, Proceedings of the V Iberian Conference of the International Permafrost Association, University of Barcelona.

Plummer, P. S. and Gostin, V. A. (1981). Shrinkage cracks: Desiccation or synaeresis? *J. Sediment. Res.*, **51**(4), 1147–1156.

Pollack, J. B., Kasting, J. F., Richardson, S. M. and Poliakoff, K. (1987). The case for a wet, warm climate on early Mars. *Icarus*, **71**, 203–224.

Ruesch, O., Poulet, F., Vincendon, M., Bibring, J.P., Carter, J., Erkeling, G., Gondet, B., Hiesinger, H., Ody, A. and Reiss, D. (2012). Compositional investigation of the proposed chloride-bearing materials on Mars using near-infrared orbital data from OMEGA/MEx. *J. Geophys. Res. Planets*, **117**, E00J13, doi: 10.1029/2012JE004108.

Salese, F., Di Achille, G., Neesemann, A., Ori, G. G. and Hauber, E. (2016). Hydrological and sedimentary analyses of well-preserved paleofluvial-paleolacustrine systems at Moa Valles, Mars. *J. Geophys. Res. Planets*, **121**, 194–232.

Scott, D. H., Dohm, J. M. and Rice Jr, J. (1995). Map of Mars Showing Channels and Possible Paleolake Basins, USGS, 2461, doi: 10.3133/i2461.

Seibert, N. M. and Kargel, J. S. (2001). Small-scale Martian polygonal terrain: Implications for liquid surface water. *Geophys. Res. Lett.*, **28**, 899–902.

Sobron, P., Wang, A., Mayer, D. P., Bentz, J., Kong, F. and Zheng, M. (2018). Dalangtan saline playa in a hyperarid region of Tibet Plateau-III: Correlated multiscale surface mineralogy and geochemistry survey, *Astrobiology*, **18**(10), 1277–1304.

Stein, N., Grotzinger, J., Schieber, J., Mangold, N., Hallet, B., Newsom, H., Stack, K., Berger, J., Thompson, L. and Siebach, K. (2018). Desiccation cracks provide evidence of lake drying on Mars, Sutton Island member, Murray formation, Gale Crater. *Geology*, **46**, 515–518.

Tanaka, K., Robbins, S., Fortezzo, C., Skinner, J. and Hare, T. (2014). The digital global geologic map of Mars: Chronostratigraphic ages, topographic and crater morphologic characteristics, and updated resurfacing history. *Planet. Space Sci.*, **95**, 11–24.

Tucker, R.M. (1981). Giant polygons in the Triassic salt of Cheshire, England; a thermal contraction model for their origin. *J. Sediment. Res.*, **51**, 779–786.

Ulrich, M., Hauber, E., Herzschuh, U., Härtel, S. and Schirrmeister, L. (2011). Polygon pattern geomorphometry on Svalbard (Norway) and western Utopia Planitia (Mars) using high-resolution stereo remote-sensing data. *Geomorphology*, **134**, 197–216.

Wang, A., Sobron, P., Kong, F., Zheng, M. and Zhao, Y.-Y.S. (2018). Dalangtan saline playa in a hyperarid region on Tibet Plateau-II: preservation of salts with high hydration degrees in subsurface, *Astrobiology*, **18**(10), 1254–1276.

Wang, A. and Zheng, M. P. (2009). Evaporative Salts from Saline Lakes on Tibetan Plateau: An Analog for Salts on Mars, Lunar and Planetary Science Conference, p. 1858.

Wang, J., Fang, X., Appel, E. and Song, C. (2012). Pliocene–Pleistocene Climate Change At the NE Tibetan Plateau Deduced From Lithofacies Variation In the Drill Core SG-1, Western Qaidam Basin, China. *J. Sediment. Res.*, **82**, 933–952.

Weinberger, R. (1999). Initiation and growth of cracks during desiccation of stratified muddy sediments. *J. Struct. Geology*, **21**, 379–386.

Weinberger, R. (2001). Evolution of polygonal patterns in stratified mud during desiccation: the role of flaw distribution and layer boundaries. *Geol. Soc. Am. Bull.*, **113**, 20–31.

Wharton, R. A., Crosby, J. M., McKay, C. P. and Rice, J. W. (1995). Paleolakes on Mars. *J. Paleolimnology*, **13**, 267–283.

Wilson, L. and Head, J. W. (2007). Explosive volcanic eruptions on Mars: Tephra and accretionary lapilli formation, dispersal and recognition in the geologic record. *J. Volcano Geothermal Res.*, **163**, 83–97.

Wray, J. J., Milliken, R. E., Dundas, C. M., Swayze, G. A., Andrews-Hanna, J. C., Baldridge, A. M., Chojnacki, M., Bishop, J. L., Ehlmann, B. L. and Murchie, S. L. (2011). Columbus crater and other possible groundwater-fed paleolakes of Terra Sirenum, Mars. *J. Geophys. Res. Planets*, **116**, E01001, doi: 10.1029/2010JE003694.

Wray, J. J., Murchie, S. L., Squyres, S. W., Seelos, F. P. and Tornabene, L. L. (2009). Diverse aqueous environments on ancient Mars revealed in the southern highlands. *Geology*, **37**, 1043–1046.

Wray, J. J., Squyres, S. W., Roach, L. H., Bishop, J. L., Mustard, J. F. and Dobrea, E. Z. N. (2010). Identification of the Ca-sulfate bassanite in Mawrth Vallis, Mars. *Icarus*, **209**, 416–421.

Xia, W., Zhang, N., Yuan, X., Fan, L. and Zhang, B. (2001). Cenozoic Qaidam basin, China: a stronger tectonic inversed, extensional rifted basin. *AAPG Bull.*, **85**, 715–736.

Xiao, L., Wang, J., Dang, Y., Cheng, Z., Huang, T., Zhao, J., Xu, Y., Huang, J., Xiao, Z. and Komatsu, G. (2017). A new terrestrial analogue site for Mars research: The Qaidam Basin, Tibetan Plateau (NW China). *Earth-Science Rev.*, **164**, 84–101.

Ye, B., Huang, J., Michalski, J. and Xiao, L. (2019). Geomorphologic Characteristics of Polygonal Features on Chloride-Bearing Deposits on Mars: Implications for Martian Hydrology and Astrobiology. *J. Earth Sci.*, **30**(5), 1049–1058.

Zhao, J. and Xiao, L. (2016). Achievements, Issues and Prospects in Study of Martian Paleolake (in Chinese). *Earth-Science: J. Chinese Geolog. Univ.*, 1572–1582.

Zhu, X. (1985). Primary discussions on the environment of the western part of the Qaidam Basin. *Environment of Qinghai*, 1–5.

https://doi.org/10.1142/9789811224027_0007

Chapter 7

Valleys

Jiannan Zhao, Yutong Shi, and Long Xiao
Planetary Science Institute, School of Earth Sciences, China University of Geosciences, Wuhan 430074, PR China

7.1 Introduction

Valleys are elongate depressions on the Earth's surface (Baker, 2019). They are usually located between ranges of mountains with steep slopes and are occupied by rivers. Valleys can be divided into different types based on their origin and morphology. For example, valleys produced by tectonic action are called rift valleys (Baker, 2019); valleys with U-shaped cross-sectional shape and formed by the process of glaciation are called glacial valleys or U-shaped valleys; very narrow, deep valleys of similar appearance are called gorges; very deep valleys with steep sides of rocks are called canyons (Fig. 7.1). The deepest valley in the world is the Yarlung Tsangpo Grand Canyon in southern Tibet, China, whose average depth is 2,268 m with the deepest part reaching 6,009 m (Song *et al.*, 2013; Nevins *et al.*, 2016). In the Qaidam Basin, valleys with different morphological characteristics have been observed, which provide a good opportunity to study the formation and morphological evolution of valleys.

Valleys are also extensively distributed on the Martian surface and have been considered as the most unexpected findings of the Mariner 9 mission to Mars (Greeley, 1994). Geomorphological analyses of the valleys show sufficient evidences of a fluvial origin, which indicates the existence of surface water at least in a period of the Martian history.

Fig. 7.1. The Grand Canyon in Arizona, USA.

With the increasing coverage of high-resolution images of Mars, more and more valleys are identified and mapped. Up to the year of 2018, a global total length of 773,559 km of valleys have been mapped (Alemanno *et al.*, 2018).

7.2 Valleys on Mars

7.2.1 *Martian valley morphology*

Martian valleys are sinuous and usually narrow with respect to their lengths (Carr, 2006; e.g. Fig. 7.2), which makes them distinguishable from the relatively linear and wide outflow channels. The widths of Martian valleys vary from less than 1 km to about 10 km and the lengths range from less than 5 km to more than 1000 km (MCWG, 1983). For example, the Marikh Vallis (Fig. 7.3) is one of the longest valleys on the Martian surface. It is about 1200 km long, extending on the Martian southern highlands from the Noachis Terra in the south to the Meridiani Planum in the north. In addition, the long regional slope to the Chryse basin in Meridiani Planum leads to the formation of several long valley networks (Carr, 2006) such as Loire Valles, Paraná Valles, and Samara Valles (Fig. 7.4).

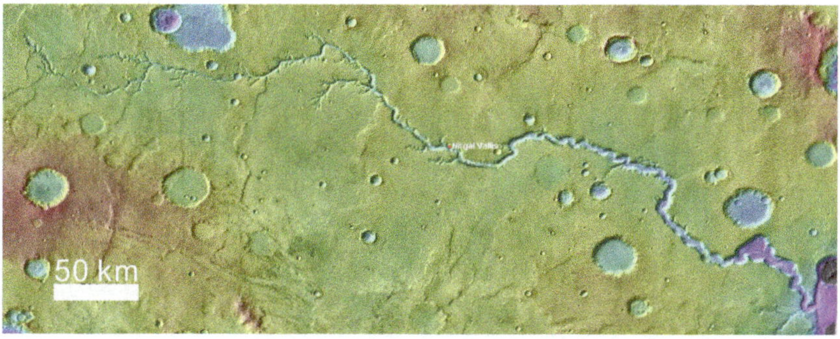

Fig. 7.2. Nirgal Vallis at 320°E, 28°S (THEMIS Daytime IR mosaic superposed by colorized MOLA topography map). It's relatively narrow compared with its length.

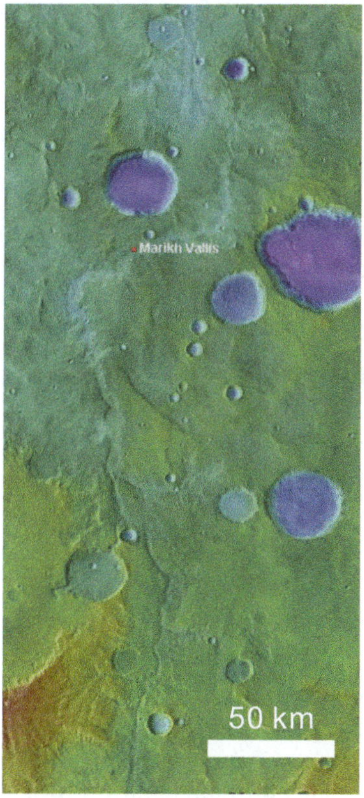

Fig. 7.3. Marikh Vallis at 4.5°E, 20.3°S (THEMIS Daytime IR mosaic superposed by colorized MOLA topography map). It's one of the longest valleys on the Martian surface.

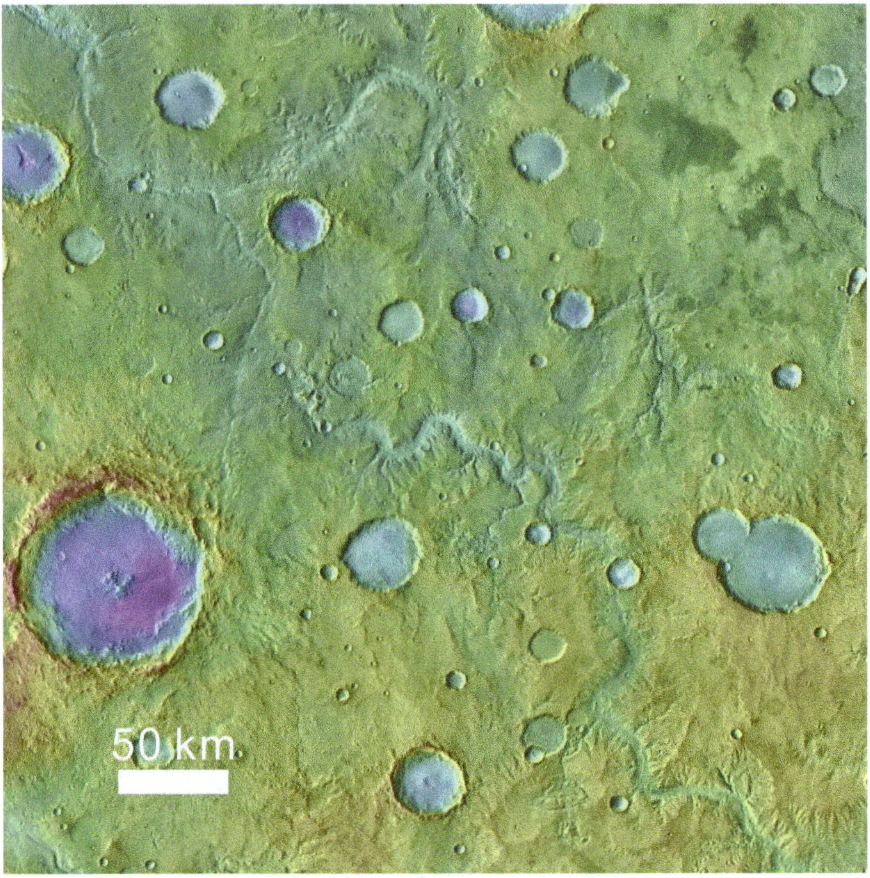

Fig. 7.4. Loire Vallis at 344°E,18°S (THEMIS Daytime IR mosaic superposed by colorized MOLA topography map). It is about 800 km in length.

In cross-section, Martian valleys usually exhibit a transformation from V-shaped in their upper reaches to U-shaped or even rectangular when extending downstream (Fig. 7.5). However, this would be easily affected by later modification processes and result in a diversity of cross-sectional shapes. The depths of Martian valleys vary in a large range from less than 1 m to more than 400 m, but most of the valleys have depths between 50 m and 200 m and the depth of incision may remain almost constant for large distances (Williams and Phillips, 2001).

Several parameters have been used to describe the geometrical and hydrological characteristics of Martian valleys. Strahler stream order is

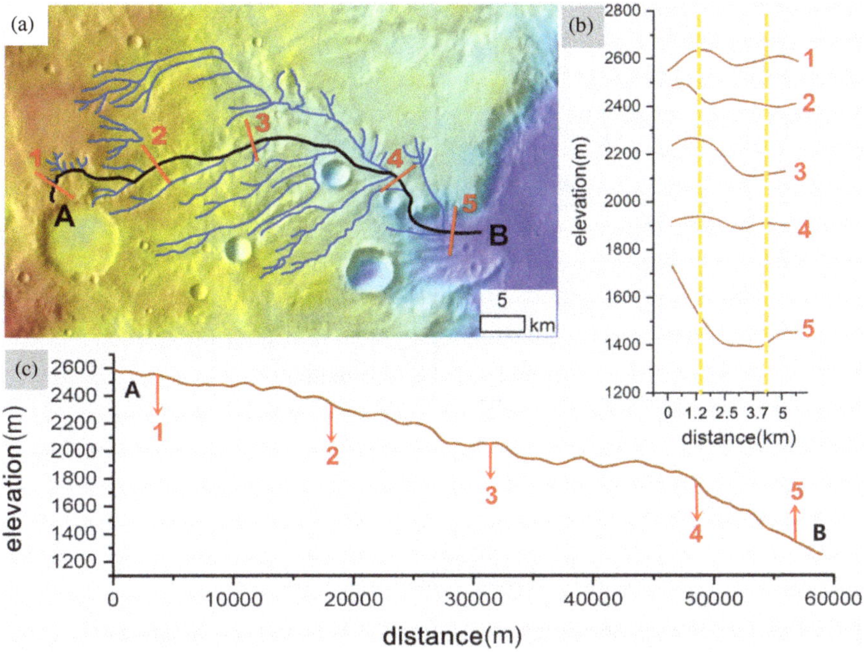

Fig. 7.5. A valley network around the Huygens Basin (59.75°E, 12.16°S). (a) Overall topography of the valley network. The background is colorized MOLA topographic map. (b) Cross-sectional profiles of different parts of the valley. (c) Longitudinal profile of the valley main stream.

used to define the branching complexity of valley networks based on the hierarchy of tributaries (Strahler, 1952, 1957). Higher orders signify higher branching complexity of a valley network. Carr (1996) found that most Martian valley networks have similar order-length relationships as the streams on Earth. Branching angle is the angle between the regression lines of two joining tributaries. Seybold *et al.* (2018) found that valley networks on Mars generally have narrow branching angles (around 40°), which is similar to terrestrial valleys found in arid landscapes.

Martian valley networks have different planimetric patterns, such as digitate, stem, parallel, rectilinear, radial, etc. (MCWG, 1983; Fig. 7.6). Although complex dendritic drainage is present on Martian surface, the overall development of Martian valleys is still immature. Alemanno *et al.* (2018) have studied the typology of Martian valleys and proposed a classification scheme that is detailed as follows:

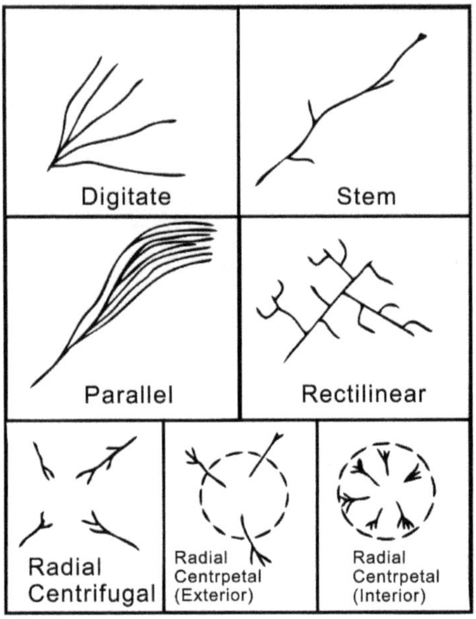

Fig. 7.6. Planimetric patterns of Martian valley networks (MCWG, 1983).

7.2.1.1 *Typical valley networks*

Typical valley networks usually have a dendritic drainage pattern, which is characterized by the presence of many short branches with one main stream, similar to common rivers on Earth. The branching structure of this type allows water to flow downstream quickly, making it a very efficient fluvial pattern. One of the most developed dendritic valley networks on Mars is the Warrego Valles (Fig. 7.7), located at the southern boundary of Thaumasia region where numerous valley networks dissected, in the area between 40°S and 44°S, and between 264°E and 269°E. Warrego Valles show an Earth-like fluvial spatial organization on the regional plateau slope. In Fig. 7.7, two mature valley networks can be observed, including the main Warrego Valley network and a secondary network.

7.2.1.2 *Longitudinal valleys*

Longitudinal valleys have narrow and sinuous main channel with lengths of hundreds of kilometers and widths of a dozen kilometers

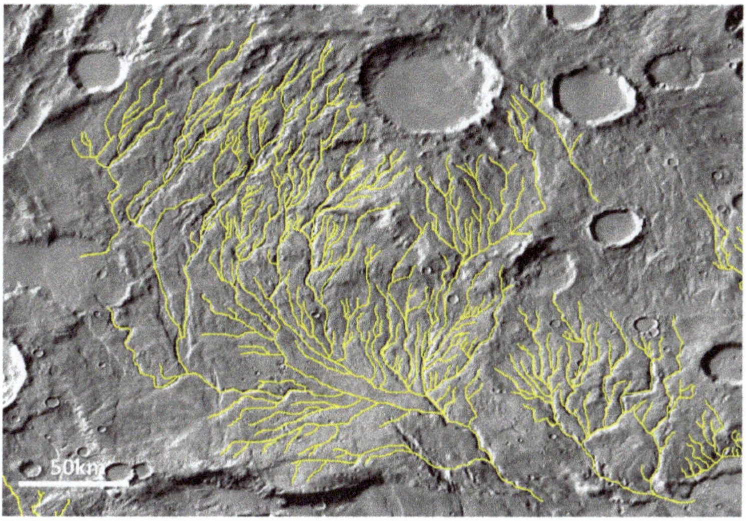

Fig. 7.7. Warrego Valles at 267°E, 42°S (THEMIS daytime IR image). It is a typical dendritic valley network.

(Baker *et al.*, 1992). Therefore, they are much larger than typical valley networks. They are also different from outflow channels as they never start from chaotic terrain areas. The Nirgal Vallis and Marikh Vallis shown, respectively, in Fig. 7.2 and 7.3 are typical valleys of this type. Each of them has a long and sinuous main stream and short and sparse tributaries connected to the main stream.

7.2.1.3 *Single valleys and valley segments*

Small single valleys and valley segments are widely distributed on the Martian surface. They are usually located close to each other but apparently disconnected. Most of them are only several kilometers to tens of kilometers in length. They could be erosional remnants of an old integrated valley system.

7.2.1.4 *Valleys associated with canyons and volcanoes*

Many Martian valleys have been observed to be located on volcanoes or near large crustal fractures. Valley systems on volcanoes (Fig. 7.8) commonly have a radial drainage pattern in which the streams flow from

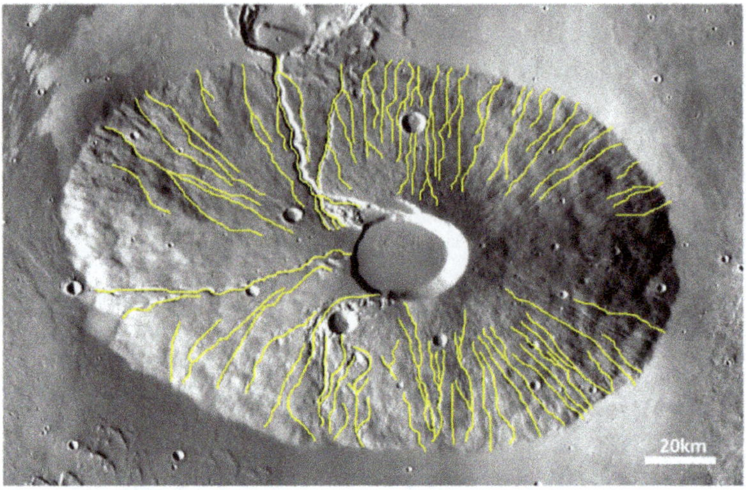

Fig. 7.8. Valleys on the Ceraunius Tholus (263°E, 24°N; THEMIS daytime IR image).

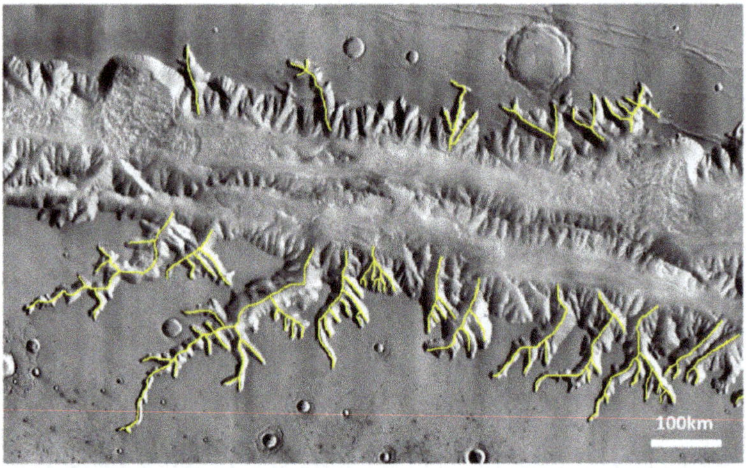

Fig. 7.9. Valleys connected to the Valles Marineris (278°E, 7°S; THEMIS daytime IR image).

a central high point near the caldera and have multiple directions. Some of the valleys may form along the pre-existing lava channels or tubes. Valleys associated with canyons (Fig. 7.9) or outflow channels are relatively small and short. They may have a theater-shaped head which indicates an origin of groundwater sapping, but small dendritic drainage patterns are also observed.

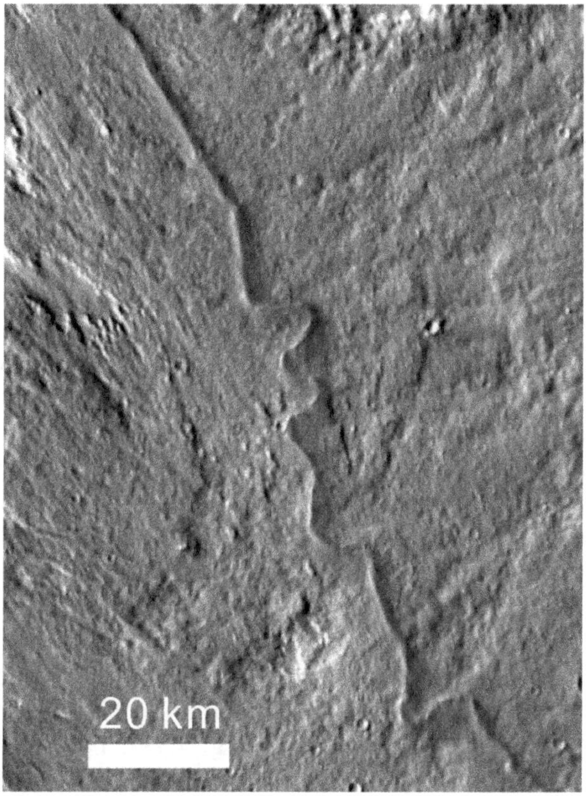

Fig. 7.10. THEMIS Daytime IR image of an inverted valley on the Martian surface (151.1°E, 2.6°S).

Besides, there are special kinds of valleys on Mars that have ridge-like appearances (Fig. 7.10). These are formed by topographic inversion of riverbeds under aeolian erosion. Riverbeds that endured cementation, armoring, or lithogenesis could become more resistant to wind erosion and when the banks of the rivers are eroded, the original riverbeds would become sinuous ridges. Now more than 200 inverted valleys have been identified on the surface of Mars (Williams, 2007, 2013; Burr *et al.*, 2010; Lefort *et al.*, 2012).

7.2.2 *Spatial and temporal distribution of Martian valleys*

Global mapping of Martian valleys started from the 1980s. The Mars Channel Working Group provided the first global distribution map of the

Martian valleys based on Viking images (MCWG, 1983). Then, Carr and Chuang (1997) thoroughly mapped the global distribution of valley networks on Mars based on imaging data from the Viking orbiter, identifying up to 11,336 tributaries. However, due to the limited resolution of image data and lack of topographic data, this mapping result is sketchy. Later, Carr *et al.* (2006; 2012) supplemented and modified the earlier mapping results based on new detection data.

Besides manual investigation of valley networks, automatic identification was conducted by Luo and Stepinski (2009). They extracted linear depressions of Mars globally based on the 463 m/pixel resolution MOLA elevation data from Mars Global Surveyor, and then manually checked the results to eliminate wrong extractions. They identified a total number of 375,550 tributaries, with the total valley length approximately 2.3 times that of the mapping results of Carr and Chuang (1997). In addition, Hynek *et al.* (2010) refined this mapping result through manual identification based on MOLA elevation data and ~200 m/pixel resolution THEMIS mosaic from Mars Odyssey, and found a strong correlation between high drainage density and reported sedimentary deposits.

The latest global mapping of valley network on Mars comes from Alemanno *et al.* (2018), based on the 463 m/pixel resolution MOLA elevation data and the 100 m/pixel resolution THEMIS mosaics, with 6 m/pixel CTX images in some areas (Fig. 7.11). This work identified more small valley networks and tributaries, which result in higher

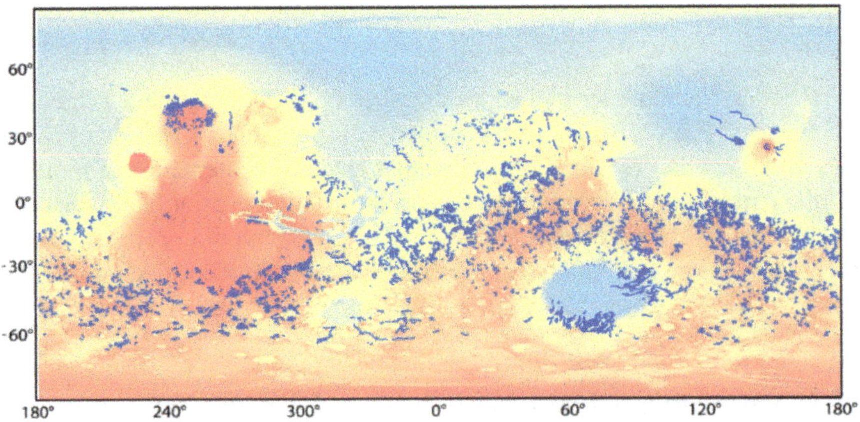

Fig. 7.11. Global distribution of valley networks on Mars based on the data from Alemanno *et al.* (2018).

drainage orders. The total length of the recognized valley network is up to 773,559 km. In addition, Alemanno *et al.* (2018) also classified the identified valley networks according to their morphological characteristics and dated the age of the formation of valley networks based on the global geological map of Mars from Tanaka *et al.* (2014).

As is shown in the mapping results above, valley networks have an inhomogeneous distribution on the Martian surface. Valley networks are concentrated in the cratered southern highlands, while sparsely distributed in the young northern plain. In the southern highlands, the distribution of the valley networks also varies between regions. Arabia Terra and west Hellas, in particular, have sparser valley distribution, while along the equator region valley networks are denser.

The age of valley networks obtained from different mapping results are slightly different, but the overall characteristics are consistent. According to the valley mapping of Carr and Chuang (1997), 90% of the valley networks are distributed on Noachian terrains, while both valley networks that incised Hesperian or Amazonian-aged terrains counts for 5%. The mapping results of Hynek *et al.* (2010) showed that 91% of the valleys are distributed on Noachian terrains, 6% on Hesperian terrains, and 3% on Amazonian terrains. Alemanno *et al.* (2018) proposed that 94% of the valley networks are distributed on Noachian terrains, 4% on Hesperian terrains, and 2% on Amazonian terrains. In general, the formation time of Martian valley networks is mostly Noachian period and may also extend to the Hesperian and Amazonian periods. Therefore, the morphological and distributional features of valley networks could also become a window to explore the changes of climate during the Late Noachian period as well as the transition from Noachian to the Hesperian.

7.2.3 *Origin of Martian valleys*

Different formation mechanisms were proposed after the discovery of Martian valley networks (summarized by Carr, 1995), including tectonic activity (Schumm, 1974), liquefaction (Nummedal, 1978), glaciation (Lucchitta, 1980), and erosion by water (Carr and Clow, 1981; Baker, 1982), lava (Carr, 1974), wind (Cutts and Blasius, 1979), liquid hydrocarbons (Yung and Pinto, 1978), or CO_2 (Sagan *et al.*, 1973). Among them, water erosion has been widely accepted as the most plausible origin of Martian valleys (Carr, 2006).

However, the water source of valley networks remains a subject of an extensive debate and it is closely related to the Noachian climate of Mars. The dendritic morphology of most Martian valley networks resembles that of terrestrial rivers, which may indicate a water supply by surface runoff. Therefore, the widely distributed valley networks on Martian southern highland, together with the identified paleolakes (Fassett and Head, 2008; Goudge *et al.*, 2012, 2015), could be formed in a warm and wet climate (Ansan and Mangold, 2006; Irwin *et al.*, 2011; Carr, 2012). This suggests that at the time of their formation, superficial temperature and atmospheric pressure were different from current conditions (Hynek *et al.*, 2010). The origin of this warm and wet climate might have been an intense volcanic activity which made the atmosphere thicker (Pollack *et al.*, 1987). This led to a greenhouse effect strong enough to produce a warming of the planet. Later, the progressive reduction of the volcanic activity and the loss of atmospheric gases, mainly due to sputtering by the solar wind (Jakosky *et al.*, 1995), caused a thinning of the atmosphere and, by consequence, a drastic reduction of the greenhouse effect.

Aside from the fluvial features, other observations also support a warm and wet Noachian scenario. While Hesperian and Amazonian terrains have well preserved impact craters, Noachian terrains are heavily eroded, and wind action alone cannot have produced such a large volume of eroded material: a likely explanation is the water flow in warm climatic conditions, very different from the present ones (Hynek *et al.*, 2010).

This climatic change is also supported by the observation of phyllosilicates in Noachian-aged areas while not in younger regions (Bibring *et al.*, 2006). Phyllosilicates are hydrated silicates, showing the presence of the hydroxyl ion OH-, which are produced only in presence of a significant quantity of liquid water (Pollack *et al.*, 1987). The presence of these minerals is a strong indication that in the past liquid water flowed on the surface of Mars. As a result, the presence of the valley networks and of these mineral deposits in the Noachian-aged terrains have frequently been associated with relatively warm and wet climatic conditions during the Noachian Era (Craddock and Howard, 2002), requiring a vertically integrated hydrologic system (Scanlon *et al.*, 2015).

There are also observations that support a cold and dry climate during which the valleys form. Sometimes the valleys observed on the Martian surface maintain their width downstream, opposed to what happens in the terrestrial case where valleys tend to become wider and less deep downstream. These valleys also have shallow tributaries of small amplitude with abrupt terminations. These features, along with U-shaped cross-sections, might suggest an origin of basal sapping (Laity and Malin, 1985), e.g. the collapse of the terrain produced by the underground ice or water.

Recent climate modeling studies (Wordsworth *et al.*, 2013) also suggest the possibility of a cold and dry Noachian where mean annual surface temperatures are below 0°C, whereas under this condition valleies could still form out of precipitation. Several mechanisms have been proposed to explain how valley networks formed under these harsh climatic conditions. Scanlon *et al.* (2013) suggest an origin of the valley networks due to snowfall rather than rainfall. Other authors explain the formation of these structures considering episodic melting of icy deposits in the highland terrains owing to temporary warm conditions produced by the volcanic activity (Halevy and Head, 2014).

The morphology of the valleys today is not original, instead, most of them have been changed by mass wasting processes, or by the failure of the side walls of the valley, giving rise to a morphology typical of groundwater sapping processes. Erosion, faulting, and dust covering could then have further modified these valley systems. Now in most cases, we can see only the remnants of the original drainage path. In addition, the valley networks seem to have formed by rainfall (Irwin, 2002; Hynek *et al.*, 2010), but it is unclear under what conditions their formation took place.

In addition, other climate scenarios have also been proposed. Seybold *et al.* (2018) found that valley networks on Mars generally have narrow branching angles (around 40°), which are similar to terrestrial valleys found in arid landscapes on Earth. Therefore, they proposed that Martian valley networks were formed in an arid continental climate with sporadic heavy rainfall events.

7.3 Valleys in the Qaidam Basin

7.3.1 *Distribution and morphology of valleys in the Qaidam Basin*

The distribution of valleys and rivers in the Qaidam Basin is mainly affected by topography and climate. The overall topography of Qaidam Basin was divided into several sub-order basins, with multiple convergent centripetal drainage systems due to the influence of neotectonic movements. Large drainage systems in Qaidam Basin include Gasikule lake system, Sugan lake system, Mahai lake system, Da Qaidam lake system,

Xiao Qaidam lake system, Dulan lake system, Taijinaier lake system, Dabuxun lake system, and Huobuxun lake system. In general, the valley networks of Qaidam Basin have an inhomogeneous regional distribution. In the southeast and northeast of the basin where there is more precipitation, valley networks are relatively dense and runoff is more abundant. While in the arid and rainless northwest, rivers are sparse. The middle of the basin is also a large area without runoff.

Most of the rivers in the Qaidam Basin are inland rivers, which are generally short in length and often originate from surrounding mountains. In the mountainous area, valley networks are relatively denser, with more tributaries which are also longer in length. After the valley extends out of the mountain pass, the water volume gradually decreases, and runoff becomes seasonal or evaporates gradually. As the slope along the river gets more moderate, frequent conversions and divergences of tributaries occur, sometimes even combined with the confluence and divergence of cross-river channels. Most of the valley networks in this area have a braid shaped planar morphology due to the densely developed tributaries, making it difficult to determine the main channel.

In the Qaidam region, more than 50 rivers have catchment areas larger than 500 km^2. Among them, relatively larger drainage systems include the Nalenggele River, Golmud River, Nuomuhong River, Qaidam River, and Bayin River.

7.3.1.1 *Nalenggele river*

The Nalenggele River, which is located in the southwest of Qaidam Basin, is the largest inland river in Qinghai Province (Fig. 7.12). It originates from Xuelian Peak in Aerge Mountain of east Kunlun Mountains, which has an elevation of 5,598 m and is covered by glaciers. The upstream called Honghe River is the major source, originating from the northern slope of Bokaleiketak Mountain, and the downstream west Kulakealagan River is sourced from the southern slope of Bukadaban Peak. Regions between 3,500–4,000 m above sea level along Nalenggele River are characterized by a large number of valleys and canyons, with large rift valleys and steep sides. The river flows to the east, then turns to the north, joins the Kulakealagan River, then turns to the north, and finally flows into East Taijinaier Lake, then into West Taijinaier Lake (Fig. 7.12). The river is 435 km long, with a drainage area of 21,898 km^2. The natural annual runoff of Nalenggele River is 1.226×10^9 m^3, and the total water resources of

Fig. 7.12. (a) The Nalenggele River originates from the Xuelian Peak in Aerge Mountain and flows northeast into the Qaidam Basin, forming a lake (panel width 250 km). (b) Xuelian Peak, the birthplace of the Nalenggele River, is more than 5,500 m above sea level (panel width 4 km); (c) Middle stream of the Nalengele River (panel width 6 km); (d) Downstream of the Nalenggele River, there are two fan deltas and an elongated lake (panel width ~140 km); (e) The end of the Nalenggele River, high relives are yardangs within the East Taijinaier Lake (panel`width 6 km).

the basin are 1.280×10^9 m^3. The river has an elevation difference of 2,458 m and an average water depth of 0.4–1.6 m. Runoff of Nalenggele River is mainly supplied by precipitation and glacier melts. Area of the upstream glacier is 572.79 km^2, with a reserve of 5.728×10^{10} m^3.

At the downstream, there are two connected deposit deltas (Fig. 7.12(d)). Due to the elevation of the terrain in front of the first delta, a meandering channel is formed to collect water from the fan deposit area, and then it continues to move downstream to enter the second fan deposit area. The end of the second fan-shaped delta forms a catchment area. This area was originally a wind-eroded yardang landform area, which was injected by water to form East Taijinaier Lake and form a spectacular semi-submerg-erd landform.

7.3.1.2 *Golmud River*

The Golmud River is located in the south of Qaidam Basin and is the second largest river in the basin (Fig. 7.13). The upper source of Golmud River is divided into two branches (Fig. 7.13(c)): the left branch Kunlun River (Naijin/Naiqigele River), which originates from the glacier covered Bokaleiketake Mountain of the Kunlun Mountain with an elevation of 4,580 m (which is also the main source of Golmud River), and the right branch Shuergan River (Xiugou/Xiugouguole River), which originates from the Tanggewula Mountains. The two tributaries meet below Nachitai, forming Golmud River. The north branch of Golmud River drains into Dabuxun Lake after passing Golmud City. The Golmud River has a length of 327 km with a drop of 1440 m, along which there is a 61-km valley in the upper reaches (Fig. 7.14). The drainage area of Golmud River from its source to Golmud hydrologic station is 18,648 km^2, and the average run-off is 2.42 m^3/s (Golmud station). Rain water and snowmelts account for 19.5% and 14.1%, respectively, as the water sources of Golmud River, while groundwater accounts for as much as 66.4%, resulting in a relatively uniform annual runoff (Li *et al.*, 2006).

7.3.1.3 *Dry valleys and inverted valleys*

Aside from large rivers such as Nalenggele River or Golmud River, there are also a large number of dry valleys distributed in the Qaidam Basin, most of which are relatively small in size (Fig. 7.15). Those dry valleys usually have widths of a few meters to hundreds of meters, lengths up to

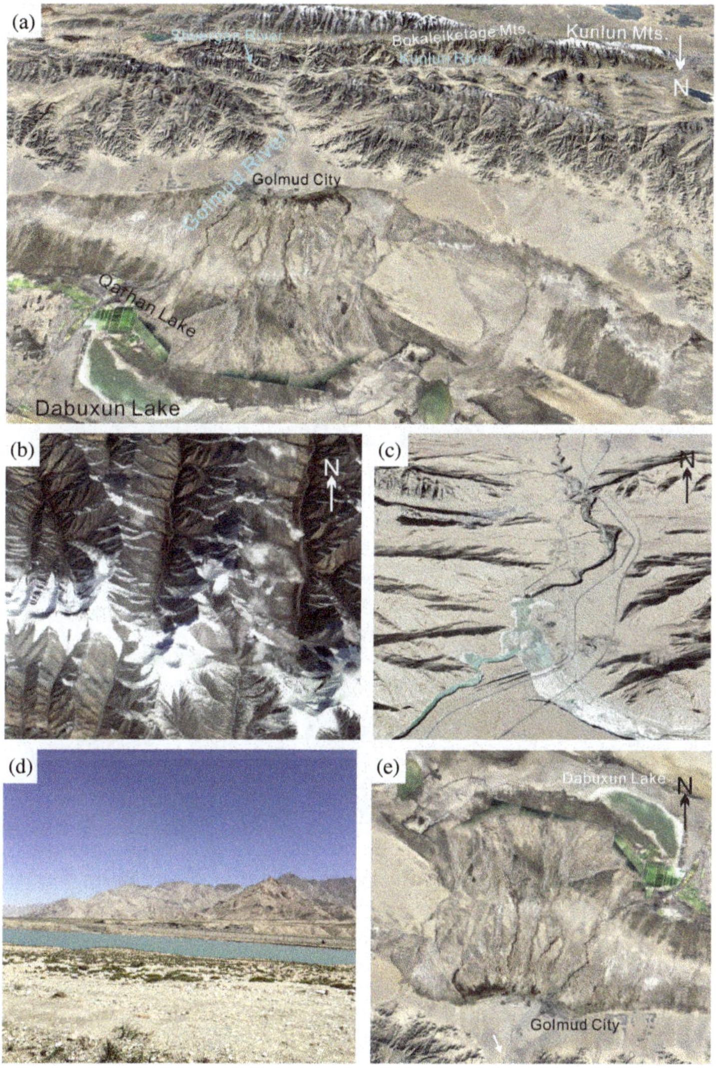

Fig. 7.13. (a) The Golmud River originates from the glaciers of Bokaleiketage Mountain in the Kunlun Mountains range, and the meltwater flows northwest from the mountains to the Qarhan (salt) Lake (panel width 220 km); (b) Bokaleiketage Mountain, the cradle of the Golmud River (panel width: 13 km); (c) Two main tributaries converge at Nachitai, forming the Golmud River, which flows out of the mountains in the northwest direction (panel width 2.5 km, center at 35°57'21.8"N, 94°48'47.54"E); (d) The Golmud River at middle stream; (e) The Golmud River comes out of the mountain pass (white arrow), passes through the Golmud City, and reaches out to Dabuxun Lake (panel width 110 km, Google Earth).

Fig. 7.14. Valley of the Golmud River.

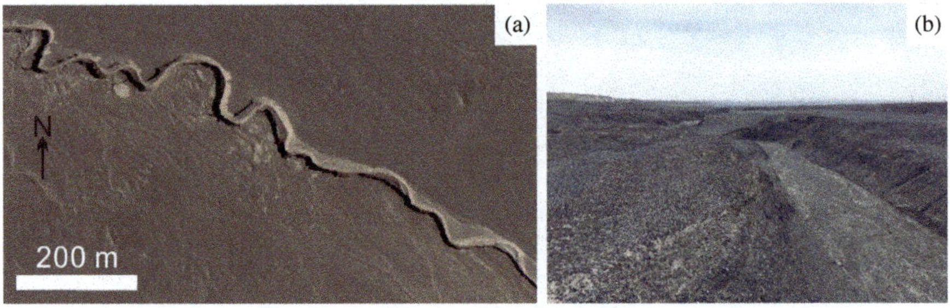

Fig. 7.15. (a) Remote-sensing image of a dry valley at the Qaidam Basin (Google Earth). (b) A dry valley in the Qaidam Basin, inside which traces of meandering rivers could be seen.

tens of kilometers, and depths up to 20 m. Most of the dry valleys extend from inside of the basin or surrounding mountains to the basin central where the elevation is relatively low. Cross beddings and binary structure could be found on the cross-section of riverbeds: riverbed sediments composed of coarse sand and gravel are deposited at the bottom and the flood plain sediments composed of fine sand or clay are at the top. Lenticular stratification, flattening stratification, and wavy stratification are observed in some riverbeds, and multiple layers of gravel, sand, and mud sediments are observed to distribute alternately, reflecting multiple changes of river hydrodynamic conditions (Fig. 7.16).

Fig. 7.16. (a) Fluvial sedimentary cross-section in the Qaidam Basin with wedge-shaped cross-bedding. (b) A riverbed profile of dry valleys near the Golmud River.

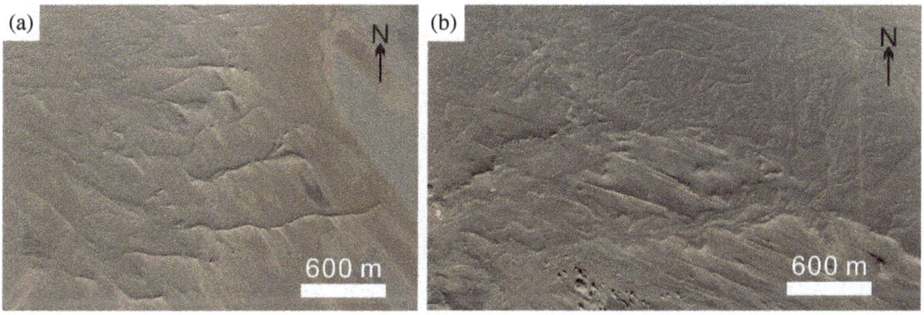

Fig. 7.17. Inverted valleys in the southern Dalangtan region (Google Earth).

Inverted valleys are also present in the Qaidam Basin, especially in the southern Dalangtan region and near the Wutumeiren Town. For example, Fig. 7.17 shows some typical inverted valleys in southern Dalangtan. They originated from the mountains in the south and extend to the northeast. Their lengths can be up to 4 km and the heights vary between 1 m to more than 10 m. In cross section, they usually have flat-topped or sharp-crested shape and interbedded boulders can be found (Fig. 7.18(a)). In addition, their surface is also covered by scattered or cemented boulders, which protect them from aeolian erosion (Fig. 7.18(b)).

7.3.2 *Origin of valleys in the Qaidam Basin*

The water sources of valleys in the Qaidam Basin are mainly snowmelt and precipitation. The surrounding mountains serve as source region for

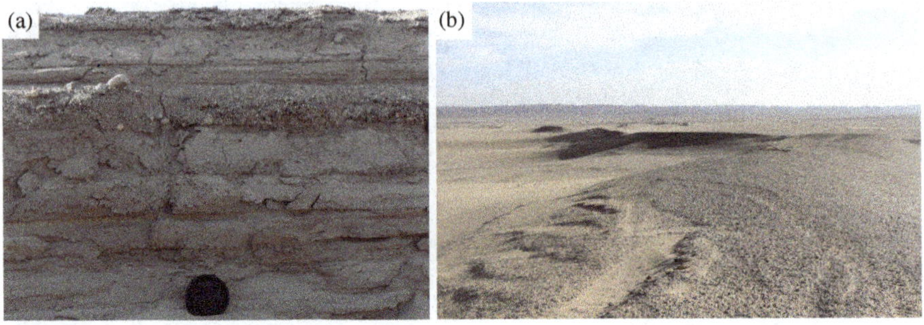

Fig. 7.18. (a) Cross-sectional profile of an inverted valley. (b) The surface of an inverted valley.

many rivers in the basin, as well as important reservoirs for groundwater. These mountain regions are usually much more humid than the basin center area, and are mostly covered by snow all year round with widespread glaciers. Rivers originating from these surrounding mountains converge towards the center of the basin, forming a radial shaped drainage pattern.

The differences in water sources of the rivers in the Qaidam Basin lead to different trends of the changing in discharge rates during different periods of the year. For rivers whose water source is dominated by atmospheric precipitation, seasonal variation of precipitation, regional topography, and hydrogeological conditions are all significant factors affecting the discharge rates of the river. Unusually, the maximum continuous discharge rate of four months could take up 50–70% of the total annual discharge, such as the Xiangride River and Chahanwusu River. For rivers mainly fed by snowmelt, air temperature is the main controlling factor of the discharge rate, which leads to much larger discharge rate in summer than in winter. Therefore, the annual distribution of river discharge is extremely uneven, with the maximum continuous discharge of four months roughly accounting for 70–90% of the total annual discharge, such as the Dahaerteng River and the Tataleng River. For rivers that have groundwater as the main water source, their discharge is evenly distributed in different periods of the year, with the maximum continuous discharge of four months roughly occupying 40–50% of the annual discharge, such as the Golmud River, Nuomuhong River, Shaliu River, Dulan River, and Bayin River. The conversion between surface water and groundwater has remarkable impact on discharge. At Zelingou Station on the upstream of the Bayin River, where the river is mostly fed by snow-melt and precipitation, the maximum continuous discharge of four months takes up to 65.3% of the total annual discharge; while at downstream after some tributaries infiltrate into the undercurrent, the maximum continuous discharge of four months has reduced to 51.5% (Tan *et al.*, 2014).

7.4 Summary

Valley networks are astonishing features extending on the surface of Mars. They can be more than 1000 km and have different planimetric patterns such as digitate, parallel, radial, and so on. In cross section, they usually exhibit a transformation from V-shaped in their upper reaches to U-shaped when extending downstream. Martian valleys have been classified into four types based on their planimetric pattern and distribution. Global mapping of Martian valleys showed an inhomogeneous distribution that most valleys are concentrated in the southern highlands, especially in the equatorial zone. Most valleys are formed in the Noachian period, while there are still about 2% that could have been formed in the Amazonian period. The morphology and water source of Martian valleys are closely related to climate. The relatively immature morphology and small branching angles may indicate that the valleys may develop in a relatively arid climate. However, this remains controversial and further studies are still needed.

In the Qaidam Basin, many inland rivers developed, among which the Nalenggele River is the largest inland river in the Qinghai Province and feeds the Taijinaier Lakes in the Qaidam Basin. Rivers in the Qaidam Basin mostly originated from the mountains surrounding the basin, including Kunlun Mountains, Altyn-Tagh Mountains, and Qilian Mountains. Atmospheric precipitation, melting snow, and groundwater provide water sources for the rivers. A comprehensive study of the valleys in the Qaidam Basin could shed light on the water supply, paleodischarge, and climatic setting of Martian valleys.

Acknowledgments

This study was supported by the Natural Science Foundation of China (42002305, 41830214), China Postdoctoral Science Foundation (2018M630889), and the Preresearch Project on Civil Aerospace

Technologies (D020101) of CNSA. We thank Prof. Long Xiao for his guidance and suggestions in improving the manuscript, and Dr. Jiang Wang for his assistance in field survey and data collection.

References

Alemanno, G., Orofino, V. and Mancarella, F. (2018). Global map of Martian Fluvial Systems: Age and total eroded volume estimations. *Earth and Space Sci.*, **5**, 560–577.

Ansan, V. and Mangold, N. (2006). New observations of Warrego Valles, Mars: Evidence for precipitation and surface runoff, *Planet. Space Sci.*, **54**, 219–242.

Baker, V. (2019). Valley. *Encyclopaedia Britannica*, https://www.britannica.com/science/valley.

Baker, V., Carr, M. H., Gulick, V. C., Williams, C. R. and Marley, M. S. (1992). *Channels and Valley Networks, Mars*. University of Arizona Press, Space Science Series, pp. 493–522.

Baker, V. R. (1982). The channels of Mars. Research supported by NASA, NSF, and Australian-American Educational Foundation Austin, TX, University of Texas Press, 1982. p. 204

Baker, V. R. and Nummedal, D. (1978). The Channeled Scabland: A guide to the geomorphology of the Columbia Basin, Washington, NASA.

Bibring, J. P., Langevin, Y., Mustard, J. F., Poulet, F., Arvidson, R., Gendrin, A., *et al.* (2006). Global mineralogical and aqueous mars history derived from OMEGA/Mars Express data. *Science*, **312**, 400–404.

Burr, D. M., Williams, R. M. E., Wendell, K. D., Chojnacki, M. and Emery, J. P. (2010). Inverted fluvial features in the Aeolis/Zephyria Plana region, Mars: Formation mechanism and initial paleodischarge estimates. *J. Geophys. Res.*, **115**, E07011.

Carr, M. (2006). *The Surface of Mars*, Cambridge University Press, Cambridge, UK.

Carr, M. H. (1974). The role of lava erosion in the formation of lunar rilles and Martian channels. *Icarus*, **22**, 1–23.

Carr, M. H. (1995). The Martian drainage system and the origin of valley networks and fretted channels. *J. Geophys. Res. Planets* (1991–2012), **100**, 7479–7507.

Carr, M. H. (2012). The fluvial history of Mars. *Philosophical Transactions of the Royal Society A: Mathematical, Physical and Engineering Sciences*, **370**, 2193–2215.

Carr, M. H. and Chuang, F. C. (1997). Martian drainage densities. *J. Geophys. Res. Planets*, **102**, 9145–9152.

Carr, M. H. and Clow, G. D. (1981). Martian channels and valleys: Their characteristics, distribution, and age. *Icarus*, **48**, 91–117.

Craddock, R. A. and Howard, A. D. (2002). The case for rainfall on a warm, wet early Mars. *J. Geophys. Res. Planets*, **107**, 21-21–21-36.

Cutts, J. A., Blasius, K. R. and Roberts, W. J. (1979). Evolution of Martian polar landscapes: Interplay of long-term variations in perennial ice cover and dust storm intensity. *J. Geophys. Res. Solid Earth*, **84**, 2975–2994.

Fassett, C. I. and Head, J. W. (2008). Valley network-fed, open-basin lakes on Mars: Distribution and implications for Noachian surface and subsurface hydrology. *Icarus*, **198**, 37–56.

Goudge, T. A., Aureli, K. L., Head, J. W., Fassett, C. I. and Mustard, J. F. (2015). Classification and analysis of candidate impact crater-hosted closed-basin lakes on Mars. *Icarus*, **260**, 346–367.

Goudge, T. A., Head, J. W., Mustard, J. F. and Fassett, C. I. (2012a). An analysis of open-basin lake deposits on Mars: Evidence for the nature of associated lacustrine deposits and post-lacustrine modification processes. *Icarus*, **219**, 211–229.

Goudge, T. A., Mustard, J. F., Head, J. W. and Fassett, C. I. (2012b). Constraints on the history of open-basin lakes on Mars from the composition and timing of volcanic resurfacing. *J. Geophys. Res. Planets*, **117**, E00J21.

Greeley, R. and Williams, S.H. (1994). Dust deposits on Mars: The "parna" analog. *Icarus*, **110**, 165–177.

Group, M. C. W. (1983). Channels and valleys on Mars. *Geol. Soc. Am. Bull.*, **94**, 1035–1054.

Halevy, I. and Head Iii, J. W. (2014). Episodic warming of early Mars by punctuated volcanism. *Nat. Geosci.*, **7**, 865–868.

Head, J. W., Marchant, D. R., Agnew, M. C., Fassett, C. I. and Kreslavsky, M. A. (2006). Extensive valley glacier deposits in the northern mid-latitudes of Mars: Evidence for Late Amazonian obliquity-driven climate change. *Earth and Planet. Sci. Lett.*, **241**, 663–671.

Hynek, B. M., Beach, M. and Hoke, M. R. T. (2010). Updated global map of Martian valley networks and implications for climate and hydrologic processes. *J. Geophys. Res.*, **115**, E09008.

Irwin, R. P. (2002). Drainage basin evolution in Noachian Terra Cimmeria, Mars. *J. Geophys. Res.*, **107**, 5056.

Irwin, R. P., Craddock, R. A., Howard, A. D. and Flemming, H. L. (2011). Topographic influences on development of Martian valley networks. *J. Geophys. Res.*, **116**, E02005.

Jakosky, B. M., Henderson, B. G. and Mellon, M. T. (1995). Chaotic obliquity and the nature of the Martian climate. *J. Geophys. Res.*, **100**, 1579.

Laity, J. E. and Malin, M. C. (1985). Sapping processes and the development of theater-headed valley networks on the Colorado Plateau. *Geol. Soc. Am. Bull.*, **96**, 203–217.

Lefort, A., Burr, D. M., Beyer, R. A. and Howard, A. D. (2012). Inverted fluvial features in the Aeolis-Zephyria Plana, western Medusae Fossae Formation, Mars: Evidence for post-formation modification. *J. Geophys. Res.*, **117**, E03007.

Li, J., Wang, H., Tan, L.-W., Liu, C.-E., H. (2006). Forcasting analysis on water resource development and programs in Ge'ermu river basin. *Plateau Earthquake Res.*, **18**, 52–61.

Lucchitta, B. (1980). Martian Outflow Channels Sculptured by Glaciers, II, Lunar and Planetary Science Conference, pp. 634–636.

Luo, W. and Stepinski, T. F. (2009). Computer-generated global map of valley networks on Mars. *J. Geophys. Res.*, **114**.

Nevins, D., Bosco, D. and Levy, P. (2016). *Tibet*. Cavendish Square Publishing, LLC.

Nummedal, D. (1978). The role of liquefaction in channel development on Mars. *Reports of Planet. Geology Prog.*, 257–259.

Pollack, J. B., Kasting, J. F., Richardson, S. M. and Poliakoff, K. (1987). The case for a wet, warm climate on early Mars. *Icarus*, **71**, 203–224.

Sagan, C. (1973). Sandstorms and eolian erosion on Mars. *J. Geophys. Res.*, **78**, 4155–4161.

Scanlon, K. E., Head, J. W., Madeleine, J. B., Wordsworth, R. D. and Forget, F. (2013). Orographic precipitation in valley network headwaters: Constraints on the ancient Martian atmosphere. *Geophys. Res. Lett.*, **40**, 4182–4187.

Scanlon, K. E., Head, J. W. and Marchant, D. R. (2015). Volcanism-induced, local wet-based glacial conditions recorded in the Late Amazonian Arsia Mons tropical mountain glacier deposits. *Icarus*, **250**, 18–31.

Schumm, S. (1974). Structural origin of large Martian channels. *Icarus*, **22**, 371–384.

Seybold, H. J., Kite, E., and Kirchner, J. W. (2018). Branching geometry of valley networks on Mars and Earth and its implications for early Martian climate. *Sci. Adv.*, **4**, eaar6692.

Shepherd, R. and Schumm, S. (1974). Experimental study of river incision. *Geol. Soc. Am. Bull.*, **85**, 257–268.

Song, Z., Zhen-han, W., Xi-tao, Z. and Ke-yan, X. (2013). Glacial dammed lakes in the Tsangpo River during late Pleistocene, southeastern Tibet. *Quat. Int.*, **298**, 114–122.

Strahler, A. N. (1952). Hypsometric (area-altitude) analysis of erosional topography. *Geol. Soc. Am. Bull.*, **63**, 25.

Strahler, A. N. (1957). Quantitative analysis of watershed geomorphology. *Eos* **38**, 913–920.

Tan, Y. (2014). Characteristics of the drainage system and water resources of the Qaidam Basin. *Water Conservancy Sci. Tech. Econ.*, **20**, 51–54.

Tanaka, K. L., Skinner Jr, J. A., Dohm, J. M., Irwin III, R. P., Kolb, E. J., Fortezzo, C. M., *et al*. (2014). *Geologic Map of Mars*.

Williams, R. M. and Phillips, R. J. (2001). Morphometric measurements of Martian valley networks from Mars Orbiter Laser Altimeter (MOLA) data. *J. Geophys. Res. Planets*, **106**, 23737–23751.

Williams, R. M. E. (2007). Global spatial distribution of raised curvilinear features on Mars, Lunar and Planetary Institute Science Conference, p. 1821.

Williams, R. M. E., Irwin, R. P., Burr, D. M., Harrison, T. and McClelland, P. (2013). Variability in martian sinuous ridge form: Case study of Aeolis Serpens in the Aeolis Dorsa, Mars, and insight from the Mirackina paleoriver, South Australia. *Icarus*, **225**, 308–324.

Wordsworth, R., Forget, F., Millour, E., Head, J. W., Madeleine, J. B. and Charnay, B. (2013). Global modelling of the early martian climate under a denser CO_2 atmosphere: Water cycle and ice evolution. *Icarus*, **222**, 1–19.

Yung, Y. and Pinto, J. (1978). Primitive atmosphere and implications for the formation of channels on Mars. *Nature*, **273**, 730.

https://doi.org/10.1142/9789811224027_0008

Chapter 8

Gullies

Yutong Shi, Jiannan Zhao and Jiang Wang

Planetary Science Institute, School of Earth Sciences, China University of Geosciences, Wuhan 430074, PR China

8.1 Introduction

Gullies are linear troughs that usually develop on slopes. On Earth, gullies can be formed due to the erosion of soils or rocks by ephemeral flowing water. Gullies have the following characteristics (Su *et al.*, 1963): firstly, the slope of a gully is larger in the upper part and gradually decreases downstream, which could be different from the slope change of the hillside; secondly, they have steep walls, resulting in V-shaped cross-sectional shapes; thirdly, there is no persistent water in gullies. In the Qaidam Basin, gullies of different morphological types can be observed. This diversity is mainly caused by differences in precipitation, topography, and substrate materials.

On Mars, gullies have been identified on crater walls, central peaks, dunes, and slopes of hills. They have similar geomorphological characteristics as terrestrial gullies and have attracted high attention as they could have been formed in recent geological epochs and may indicate recent water activities (Carr, 2006). However, the origin of gullies on Mars is still controversial and many hypotheses have been proposed, including mass wasting, liquid CO_2, groundwater seepage, melting of subsurface ice, etc.

Another type of young landform on Mars that seems to have a water-related genetical connection with gullies is recurring slope lineae (RSL).

RSLs are dark narrow linear features distributed widely over the Martian surface and were first discovered by McEwen *et al.* (2011) based on the observation of data from High Resolution Imaging Science Experiment (HiRISE) on board Mars Reconnaissance Orbiter (MRO). The occurrence of RSL differs with seasons, showing increasing of length down steep slopes during warmer seasons and fading during cold seasons, yet its formation process remains controversial, possible source of material including atmospheric water vapor (McEwen *et al.*, 2015; Gough *et al.*, 2019), groundwater or ground ice (Stillman *et al.*, 2014; Abotalib and Heggy, 2019), briny fluids (Chevrier and Rivera-Valentin 2012; Ojha *et al.*, 2015) and even granular flows (Dundas *et al.*, 2017; Schaefer *et al.*, 2019).

The occurrence of RSL on the Mars surface is often associated with gullies, but there are still significant differences between RSL and gully landforms. The seasonality of RSL, for example, is one of the most important characteristics of RSL which differs from other long-lasting fluvial landforms, which may suggest a unique formation process for RSL that differs from that of gullies, although their commonly associated appearance seems to suggest a shared origin background. Furthermore, unlike gullies which are also common on Earth, RSL seems to be lacking in terrestrial analog sites. Levy (2011) once took the water/brines flow track through shallow subsurface over impermeable ice layer in Antarctica as analog to RSL on Mars based on their shared features, including downslope propagation of linear or branching patterns of surface darkening, growth during spring/summer seasons, and fading during winter, but the terrestrial sites would generate more sinuous flow with broad channels in the ice table that serve as preferred flow paths over interannual timescales (Levy, 2011). Aside from this example, no other terrestrial analog sites of RSL have been identified yet, further studies concerning the formation of this unique gully-like feature is yet to be conducted. This chapter will mainly focus on the distribution and morphology of gullies on Mars and its widely distributed analogs in the Qaidam Basin.

8.2 Gullies on Mars

8.2.1 *Global distribution of Martian gullies*

Gullies are relatively young landforms on the Martian surface. They were first reported by Malin and Edgett (2000) using Mars Orbiter Camera (MOC) on board Mars Global Surveyor. They identified gullies in ~120 locations and most of the gullies are on interior walls or central peaks of impact craters, others are on walls of martian valley systems (Nirgal Vallis and Dao Vallis) and pits in the south polar region. They also calculated the orientations of the gullies and found about half of the gullies are located on south-facing slopes, while only about 20% face north. What's more, they occur more often on poleward facing slopes than on equator-facing slopes.

Thousands of gullies have been identified on the surface of Mars (Harrison *et al.*, 2015). These gullies are mainly distributed on steep slopes at north and south high latitudes, with higher concentrations in the southern hemisphere (Fig. 8.1), and an obvious zonal distribution in the southern hemisphere, 30–60°. The high correlation between gully distribution and latitude may indicate the close relationship between gully genesis and climate. In terms of gully scale, the total length of gullies varies from several hundred to several thousand meters. The average length of gullies is different with different gully morphological types, with a range of 1,600–2,000 m. The longest gullies can reach to nearly 8,000 m long. The extension direction of gullies has a preference of poleward rather than eastward or westward. Gullies develop on different kinds of landforms, including impact crater walls (Fig. 8.2(a)), dunes (Fig. 8.2(b)), gully flanks, slopes, and crater central uplift (Fig. 8.2(c)).

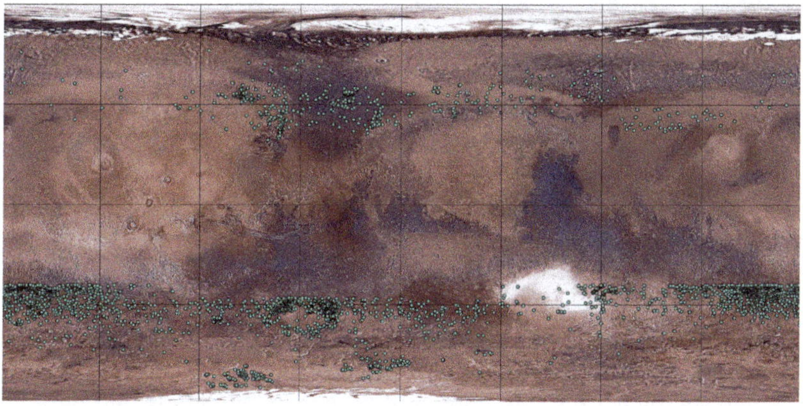

Fig. 8.1. Gully distribution on Mars based on >54,000 images from the Mars Reconnaissance Orbiter (MRO) Context Camera (CTX) covering ~85% of the martian surface at a resolution of ~6 m/pixel.

Source: Harrison *et al.* (2015).

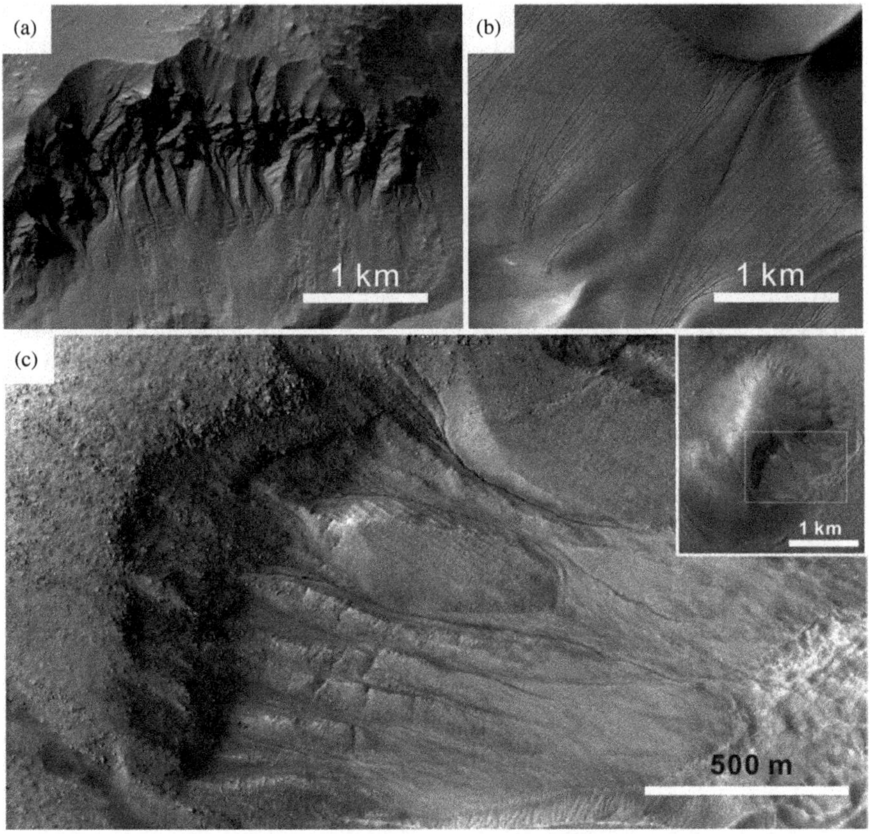

Fig. 8.2. (a) Small gullies on crater walls (CTX: D15_033115_1430_XI_37S036W); (b) Linear gullies on dunes in Russell crater (12.9°E–54.3°N) (HiRISE: PSP_007229_1255_RED); (c) Gullies on the central peak of Lohse crater (343.2°E, –43.2°N). The up-left image shows a general view of the central peak of Lohse crater, regions in the yellow box are zoomed in to the down-right. Gullies have a length of 1 km, width of 100 m and depth of tens of meters (HiRISE: ESP_022697_1365_RED).

8.2.2 *Martian gully morphology*

Martian gullies were generally constituted of three parts: the head alcove, which acts as the source, the main and secondary channels that define the transportation conduits, and the depositional apron (Fig. 8.2) (Malin and Edgett, 2000). Based on the development and morphological features of the three components, gullies on Mars are classified into two main groups with several types. The classification and characteristics of each gully type are listed in Table 8.1 (see Fig. 8.3).

Table 8.1. Morphology classification of Martian gullies.

Gully classification		Alcove	Channel	Apron
I-type gully: Mainly distributed on carter walls, or on the central peak of craters.	Classic gully	Single alcove narrowing downslope	V-shaped erosional channel that may divert around the topography	Alluvial apron or lobate fan
Gullies constitute of alcove, channel, and apron.	Complex gully	One or more alcoves feeding into a single channel	One or more-shaped erosional channels that may exhibit braiding. Channel often dissects the developing apron	One or more alluvial aprons that often exhibit braiding
	Channel gully	No alcove	V-shaped channel that exists only in the colluvium above any local bedrock layers	Small alluvial aprons or lobate fans
	Mantled gully	Alcove often filled by collapse or aeolian deposition	Channel obscured by infilling or aeolian deposition	Aprons that have been covered by subsequent deposits or mantling or that have edges that have been eroded
II-type gully: Mainly distributed on slopes on dunes, generally	Alcove-apron feature	Single alcove narrowing downslope	No V-shaped channel	Alluvial apron or lobate fan
manifested as a slender ditch with no significant collapse at gully head and no obvious accumulation of apron.	Abbreviated alcove gully	Single alcove narrowing downslope and abbreviated upslope by an obvious layer within the subsurface	V-shaped erosional channel that may divert Around topography	Alluvial apron or lobate fan
A semi-circular uplift is often shown at channel termination.	Linear gully	No alcove or extremely small triangular alcove	Narrow U-shaped channel that maintains its width and will divert around the topography	No apron

Source: Auld and Dixon (2016).

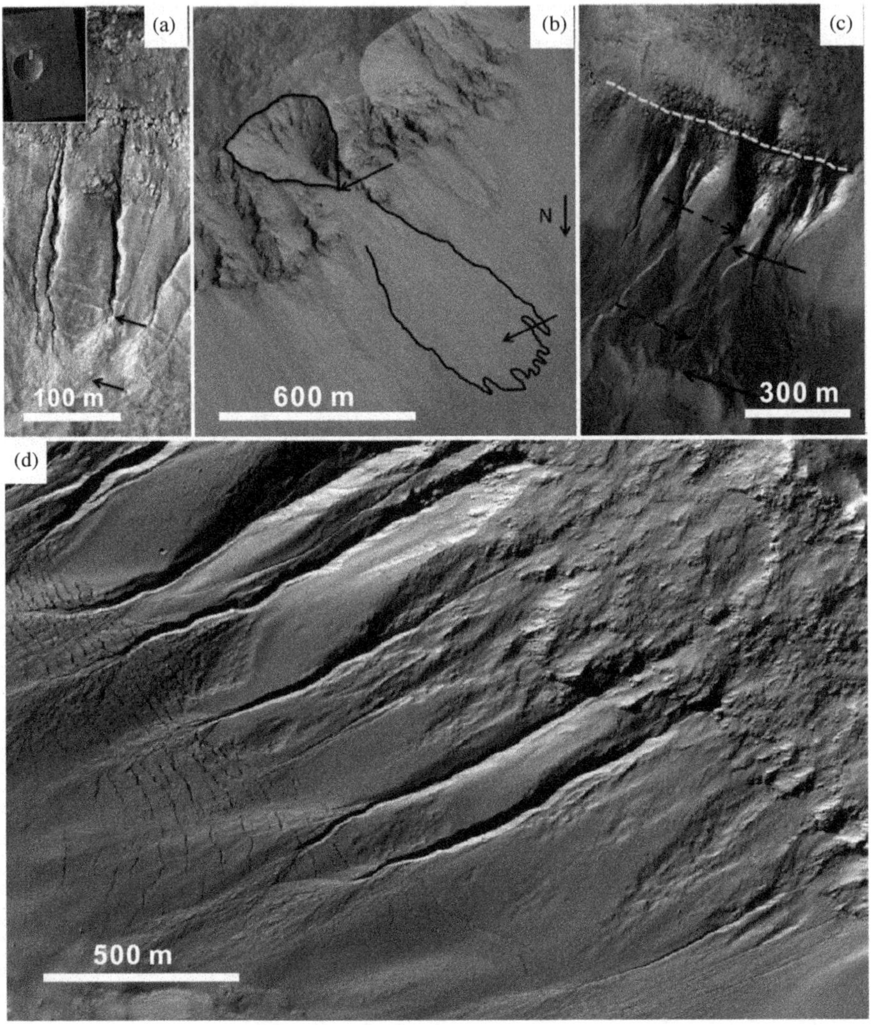

Fig. 8.3. Special types of gullies. (a) Channel gullies; (b) alcove–apron gullies; (c) alcove gullies; (d) gullies on crater walls with undistinguishable alcove (176.5°E–37.4°N), soft-sediment deformation structures can be seen at the down-right corner of the image to be cut by gullies (HiRISE: PSP_008739_1425_RED).

Source: Modified from Auld and Dixon (2016).

- **Classical gullies:** Classical gullies have all three of the main components of gullies: an alcove, a channel, and an apron. Classic gullies occur widely across Mars, especially concentrated in two bands around the planet: the more prominent band occurs between 31° and 51°S with another band occurring between 23° and 57°N. About 75% of the classic gully locations occur in the southern band, while 12.5% fall in the northern band. The remaining classic gullies are scattered across the planet with a minor concentration on a scarp in Noachis Terrain 71°S. Classic gullies appear to occur more often on pole-facing slopes than on east- or west-facing slopes.

- **Complex gullies:** Complex gullies also have all three of the main gully components, they are termed complex because either they have multiple alcoves that converge to a single channel or they have multiple channels and/or aprons that diverge from one alcove. Complex gullies are among the largest of the gully forms, usually displaying large aprons. Complex gullies mainly occur in a band from 32° to 49°S in a narrow strip.
- **Channel gullies:** Channel gullies include gullies without alcove or gullies without apron, which exist primarily as channels. Channel gullies occur principally in the southern hemisphere between 31° and 46° S, but also occur in the northern hemisphere, associated with classic gullies.
- **Alcove–apron:** Alcove–apron type gullies do not exhibit all three main components of gullies, instead they are gully-like features that need to be examined as they may have formation processes related to the development of classic gullies. Alcove–apron type gullies usually display a typical alcove narrowing downward, and also have a typical apron at the bottom of the slope. The occurrence of alcove–apron features ranges between 47°S and 43°N, and they also occur across the equatorial region.
- **Mantled gullies:** Mantled gullies are gullies that have been either mantled or eroded, which occurs over time as a consequence of a variety of surface processes, which redistribute the sediments on the slope so that the alcove, channel, and apron are difficult to distinguish in available resolution.

8.2.3 *Origin of Martian gullies*

Martian gullies were firstly interpreted as the results of fluid seepage and surface runoff based on their morphology that is consistent with fluid-mobilized mass movement processes (Malin and Edgett, 2000). Musselwhite *et al.* (2001) proposed that the action of a CO_2 suspended flow could have produced the recent small gullies on Mars, and therefore that liquid water is not required. The model involves the build-up of a liquid-CO_2 aquifer behind and below a dry-ice barrier (dam/cap rock) in the pore spaces a few meters into the rock from the cliff face and on order of a hundred meters below the top of the cliff brink surface. Seasonal (or obliquity-cycle-seasonal) heating causes pinching out of the dry-ice barrier and rapid release of the liquid CO_2. Erosion of the gullies occurs as the rapid vaporization of the liquid CO_2 with entrainment of rock and clathrate-hydrate ice produces a density flow analogous to a terrestrial *nuée ardente* turbulent.

The origin of Martian gullies has not been concluded yet. Some gullies located on the sidewalls of impact craters and grabens are thought to have been caused by melting of snowpacks which have existed for thousands of years in the early Martian climate or near-surface water eruptions, while gullies on dunes are thought to be related to the melting of snow or ice, and the annual change of their morphology indicates the existence of some intense geological processes. Gullies tend to occur where there is seasonal carbon dioxide ice formation, suggesting that gullies may also be driven by anhydrous processes, such as the melting of carbon dioxide ice. In addition, debris flow, mass slope movement, melting of atmospheric ice or snow, and erosion of brine have also been proposed to explain the formation of gullies on Mars.

Proposed formation mechanism for Martian gullies are mainly as follows:

1. Groundwater seepage (Malin and Edgett, 2000; Mellon and Philips, 2001; Heldmann and Mellon, 2004; Heldmann *et al.*, 2005).
2. Landslides caused by water and ice-rich deposits (Malin and Edgett, 2000).
3. Melting of atmospheric snow or ice (Costard *et al.*, 2002; Christensen, 2003).
4. Sublimation–condensation cycle of solid carbon dioxide and its induced surface landslide (Pilorget and Forget, 2015) (see Fig. 8.4).
5. Intermittent bursts of underground solid carbon dioxide (Hoffman, 2002; Balme *et al.*, 2006; Cedillo-Flores *et al.*, 2011; Sylvestetal, 2015).
6. Erosion caused by brine flows (Anderson *et al.*, 2002; Chevrier and Altheide, 2008; Chevrier *et al.*, 2009; Chevrier and Rivera-Valentin, 2012).

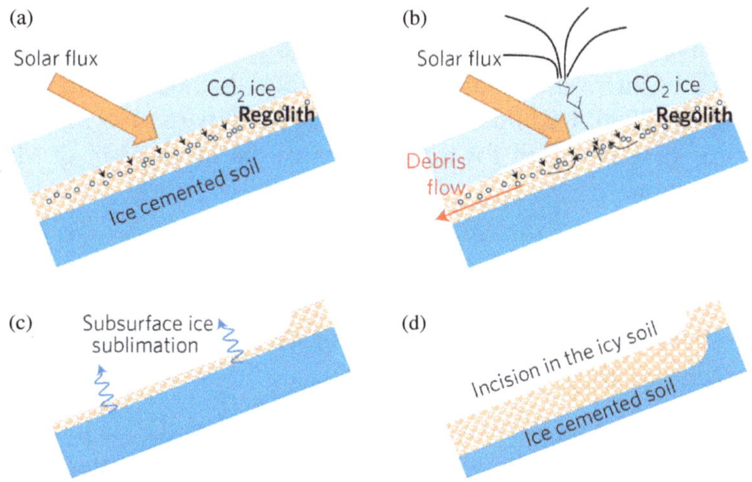

Fig. 8.4. Sublimation–condensation cycle of solid carbon dioxide and its induced surface landslide.

Source: Pilorget and Forget (2015).

According to recent observations, new gullies are still forming on Mars. To explain this continuously observed formation of new gullies on Mars, scholars who support the fluvial action (e.g. groundwater seepage, snow or ice melts) as the dominant mechanism of gully formation take the absence of evaporite or water-bearing minerals as evidence, claiming that these observed changes are only the degradation of old gullies, and the real gully formation process is no longer active (Kolb *et al.*, 2010; Schon and Head, 2011), while the source of water still remains controversial. Those who support the solid carbon dioxide sublimation and other non-fluvial processes as the main cause of gully formation believe that the current environment on Mars is not suitable for the existence of liquid water, and the continued evolution of gullies could be dominated by factors outside of fluvial water. The timing of gully formation is consistent with the time of year when seasonal frozen carbon dioxide is exposed to the surface and experiences sublimation (Pilorget and Forget, 2015). Through the simulation experiment, they confirmed that the solid carbon dioxide layer would rupture in a certain angle of sunlight, and its sublimation can indeed trigger a large-scale landslide of the underlying loose sediments. This hypothesis of sublimation–condensation cycle of carbon dioxide causing landslides does not involve the participation of liquid water, and could also solve the problem of contradiction between a small number of matter sources and a large-scale gully that occurred in the hypothesis of intermittent burst of solid carbon dioxide (Hoffman, 2002; Balme *et al.*, 2006; Cedillo-Flores *et al.*, 2011; Sylvest *et al.*, 2015), and could provide a new idea for the question of the mechanism of Martian gully formation.

8.3 Gullies in the Qaidam Basin

8.3.1 *Distribution and morphology of gullies in the Qaidam Basin*

Gullies in the Qaidam Basin are widely distributed on the slopes or valley walls around the basin. Shorter gullies are usually several meters long and dozens of centimeters wide, while the longer ones can reach more than 300 m and the width can reach 20 m.

The morphology of gullies is obviously affected by substrate materials. Long-linear gullies (Fig. 8.5) are developed in the Qianfoya region where semi-consolidated mudstone of Tertiary system is the dominant substrate material. These materials are heavily eroded and the depth of the gullies can be up to 10 m. The lengths of the gullies vary between 50 and 400 m and the widths are up to 30 m.

Narrow and densely distributed gullies occur on the slopes of low hillocks with unconsolidated deposits of Quaternary system. This kind

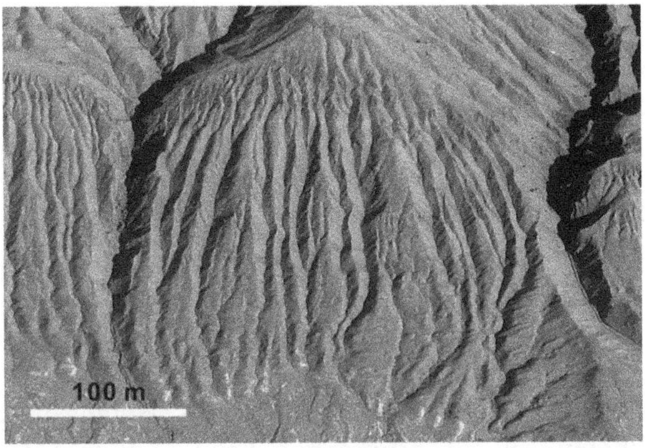

Fig. 8.5. Linear gullies at the southwest of Qaidam Basin. Gullies could extend up to 400 m, with widths of around 20 m.

of gully forms quickly after heavy rainfall and is the most common gully type in the Qaidam Basin (Fig. 8.6(a)). They are usually less than 20 m in length and have a linear or dendritic pattern (Figs. 8.6(b)–8.6(d)). Some of the linear gullies could be found on near-vertical slopes (Fig. 8.6(e)).

Gullies in the Qaidam Basin that are analogous to typical gullies on Mars are mainly distributed on slopes of large mountains around the Qaidam Basin (e.g. Kunlun Mountains, Qilian Mountains, and Altyn Mountains), most of which have developed on rocks of different lithology and strata. Their sizes vary across a large range, with the alcove in the upstream being the widest part. Similar to the most typical type of Martian gullies, small fan-shaped or conical deposits occur at the end of some gullies in the Qaidam Basin (Fig. 8.7).

8.3.2 *Origin of gullies in the Qaidam Basin*

Gully formation on Earth is controlled by three main conditions (Su *et al.*, 1963):

1. **Suitable strata or soil layers:** Gullies tend to occur in loose and easily disintegrating, weakly consolidated rocks (mainly argillaceous rocks) and cohesive soil.
2. **Appropriate surface topography:** Gullies tend to form on the steep slopes where the water erosion force is larger and easily wash away part of the soil materials, to form small grooves and then gullies, especially where the slope changes. This is because when water flows through this salient point, the speed of flow will suddenly increase, leading to an increase in the scour force, and resulting in the formation of a small length of trench from this point, which gradually develops into a gully.
3. **Appropriate climate conditions:** The climate conditions here mainly refer to the characteristics of precipitation, that is, the amount of precipitation and rainfall intensity. In general, it is easier to develop gullies in areas with more precipitation, especially in areas with more heavy rain. In addition, in areas where there is a large amount of melting snow, gully development is also easy. Aside from the three conditions mentioned above, the development of vegetation on the slope and man-made changes in the ground slope are also important factors affecting the formation of gullies.

Fig. 8.6. (a) Newly formed small gullies after temporal rainfall; (b) small linear gullies developed on slopes; (c) small gullies with dendritic morphology; (d) linear and dendritic gullies on slopes; (e) linear gullies developed on steep slopes.

Fig. 8.7. A gully with newly formed fan-shaped deposits.

The conditions mentioned above also affect the distribution of gullies directly. For example, the arid and semi-arid climate areas in northwest China have a low annual precipitation rate, which differs greatly among regions, with most of the precipitation falling in the form of heavy rain, and this climate is conducive to the development of gullies. In addition, gullies are very well developed in the loess regions which are widely distributed in China, because loess easily collapses and has vertical fractures. Gullies are also easily formed where there is less vegetation.

On Earth, the development of gullies can be roughly divided into four stages (Su *et al.*, 1963):

1. **Water cave stage:** A point protruding on a steep slope, a part of it is first washed away, and a channel of small length and area (namely water cave) is formed. At this stage, as the initial erosion cuts into the surface, the bottom of the longitudinal section is still convex upward, and the slope is still steep. If the environment condition does not change, the downward cutting will continue and gradually enter the next stage.

2. **The cutting stage:** The downstream part of the water erosion cave continues to incise the surface, the bottom of the gully deepens, and extends upstream. Deposits began to accumulate at the mouth of the gully. The slope of the valley bottom is still steep and the longitudinal section is still convex.
3. **Balanced section stage:** The steeper valley bottom constantly cuts into the depths, the upstream extension continues, the slope of the valley bottom gradually becomes gentle and the longitudinal section turns into concave. The down-cut effect begins to weaken, and even stops completely, and the equilibrium section begins to form. However, the erosion of water on both sides does not stop, and the width of gully continues to increase. At the bottom of the gully downstream, the gully alluvium is accumulated.
4. **Ditch stage:** Under the impact of lateral erosion, the width of the gully increases. The slope of the two gully walls gradually becomes gentle, and apron begins to accumulate at the bottom of the gully. At this point, the gully activities generally stop. Such inactive gullies are called ditches.

In the Qaidam Basin, the climate is arid and precipitation is limited. Gullies, as one of the erosion landforms caused by rapid surface flow, are widely distributed on the steep mountain slopes of the Qaidam region. Gullies are mainly formed from intermittent rainfall that flows over the surface, and most of them come from the local rainstorm in a short time, rather than from the sparse precipitation in the large area. In the short period when the precipitation rate is relatively high, the fast-gathering water incises the surface to form a gully and elongates source-ward through headward erosion. On slopes with homogeneous rock compositions and similar topographical features, gullies are evenly distributed at roughly similar intervals, and their plane shape changes from the common dendritic shape on slopes less than 30° to the flat shape on slopes that are even steeper. The difference of surface lithology and the distribution of tectonic lines can affect the morphology and distribution of gullies. Most gullies in the Qaidam Basin do not have water all year round or only have a very small flow of water, which contributes little to shaping the gully beds or transporting detrital materials. The large flow of water in gullies usually only occurs when there are heavy rains or a large amount of snow melts, during which time the amount of water in the gully is relatively concentrated, and the erosion rate is significantly boosted to further shape the gully morphology, and finally produces the dense and deep-cut gully landform now commonly observed in the Qaidam Basin.

The study of gullies in the Qaidam Basin are helpful for us to understand the formation of Martian gullies. Most Martian gullies have their counterparts in the Qaidam Basin. However, it should be noted that water participates in the formation of all types of gullies in the Qaidam Basin, while on Mars the role of water in gully formation is still unclear, and maybe future *in situ* explorations of Martian gullies can shed light on this problem.

8.4 Summary

The widely distributed Martian gullies are linear troughs developed on steep slopes at north and south high latitudes, their formation mechanism remains debatable. Typical gully consists of three parts: alcove, channel, and apron, while other kinds of gullies that lack one or two of these parts also exist. On earth, the origin of gullies are mostly related to precipitation. In the Qaidam Basin, many water-carved gullies can be observed on slopes or valley walls, whose morphology and distribution pattern is similar to Martian gullies. The analog study of gullies in these two sites may bring significant clues for the Martian gully formation mechanism and climate background.

Acknowledgments

This study is supported by the Natural Science Foundation of China (42002305, 41830214), Postdoctoral Science Foundation (2018M630889), and the Pre-research Project on Civil Aerospace Technologies (D020101) of CNSA. We thank Prof. Long Xiao for his guidance and suggestions in improving the manuscript.

References

Abotalib, A. Z. and Heggy, E. (2019). A deep groundwater origin for recurring slope lineae on Mars. *Nat. Geosci.*, **12**(4), 235–241.

Andersen, D. T. (2002). Cold springs in permafrost on Earth and Mars. *J. Geophys. Res.*, **107**(E3), 5015, doi:10.1029/2000JE001436.

Auld, K. S. and Dixon, J. C. (2016). A classification of martian gullies from HiRISE imagery. *Planet. Space Sci.*, **131**, 88–101.

Balme, M. and Greeley, R. (2006). Dust devils on Earth and Mars. *Rev. Geophys.*, **44**, RG3003, doi:10.1029/2005RG000188.

Cedillo-Flores, Y., Treiman, A. H., Lasue, J. and Clifford, S. M. (2011). CO_2 gas fluidization in the initiation and formation of Martian polar gullies. *Geophys. Res. Lett.*, **38**, L21202, doi:10.1029/2011GL049403.

Chevrier, V. F. and Altheide, T. S. (2008). Low temperature aqueous ferric sulfate solutions on the surface of Mars. *Geophys. Res. Lett.*, **35**, L22101, doi:10.1029/2008GL035489.

Chevrier, V. F., Hanley, J. and Altheide, T. S. (2009). Stability of perchlorate hydrates and their liquid solutions at the Phoenix landing site, Mars. *Geophys. Res. Lett.*, **36**, L10202, doi:10.1029/2009GL037497.

Chevrier, V. F. and Rivera-Valentin, E.G. (2012). Formation of recurring slope lineae by liquid brines on present-day Mars. *Geophys. Res. Lett.*, **39**, L21202, doi:10.1029/2012GL054119.

Christensen, P. R. (2003). Formation of recent martian gullies through melting of extensive water-rich snow deposits. *Nature*, **422**(4), 45–48.

Costard, F., Forget, F., Mangold, N. and Peulvast, J. P. (2002). Formation of recent Martian debris flows by melting of near-surface ground ice at high obliquity. *Science*, **295**, 110–113.

Dundas, C. M., McEwen, A. S., Chojnacki, M., Milazzo, M. P., Byrne, S., McElwaine, J. N. and Ursob, A. (2017). Granular flows at Recurring Slope Lineae on Mars indicate a limited role for liquid water. *Nat. Geosci.*, **10**, 903–907.

Edwards, C. S. and Piqueux, S. (2016). The water content of recurring slope lineae on Mars. *Geophys. Res. Lett.*, **43**, 8912–8919.

Gough, R. V., Nuding, D. L., Toigo, A., Guzewich, S. and Tolbert, M. A. (2019). An examination of atmospheric water vapor as a source for recurring slope lineae on Mars, *Ninth International Conference on Mars*, LPI Contrib. No. 2089.

Harrison, T. N., Osinski, G. R., Tornabene, L. L. and Jones, E. (2015). Global documentation of gullies with the Mars Reconnaissance Orbiter Context Camera and implications for their formation. *Icarus*, **252**, 236–254.

Heldmann, J. L. (2005). Formation of Martian gullies by the action of liquid water flowing under current Martian environmental conditions. *J. Geophys. Res.*, **110**, E05004, doi:10.1029/2004JE002261.

Heldmann, J. L. and Mellon, M.T. (2004). Observations of Martian gullies and constraints on potential formation mechanisms. *Icarus*, **168**, 285–304.

Hoffman, N. (2002). Active polar gullies on mars and the role of carbon dioxide. *Astrobiology*, **2**(10), 313–323.

Kolb, K. J., Pelletier, J. D. and McEwen, A. S. (2010). Modeling the formation of bright slope deposits associated with gullies in Hale Crater, Mars: Implications for recent liquid water. *Icarus*, **205**, 113–137.

Levy, J. (2011). Hydrological characteristics of recurrent slope lineae on Mars: Evidence for liquid flow through regolith and comparisons with Antarctic terrestrial analogs. *Icarus*, **219**(1), 1–4.

Malin, M. C. and Edgett, K. S. (2000). Evidence for recent groundwater seepage and surface runoff on Mars. *Science*, **288**, 2330–2335.

McEwen, A., Chojnacki, M., Dundas, C., Ojha, L., Masse, M., Schaefer, E. and Leung, C., 2015. Recurring Slope Lineae on Mars: Atmospheric Origin? *European Planetary Science Congress*, EPSC Abstracts, Vol. 10, EPSC 2015-786-1.

McEwen, A. S., Dundas, C. M., Mattson, S. S., Toigo, A. D., Ojha, L., Wray, J., Chojnacki, M., Byrne, S., Murchie, S. L. and Thomas, N. (2014). Recurring slope lineae in equatorial regions of Mars. *Nat. Geosci.*, **7**, 53–58.

McEwen, A. S., Ojha, L., Dundas, C. M., Mattson, S. S., Byrne, S., Wray, J. J., Cull, S. C., Murchie, S. L., Thomas, N. and Gulick, V. C. (2011). Seasonal flows on warm Martian slopes. *Science*, **333**, 740–743.

Mellon, M. T. and Phillips, R. J. (2001). Recent gullies on Mars and the source of liquid water. *J. Geophys. Res. Planets*, **106**, 23165–23179.

Ojha, L., McEwen, A., Dundas, C., Byrne, S., Mattson, S., Wray, J., Masse, M. and Schaefer, E. (2014). HiRISE observations of recurring slope lineae (RSL) during southern summer on Mars. *Icarus*, **231**, 365–376.

Ojha, L., Wilhelm, M. B., Murchie, S. L., McEwen, A. S., Wray, J. J., Hanley, J., Massé, M. and Chojnacki, M. (2015). Spectral evidence for hydrated salts in recurring slope lineae on Mars. *Nat. Geosci.*, **8**, 829–832.

Pilorget, C. and Forget, F. (2015). Formation of gullies on Mars by debris flows triggered by CO_2 sublimation. *Nat. Geosci.*, **9**, 65–69.

Schaefer, E. I., McEwen, A. S. and Sutton, S. S. (2019). A case study of recurring slope lineae (RSL) at Tivat crater: Implications for RSL origins. *Icarus*, **317**, 621–648.

Schon, S. C. and Head, J. W. (2011). Keys to gully formation processes on Mars: Relation to climate cycles and sources of meltwater. *Icarus*, **213**, 428–432.

Stillman, D. E., Michaels, T. I. and Grimm, R. E. (2017). Characteristics of the numerous and widespread recurring slope lineae (RSL) in Valles Marineris, Mars. *Icarus*, **285**, 195–210.

Stillman, D. E., Michaels, T. I., Grimm, R. E. and Harrison, K. P. (2014). New observations of martian southern mid-latitude recurring slope lineae (RSL) imply formation by freshwater subsurface flows. *Icarus,* **233**, 328–341.

Sylvest, M. E., Conway, S. J., Patel, M. R., Dixon, J. C. and Barnes, A. (2016). Mass wasting triggered by seasonal CO_2 sublimation under Martian atmospheric conditions: Laboratory experiments. *Geophys. Res. Lett.*, **43**(12), 363–370.

Zurek, R. W. and Smrekar, S. E. (2007). An overview of the Mars Reconnaissance Orbiter (MRO) science mission. *J. Geophys. Res.*, **112**, E05S01, doi:10.1029/2006JE002701.

https://doi.org/10.1142/9789811224027_0009

Chapter 9

Habitability and Astrobiological Significances

Ting Huang[*,†], Ziye Cheng[‡,§], and Long Xiao[*,†,§]

[*]*State Key Laboratory of Lunar and Planetary Sciences, Macau University of Science and Technology, Macau, PR China*

[†]*CNSA Macau Center for Space Exploration and Science, Macau, PR China*

[‡]*Guangdong University of Finance & Economics, Guangzhou, PR China*

[§]*Planetary Science Institute, School of Earth Sciences, China University of Geosciences, Wuhan 430074, PR China*

9.1 Introduction

Mars has been a hot subject for explorations because it might be a habitable planet. Survival of terrestrial life has always relied on water. Searching for life traces on Mars is in fact following the evidence of water, especially liquid water. Many studies suggest it has a warm and wet past with a long time of liquid water on its surface (Cabrol and Grin, 1999; Fassett and Head III, 2008; Poulet *et al.*, 2005) and probably presently liquid water is underneath (Martín-Torres *et al.*, 2015; Orosei *et al.*, 2018). The existence of water, the most important agent for life to form and survive, has become a key clue to assess the habitability and search for life on Mars.

9.2 Water on Mars

Multi-source remote sensing data show that there was a large amount of surface water activity on the surface of Mars. These ancient surface waters carved many gullies with different forms and preserved the remains of ancient lakes and even ancient oceans. At present, water ice mainly occurs in polar ice caps on the surface (Bibring *et al.*, 2005; Byrne *et al.*, 2009), buried glaciers in the mid-latitudes (Holt *et al.*, 2008), and subsurface ice beneath the topsoil (Byrne *et al.*, 2009; Carr, 1986; Dundas *et al.*, 2014; Mellon *et al.*, 2004; Mellon and Jakosky, 1995). While the direct observation of fluvial activity is relatively rare, the recurring seasonal slope pattern indicates that probably there is still small-scale surface water activity in the present cold and dry Mars.

In 1971, image data from the Mariner 9 spacecraft revealed several canyon systems on the Martian surface. By comparing the topography of the canyon system on Earth's surface and combining the discovery of water ice and evaporites on the surface of Mars, it was believed that the canyon formation was the result of abundant flowing water activity on the surface. It also indicated that the climate of Mars was warm and humid in a certain period in the past.

More geomorphical evidence and rock record from Mars mission revealed that there are different types of surface water in the various geological periods of Mars. Crater dating results show their age with the following order: valley network (Noachian), paleolakes (Later Noachian–Early Hesperian), the ancient oceans (Noachian–Hesperian), outflow channels (Hesperian), gullies (Amazonian), recurring seasonal slope (today). At present, the judgment of the activity of ancient surface water is mainly based on a large amount of fluvial landforms and the rich area of water-bearing minerals.

The valley network is the oldest form of flowing water, probably caused by the slow erosion of flowing water. The network of valleys that are widely distributed over the Noachian-aged region indicates that Mars had a warm, wet climate and extensive water flow during its early life, possibly sourced from meteoric water, groundwater ice, or meltwater. The outflow channel is a linear landform formed by large-scale flowing water scouring the ground. The formation of most outflow channels is related to impact, volcanic activity, and regional tectonic activity, which can cause groundwater eruption or lake discharge. And the melting of shallow surface water ice in the polar region is also one of its causes. Gullies are young erosion marks that develop on steep slopes on the surface of Mars.

They are usually small in size, tens of meters wide and hundreds of meters long. The gullies may have been mainly caused by erosion by flowing water during the Amazonian Epoch, providing clues to recent liquid water activity on the surface of Mars. The recurring seasonal slope is a kind of geomorphology that reveals trace of curving fluvial activity, generally appears on the slope that faces the equator, showing the dark color stripe that has seasonal change. Some studies suggest that it is caused by the intermittent flow formed by the process of brine thawing (Cabrol and Grin, 2010; Fassett and Head III, 2008; Malin *et al.*, 2006; Stillman *et al.*, 2016). Mars's wet past is reflected not only in the many flowing water features on the surface but also in the special minerals and rocks formed by the presence of water. Orbiters and landers have identified a variety of salt minerals on Mars, including carbonate (Ehlmann *et al.*, 2008), hydrous sulfate (Gendrin *et al.*, 2005), phyllosilicates (Poulet *et al.*, 2005), and chlorine salts (Osterloo *et al.*, 2008), of which phyllosilicates are the most widely distributed. Phyllosilicates (clay minerals) widespread in Mars's Noachian terrains indicate long-duration interaction between water and rock over 3.7 billion years ago. Their formation suggests involvement of hydrothermal groundwater circulation (Ehlmann *et al.*, 2011). Data from the Thermal Emission Imaging System on the Odyssey (NASA/JPL) showed signals of chloride in the low-albedo region of the Noachian and Hesperian periods. Most of them are located in low-lying areas. The formation of the chlorides is likely due to the overflow of water-bearing depressions or groundwater upwelling, followed by a process of evaporation, and this procedure may have lasted for one billion years (Osterloo *et al.*, 2008, 2010). Moreover, sulfate-rich deposits were also detected in different regions of the Martian surface, some of which may have resulted from chemical precipitation through evaporative processes (Gendrin *et al.*, 2005; Grotzinger *et al.*, 2014). These evaporites suggest a watery past on Mars, which could have been favorable for the occurrence of life (Marlow *et al.*, 2011).

Aside from geomorphological and mineralogical evidence of ancient water activities identified on Mars, direct proof from space missions also provides water-content data of the Mars atmosphere. Martian atmospheric water vapor is condensed closely to the surface of Mars and its vertical distribution is variable with season and location (Clancy *et al.*, 1996). Mars atmospheric water detector on Viking orbiter measured the water vapor content of the Mars atmosphere. The column abundances vary between 0 and about 100 precipitable microns (pr-μm) (Jakosky and Farmer, 1982). Large differences in water-vapor behavior between hemispheres and seasons were detected. A maximum value of ~100 pr-mm of water vapor abundance was observed at high latitudes during midsummer in the north, and ~50 pr-mm in the south. Whereas low water vapor abundance (<5 pr-mm) was observed at middle and high latitudes in the fall and winter of both hemispheres (Smith, 2002).

Another important pattern of water on Mars is water ice, which indeed conserves most abundant Mars water. Water ice is visible in polar regions of Mars (Bibring *et al.*, 2004; Grima *et al.*, 2009). This includes water ice in residual ice sheets and polar layered deposits. The residual ice sheet is a large amount of water ice lasting at least several centuries after the sublimation of Mars's two-stage seasonal CO_2 ice sheet (Bibring *et al.*, 2004; Byrne *et al.*, 2009). The polar layered deposits are mainly composed of atmospheric settling dust and water ice. Just as the ice cores on Earth record the Earth's climate change, the polar layered deposits also record the climate change entirely, which is an important geological feature to predict the recent climate change on Mars. The north polar layered deposits contain extremely fine stratified structures of water ice stacked with mixed dust more than 2 km thick. They indicate a climate change ranging from a 10^5 to 10^9 years-period. Underground ice on Mars formed in a dry and cold climate similar to the current Martian surface condition. It has a relatively recent geological age, less than 50,000 years old (Mellon *et al.*, 2004). In the current Martian climate, temperature, and pressure conditions, subsurface ice is stable at depths ranging from a few centimeters to several tens of centimeters below in mid-high latitudes above 40°. Subsurface ice was proved in several places such as Pavonis and Arsia Mons and Medusae Fossae Formation (Feldman *et al.*, 2002; Weiss and Head, 2014; Wilson *et al.*, 2018).

As of now, there is no long-term storage of liquid water found on the modern Mars surface, but temporary liquids exist. One scenario is when normal-incidence insolation illuminates thin ice accumulated in depressions or gullies, liquid water might exist transitorily on Mars (Hecht, 2002). Thirty-two gullies located in the southern crater wall might form as the result of saturation of near-surface regolith by a liquid (Lanza *et al.*, 2010). And liquid water signal (20-km wide zone) was detected by radar 1.5 km below the ice of south polar layered deposits (Orosei *et al.*, 2018). In another scenario, salts (e.g. perchlorate) on Mars could decrease the freezing point of water allowing the brine to form. As salts easily absorb atmospheric water vapor to form hydrated compounds or liquid solution, it is predicted that under optimum condition, the temporal formation of briny water is possible at the shallow subsurface (5 cm deep) of

Mars (Martín-Torres *et al.*, 2015). Actually, direct liquid water has been detected during the Phoenix landing mission, and probably was briny water saturated by perchlorate (Fischer *et al.*, 2014; Smith *et al.*, 2009).

Based on research data to date, "following the water" has seemingly provided positive support for searching for life on ancient and modern Mars.

9.3 Organic Compounds Found on Mars

Besides liquid water, chemical energy, and essential elements, organic compounds are critical for life's survival. Organic compounds could come from both biotic and abiotic processes. Complex organics are prevalent in the Universe as they are detected in Solar System objects, stars, and galaxies. The common view is that these organics were synthesized abiotically on a large scale (Kwok, 2017). Whereas many organic compounds are produced by living organisms, the remnants of past lives also contribute to their accumulations. If been protected well, their identification afterward will be one of the pieces of evidence of everlasting life. Considering their possible origination, organic molecules, if been detected on Mars, have significant astrobiological implications. Data from Martian meteorites and space missions have provided evidence of organic molecules in ancient Martian rocks and in the atmosphere. The key question is how to distinguish if they were produced from living organisms.

9.3.1 *Organics in Martian meteorites*

The first evidence of extraterrestrial organic matter paraffin was discovered from meteorite "Orgueil" in 1961 (Nagy *et al.*, 1961). In the 1970s, other organic compounds including aromatic and aliphatic hydrocarbons, amino acids, carboxylic acids, amines, alcohols, and carbonyls were discovered in meteorites successively. Since then, most molecular building blocks of life were identified in the soluble components of carbonaceous chondrites (Schmitt-Kopplin *et al.*, 2010). Moreover, compared to 20 amino acids used by terrestrial protein, over 100 amino acids were found in meteorites (Pizzarello and Shock, 2010).

Several pieces of research have been focusing on organics in Martian meteorites. Polycyclic aromatic hydrocarbons (PAHs) were found in Dar al Gani 476 Martian meteorite, EETA79001, Nakhla, and Tissint, including phenanthrene (and alkylated derivatives), pyrene, perylene, and anthracene. As the PAHs in the meteorite lack dibenzothiophene, a ubiquitous organic on Earth, PAHs found in Dar al Gani 476 are thought indigenous to this meteorite. The dominant organic compounds found in EETA79001 and Nakhla include aromatic and alkyl-substituted aromatic hydrocarbons (benzene, C1-C2 alkylbenzenes, and naphthalene) along with oxygen-containing (phenol) and nitrogen-containing (benzonitrile) aromatics (Sephton *et al.*, 2002).

Tissint is a bit different in organic composition. It is a depleted picritic shergottite that fell in the dry Morocco desert on July 18, 2011, as the fifth Martian meteorite ever witnessed falling to Earth. Most of the pieces were collected over months, thus Tissint is regarded as a perfect Martian sample minimally contaminated by terrestrial materials. Tissint's overall composition corresponds to that of aluminum-poor ferroan basaltic rock (Aoudjehane *et al.*, 2012). And it appears to be derived from a deep mantle source region (Brennecka *et al.*, 2014). The updated research reported that the organic profile in Tissint meteorite is dominated by aromatic hydrocarbons, including oxygen and nitrogen-containing aromatics, as well as organic compounds containing sulfur, such as thiophenes. Meanwhile, the organic composition of the soil in the meteorites' strewn field has been studied as a comparison to that of meteorites. They found that the organic compounds in the soil are dominated by typical breakdown products of biopolymers, including benzofuran from carbohydrate decomposition, nitrogen-bearing compounds from amino acids or proteins, and midchain length alkene/alkane dominated aliphatic doublets up to C18 from the pyrolysis of an aliphatic biopolymer. Organics in the soil indicate that they come from an extant or recent biological source, probably algal, at the recent wetter localities which are quite different from those in the meteorites (Jaramillo *et al.*, 2019).

In addition, since macromolecular carbon in meteorites is always associated with the igneous phase, it could be a production of crystallization of the host magma. This kind of textural relationship doesn't support any biotic origination of the macromolecular carbon and PAHs. It is indicated that primary organic carbon is ubiquitous in Martian basaltic rocks and it formed through abiotic igneous processes (Steele *et al.*, 2012).

9.3.2 *Organics detected by Mars missions*

The Mars missions which were able to detect organic compounds include the Viking landers (1976), the Phoenix lander (2008), and Mars Science

Laboratory with Curiosity rover (Sobron *et al.*, 2013). They were equipped with gas chromatograph–mass spectrometer (GC–MS), Thermal and Evolved Gas Analyzer (TEGA), and Sample Analysis at Mars (SAM), respectively, in order to analyze organic components on Mars. However, there is a challenge to look for organics on Mars which come from soluble inorganics. The Martian origin ionic compounds, ClO4-, ClO3-, and NO3-, all are oxidizing agents, may contribute to the oxidization of organic compounds. Consequently, only highly refractory and/or well-protected organics are likely to be preserved and detected (Kounaves *et al.*, 2014).

At the beginning of Vikings' landing, the gas chromatograph–mass spectrometer equipped on the lander did not detect any convincing organics. Later perchlorate was discovered on Mars and it was proved that this oxidizing salt is widespread on the planet. During the pyrolysis process, perchlorate could chlorinate and oxidize organic compounds, degrading or removing their signals. As a result, chloromethane and dichloromethane are easily produced. Thus indigenous organic content in Vikings' landing sites is reevaluated to be 0.7–6.5 ppm (Navarro-González *et al.*, 2010). The hypothesis has been challenged (Biemann and Bada, 2011) and then debated (Navarro-González and McKay, 2011). A dependent analysis from Guzman *et al.* (2018) proved chlorobenzene was indigenous to Mars, even the content was low (0.08–1.0 ppb) and instrumental contamination could not be totally excluded.

Similarly, carbon source for chlorinated hydrocarbons was detected by Curiosity in samples from Rocknest (Leshin *et al.*, 2013) and mudstone of Yellowknife bay (Ming *et al.*, 2014). Chloromethane, dichloromethane, and trichloromethane were found in both sites, while chloromethylpropene and carbon tetrachloride were detected in Rocknest and Yellowknife bay, respectively. Even they are not definitively of Martian origin, part of them are from the Mars surface. And indigenous Martian or meteoritic organic carbon sources may be preserved in the mudstone. Research from Eigenbrode *et al.* (2018) proved that they reported organic matters (including thiophenic, aromatic, and aliphatic compounds) preserved in a 3-billion-year old lacustrine mudstone at Gale Crater. The results convincingly showed the detection of organic compounds on Mars's surface. It is not clear if the organics have an indigenously biological origin yet, the discovery shed light on future *in situ* tests with deep samples in order to identify life traces (if any existed) conserved underground.

9.3.3 *Methane in Mars atmosphere*

The occurrence of methane on Mars is of remarkable astrobiological significance. Because it is related to geochemical and biotic processes on Earth, such as hydrothermal activity and methanogenic archaebacteria metabolism. Since Formisano *et al.* (2004) firstly reported methane detected in Mars atmosphere ranging from 0 to 30 parts per billion by volume (ppbv), trace amounts of methane have been found by various missions and observational studies (Korablev *et al.*, 2019; Mumma *et al.*, 2009; Webster *et al.*, 2015, 2018). Signs of methane in the Martian atmosphere suggest that the process of exchanging carbon between the planet's crust and atmosphere may have been active to some extent.

However, neither the source of that methane nor the driving force of its release is understood on Mars. On modern Earth, the volcanic processes have been attributed as a minor source of methane emission. While on Mars, this kind of geological methane is not determined due to the lack of recent volcanism, hydrothermal activity, or hotspots (Etiope *et al.*, 2007). How about living microorganisms, such as methanogens? When the H_2/CH_4 ratio is lower than 40, it is indicated that there is a biological process producing atmospheric methane. Whereas in the Martian atmosphere, this ratio is much higher, which is not favorable for methane's biological origin (Oze *et al.*, 2012). On the other hand, the laboratory observed substantial levels of methane produced by methanogenic bacteria growing on simulated Martian soil (Kral *et al.*, 2009). Methanogens use hydrogen as their energy source, and carbon dioxide (CO_2) as their carbon source, and the subsurface of Mars seems habitable for them (Blaut, 1994). But to date, no evidence of the presence of such organisms has been found on Mars. Future missions should dig deep enough to search if microscopic life is the seasonal source of Martian methane.

Early Mars might be habitable, this does not mean that Mars was inhabited. Currently, neither organic matter found in Martian meteorites, rocks nor in the atmosphere can prove their biological origin. Therefore, the assistance of other means is needed to determine the possible biological origin of organic matter on Mars in future surveys. The terrestrial analogs to Mars are good targets for investigation. Works on life-searching in terrestrial old samples are aiding our understanding of how biomarkers are preserved in rocks. This, in turn, provides guidance for Martian life investigation.

9.4 Habitability Assessment of Mars Analog Sites

The mission to search for life on Mars focused on looking for possible water, chemical, and biological signatures in the soil and rocks on the planet's surface, and biomarker gases in the atmosphere. However, existing shreds of evidence from Martian meteorites and probes of Mars have not yet been able to determine whether there is life on the planet.

More update researches are needed to (1) increase our understanding of possible traces of life on Mars, (2) to conduct more experiments to select the most suitable instruments for life detection on Mars, and (3) to rehearse the adequate processing of Mars samples in advance. Consequently, future probes could be more targeted. Fortunately, all of these can be learned by studying the terrestrial analogs to Mars with several advantages: (1) the existence of life forms present in the analog area can provide a reference for the detection of life signals near the surface of Mars, (2) compared with space exploration, it is easier to obtain samples, and (3) the test of life detection method and equipment can be carried out in the analog region. Hence, the study of terrestrial analogies is urgent to help assess Mars's past and present habitability.

In view of Mars's past and current geological and climatic characteristics, the extreme environments analogous to Mars include permafrost (Doran *et al*., 2010), desert (Navarro-González *et al*., 2003), hypersaline playas (Xiao *et al*., 2017), hydrothermal system (Brown *et al*., 2010), and acidic watery condition (Amils *et al*., 2007; Fernández-Remolar *et al*., 2005). Some environments can be compared to one aspect of Mars, and some environments can be compared to multiple aspects.

Three kinds of Martian environments have been proposed as potential systems of astrobiological interest, including near-surface hydrothermal system, delta, and lacustrine environments (Summons *et al*., 2011). Deltaic and perennial lacustrine environments are ranked high due to their potential to concentrate those organics through hydrodynamic processes. Paleolakes identified on Mars are of great astrobiological importance as they might preserve biomarkers of extinct life if it existed. What's more, the surviving potential of life in the shallow subsurface of Mars, especially accompanied by evaporites, is not excluded. Considering this, Martian paleolakes are candidates for life exploration. And terrestrial deposits of paleolakes, including salt flats, are ideal analog targets for Martian life investigation and assessment.

Generally, salt flats on Earth have evolved from the lacustrine system, they have recorded a series of sediments and life forms over geological time. Salt crystals have unique characteristics distinguished from other minerals: (1) evaporite is deliquescent contributing to form a humid microenvironment suitable for the reproduction and growth of microorganisms, (2) most salt crystals are translucent, which can filter out harmful radiation at the same time, favor photosynthesis, and (3) the salt minerals preserved in extreme drought conditions can prevent the degradation of the life materials inside, which can preserve the life traces for a long time (Wierzchos *et al.*, 2006). The accumulated abundant evaporites are challenges as well as providing a chance for the survival of microorganisms.

To understand and identify life on other planets, one of the keys is to study the preservation of organic compounds and their precursor microorganisms on Earth (Preston *et al.*, 2011). Biomarkers, e.g. lipids, esters, fatty acids, pigments, are complex organic compounds composed of carbon, hydrogen, nitrogen, and other elements. They occur in sediments, rocks, and crude oils and show little or no change in structure from their parent organic molecules in living organisms (Peters *et al.*, 2005). Hexafunctionalized hopanoids were most abundant in lacustrine sediments, possibly due to greater contribution to the organic matter by methanotrophic bacteria (Peters *et al.*, 2005). Biomarkers detected from evaporites offer ancient life information in exact geological and climatic history. The analysis of biological molecules preserved in evaporites and their controlling factors gives clues for searching for extinct life on Mars, especially in paleolake sites.

The other key is the study of microbes that have survived in extreme environments on Earth. For a long time, the limits for life's surviving were thought to be very narrow, until 1969, when *Thermus aquaticus* was found in Yellowstone National Park in the USA (Brock and Freeze, 1969). The new discovered thermophilic microorganism opens a new study-field that is "Extremophile". Scientific expeditions have discovered an astonishing diversity of extremophiles (prokaryotes, eukaryotes, and viral particles) in a variety of extreme environments such as deep-sea hydrothermal vents, sub-glacier lakes, acid river system, deserts, and saline lakes. They represent the most extreme life forms on Earth. Considering the similarity between terrestrial extreme environments and other planets, including Mars, the study of extremophile has significant importance in searching for life out of Earth.

In total, terrestrial paleolakes are good analogs in order to promote the investigation of Martian life. And it will be an excellent strategy when combined with searching for existing extremophiles and past biomarkers in this region. The Qaidam Basin once had a giant ancient lake and has widespread lacustrine deposits at present. Its arid, hypersaline salt flats pose great challenges for the existence of life on the one hand. The surface evaporite crusts and the subsurface ancient sediments are habitats or life-preservation sites in such extreme environments on the other hand. Therefore, we choose the Qaidam Basin as a study site for astrobiology investigation, in order to analyze the preservation possibility of life in the evaporative deposits of Mars.

9.5 The Habitability of the Qaidam Basin

Life can be preserved in the form of living organisms or traces of past life in geological forms. The life system in the Qaidam Basin is investigated by studying the existing extremophile and the past biomarkers in its lacustrine system. This research provides a basis for evaluating the possible traces of life in the sediments of paleolakes on Mars.

9.5.1 *Microorganism studies of saline lakes and evaporative sediments in the Qaidam Basin*

The salt flat (i.e. playa) is the final product of the evolution of the saline lake, a typical extreme environment for life on earth. Microorganisms in modern lakes (including lake water and sediment) partly reflect the past ecosystem of the salt flats. The traces of life preserved in the salt flats can reveal the easily preserved remains of life and its preservation medium. Both of them reveal the development and evolution of the entire lake ecosystem and provide references for understanding the possible life forms in the paleolakes and evaporative deposits on Mars.

Compared to microbiology of dried playas, that of watery saline lakes of the Qaidam Basin has been well studied. Fungi and cyanobacteria are dominant primary producers of the lakes. According to the study of six lakes (Table 4.2, except Erhai and Dabuxun Lakes, including another Lake Geermu) in the Qaidam Basin, salinity is the main factor for algal community composition. In the lakes with high salinity ($>50\ gL^{-1}$), eukaryotic algae including *Chlorophyta* (Chlorella, Dunaliella),

Bacillariophyta (Fistulifera), *Streptophyta* (Chara), and *Dinophyceae* (Kryptoperidinium foliaceum) are the dominant species. In the lower salinity lakes (salinity: 0.5–50 gL^{-1}), Cyanobacteria (*Cyanobium* and *Synechococcus*) is dominant. Algae here shows tolerance to different salinity, which may be an adaptation to saline lake's evolution. What's more, the microbial communities in lakes' sediments may be derived from the brine water, as the community structure is very similar to that of the water (Liu *et al.*, 2016). This indicates the significance of modern lake's ecology, which helps to understand the preservation of ancient life in the sediments of playas.

As to surface sediments of the seven lakes (Table 4.2, except Keluke Lake) in the Qaidam Basin, 646 strains of bacteria were obtained (Fig. 9.1). They were affiliated with 4 bacterial phyla (i.e. 43.2% Firmicutes, 30.9% Proteobacteria, 18.3% Actinobacteria, and 7.6% Bacteroidetes), 7 classes, 18 orders, 39 families, 89 genera, and 210 species. There are differences in the community composition (dominant and unique groups) of culturable bacteria between lakes with different salinities. For example, in the fresh lake Erhai, 47 strains were dominated by *Paenibacillus*, *Bacillus*, *Pseudomonas*, and *Dietzia*, without halophilic taxa. While in the slight saline Qinghai Lake, *Bacillus*, *Aurantimonas*, *Halomonas*, and *Micrococcus* are dominant. Isolates from the mid-salinity hypersaline lakes Gaihai and XiaoQaidam were dominated by *Halobacillus*, *Bacillus*, *Halomonas*, *Planococcus*, *Nesterenkonia*, *Roseovarius*, and *Salegentibacter*. For hypersaline lake Chaka, strains were obtained with the dominance of *Bacillus*, *Halobacillus*, *Halomonas*, and *Virgibacillus*. In total, halophiles (such as *Halomonas* and *Halobacillus*) are becoming dominant among culturable bacteria with increasing salinities. Moreover, microorganisms in saline lakes with moderate salinity have high culturability (Jiang *et al.*, 2018).

Not only microbial community composition of water and surface sediments is controlled by salinity gradient, but also subsurface sediments (Jiang *et al.*, 2007). Besides, the total organic carbon (TOC) content of sediments affects bacterial community in lakes' subsurface sediments. According to a six-meters long core, the content of TOC was highest in the 1.65–4.20 m depth section (>3.5%), lowest (<3%) was lower than the 4.20 m depth section, and medium (3–3.5%) was higher than the 1.65 m depth section. In the sediments with the highest TOC content, there is more archaea than that of other samples. In the sediments with the

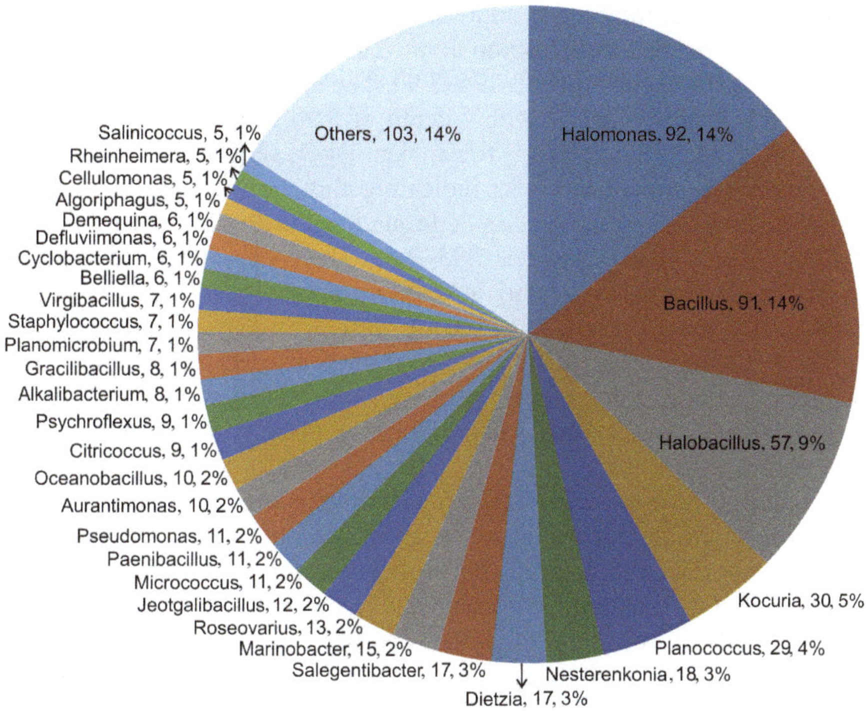

Fig. 9.1. Genus affiliation of the isolates obtained from seven lakes in the Qaidam Basin (Jiang *et al.*, 2018).

lowest TOC content, Firmicutes and proteobacteria are more abundant than others (Jiang *et al.*, 2018). However, no obvious correlation between microbial changes and mineralogy has been revealed.

Investigation of living forms in hypersaline playa is one of ways to testify life's tolerance to terrestrial extreme conditions including the mechanism of life surviving in hypersaline, arid, and elevated UV conditions.

Dalangtan Playa is one of the oldest and driest salt flats in the Qaidam Basin. As a high plateau playa, Dalangtan is exposed to strong ultraviolet radiation and suffers from extreme drought as well as extremely salinity. There are two main types of sedimentary compositions on the surface of Dalangtan: (1) crusts composed of halite, gypsum, and a small amount of mirabilite, and (2) mixtures composed of evaporites and feldspar, quartz, and other debris particles (Sobron *et al.*, 2018). Surface sediments in the playa are alkaline with a pH of 7.95–8.90. The mineral composition of the studied two profiles (one is with a depth of 5.90 m and the other one is with a depth of 6.85 m) reflects the ancient lake's sedimentary sequences with multiple sedimentary cycles indicating climatic and tectonic changes during geological history. The underground sediments are more alkaline than the surface, with pH values 7.93–9.48. The water content of samples below the surface varies greatly at different depths. It is quite dry at the surface with water content from 0.08–0.33%, with a more moisture subsurface with water content from 5.14–12.45%. Due to the different proportion of soluble salts and evaporative rates of sediments at different depths, salinity of subsurface samples varies greatly, which is in the range of 0.4–50%.

Microbial analysis in Dalangtan Playa helps enrich the diversity of culturable microorganisms in the evaporites of the basin and provides valuable experimental materials for further research on the adaptability of microorganisms to extreme environments.

By means of culturing, 50 isolates of bacteria and 7 fungal isolates were obtained from both surface and subsurface samples using three media (R2A, Marine, and MGM). A series of salinity gradients of the media was set by adding NaCl. The bacteria isolated from underground samples mainly belongs to the genus *Bacillus* of phylum Firmicutes, which is a possible dominant species. Other isolated genera include *Oceanobacillus*, *Halobacillus*, *Gracilibacillus*, *Thalassobacillus*, *Sediminibacillus*, *Ornithinibacillus*, *Pontibacillus*, *Microbacterium,* and *Nocardiopsis*. Among them, three moderate halophiles (*Oceanobacillus*, *Gracilibacllus*, *Thalassobacillus*) are isolated from the surface samples. The fungal strains isolated from underground samples belong to the genera *Aspergillus* and *Penicillium* of Eurotiales and *Cladosporium* of Capnodiales. Among them, *Aspergillus* was isolated from the surface samples (Figs. 9.2 and 9.3).

Besides traditional culturing method, CARD-FISH (Catalyzed Reported Deposition-Fluorescence *In Situ* Hybridization) and laser confocal microimaging were used to analyze subsurface samples of Dalangtan Playa. Firmicutes accounted for about 76% of the total bacteria, respectively. CARD-FISH is a supplement and verification to the culturing results. The results show that subsurface samples have bacteria abundance of (1.05–4.17) $\times$ $10^5 g^{-1}$. Confocal microscopy obtained clear images of bacterial colonies in the sediments (Fig. 9.4).

Fig. 9.2. Phylogenetic dendrogram of culturable bacteria from Dalangtan Playa based on 16S rRNA gene sequences.

Because by means of culturing only a few microorganisms can be recovered in the natural environment, the vast majority cannot be cultured and isolated under laboratory conditions. As a supplement, high-throughput sequencing was conducted by extracting total DNA from the environment to analyze the genetic composition and function of the microbial community in Dalangtan Playa. Genomes of the microbial community in environmental sample, i.e. the sum of all microbial genetic materials in the habitat, were expected. However, after different DNA extraction techniques were tried, the total DNA materials extracted from

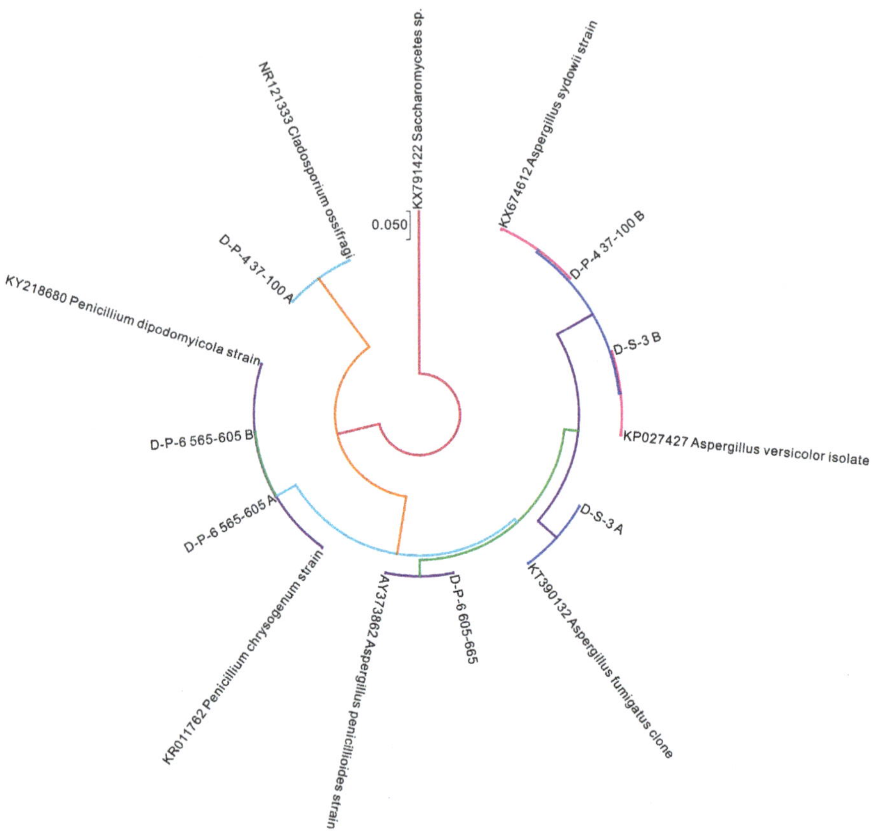

Fig. 9.3. Phylogenetic dendrogram of culturable fungi based on rRNA-ITS gene sequences in Dalangtan Playa.

samples revealed very low concentrations, which range from 1.0 to 5.0 ng/μl. After the amplification of DNA using bacterial primers (27F/1492R, s-d-bact-0341-b-s-17/s-d-bact-0785-a-a-21), archaea primers (Arch1R/Arch1F), and fungal primers (563F/1132R), no effective amplification products were obtained, which also proved that the biomass in the dried saline playas of the Qaidam Basin was extremely scarce.

In general, the very low counts of microbes are affected by the low TOC values (with an average of 0.03%). Heterotrophic microbes make use of organic carbon source as their source of energy. Therefore, the rare organic carbon limited their growth and metabolism.

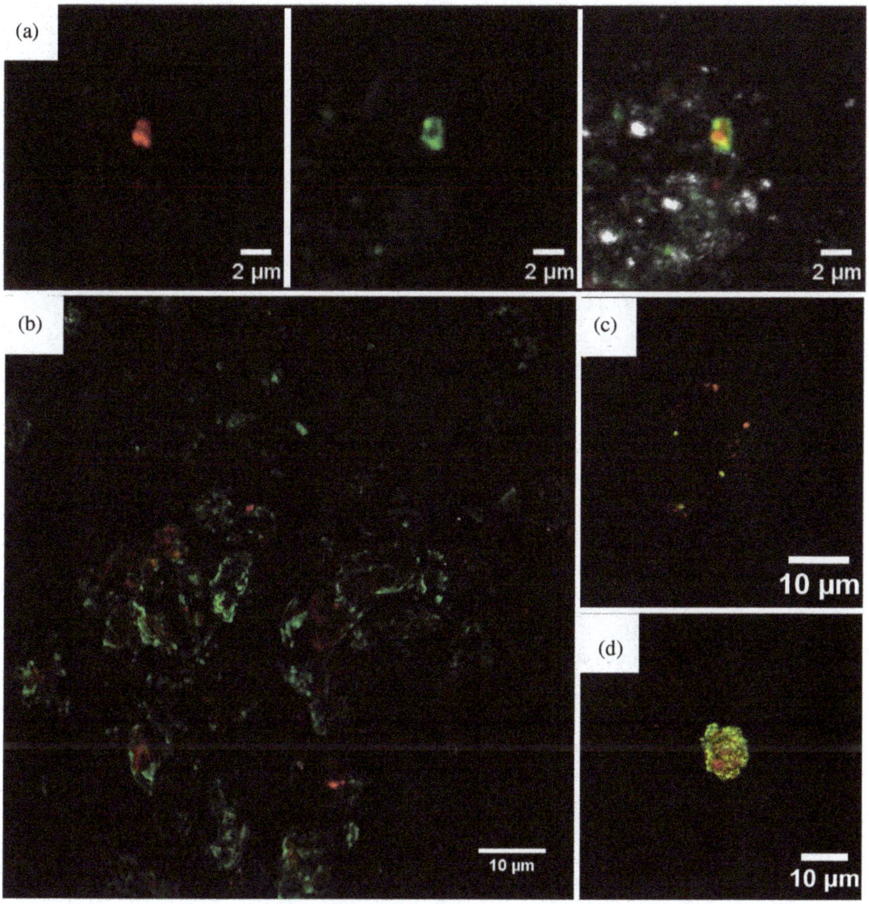

Fig. 9.4. CARD-FISH hybridization analysis of subsurface samples from Dalangtan Playa in the Qaidam Basin. (a) From left to right: epifluorescence micrographs showing bacterial cells hybridized with the universal probe EUB388I-III mix (left column), the same field stained with Syto9 (middle column), and the overlap of stained field (right column). (b), (c), and (d) more examples of the hybridized as well as stained field of samples in the Qaidam Basin.

Different from lakes' samples, where salinity gradient has significant effect on biological diversity, the availability of liquid water in evaporative deposits is the key factor for microbial life in Dalangtan Playa. Moisture content of the subsurface samples, calculated as moisture content (%) = (Mass $_{wet}$ -Mass $_{dried}$)/Mass $_{wet}$, varied greatly along the depth.

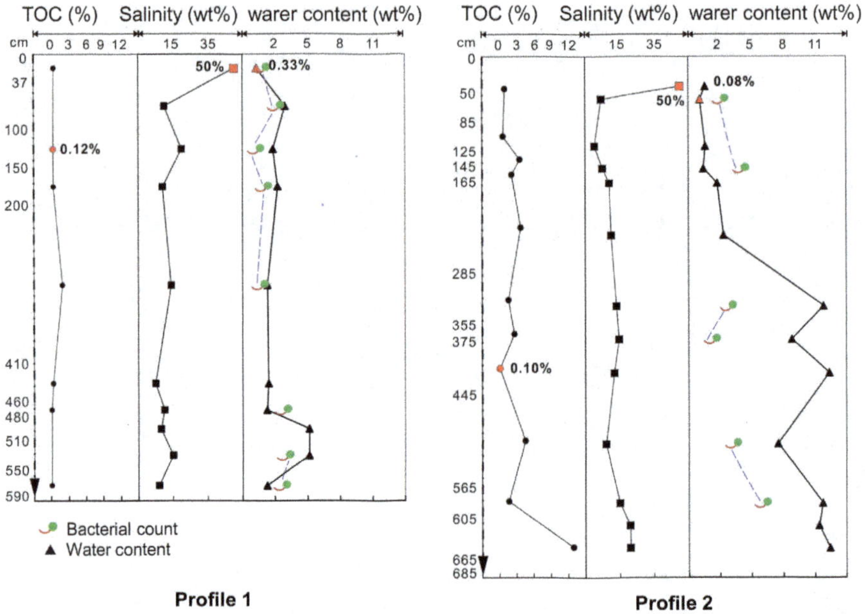

Fig. 9.5. Relationship between bacterial count and environmental physical and chemical properties, data recovered from two profiles in Dalangtan Playa. Red refers to the lowest or highest values.

The lowest moisture content was 0.08% close to the surface (0.5 m to the surface), and the highest value 12.45% was observed in the bottom (at a depth of 6.85 m). The number of cells in sediments in the Dalangtan Playa is positively correlated with water content (Fig. 9.5).

Most microorganisms on Earth live in the water activity range from 1 to 0.90 (Grant, 2004). Stevenson *et al.* (2017) reported the minimum water activity of 0.585 for the reproduction of the fungus *Aspergillus penicillioides*. Although water activity was not measured in Dalangtan Playa, water content to a certain extent can also indicate differences in water activity in samples (Achille and Hynek, 2010). In this case, the microbes (here are *Bacillus* and *Oceanobacillus*) that survived in the driest sediments of Dalangtan Playa represent the lifestyles which may be tolerant to lower water activity.

In the absence of sufficient water, microbial diversity might have a negative correlation with salinity. While other factors such as mineralogy and pH values did not show evident correlation with microbial distribution in various samples. Previous studies about sediments in saline lakes support this result. Regarding the small sample size of microbes investigated in Dalangtan Playa, more researches should be done in order to have a further explanation about the mechanism of microbial survival strategy in the super arid and saline condition.

From this investigation, it is impressed that the life diversity of hypersaline and arid sediments of the Qaidam Basin is quite low. *Bacillus* dominates the culturable isolates, which indicate its tolerance to such extreme conditions. Whereas the microbial diversity in the watery lakes is much higher. On the one hand, the playa is in a state of long-term drought due to the lack of liquid water resources; on the other hand, the available organic carbon content is low, which limits the metabolic activities of microorganisms. The similarity between the two habitats is that hypersaline sediments (both in the surface of lakes and in the playa) are inhabited with Firmicutes which suggests its tolerance to harsh conditions.

The existing microorganisms in the saline lake represent the active life forms in the lake evolution system. And the extremely dry playas, with their high ultraviolet radiation, provide a challenge for most life forms to survive. Combined with the study of the traces of past life (such as lipids) preserved in evaporite, it is of great significance to reveal the evolution law of the lake ecosystem in a different time and different space.

9.5.2 *Lipids recovered from evaporites of the Qaidam Basin*

Continental evaporites are deposits that originate from the evaporation of saline waters in the low areas of saline lakes from all continents, and mainly consist of chloride, sulphate, and potash minerals. On Earth, terrestrial hypersaline environments of arid regions are the most likely analogs of Martian regions in terms of biosignatures. In the extremely arid environment, the high solar and UV radiation, extreme dryness, and the accumulated abundant evaporates are the challenges to the survival of microorganisms. Therefore, the analysis of biological molecules preserved in evaporites of hyperarid environments, and the study of the differences of these preserved biomolecules and their controlling factors, have important significance for the exploration of traces of life on Mars.

Because of their refractory characteristics, lipids are widely used as a useful biomarker to identify microorganisms throughout Earth's history.

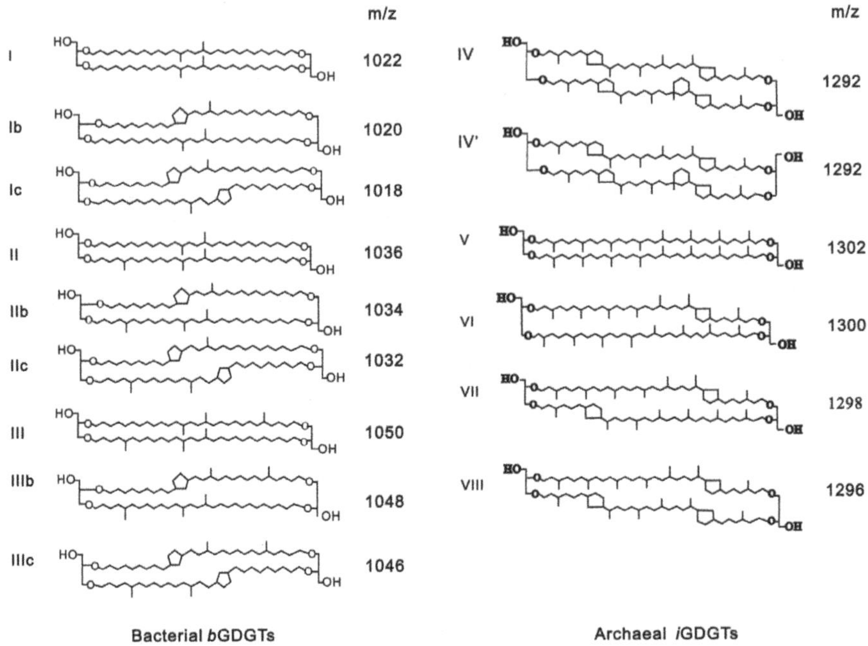

Fig. 9.6. Structures of bacterial and archaeal GDGTs.

Lipids are a structurally diverse group with high biological specificity, and thus are very valuable for assessing sources of organic matter and investigating microbial communities in sediments (White *et al.*, 1979; Zelles, 1999). In general, methyl-branched fatty acids are common constituents of bacterial cell walls. And archaeal tetraether membrane lipids, i.e. glycerol dialkyl glycerol tetraethers (GDGTs; Fig. 9.6) have been recently recognized to be sensitive to the environment (Grice *et al.*, 1998; Kaneda, 1991; Kates, 1996).

We have examined the concentration and distribution of lipids biomarkers (branched fatty acids, GDGTs, and archaeol) preserved in evaporite deposits from different sites in Qaidam Basin as well as their relationships with the associated evaporates (Figs. 9.7–9.9).

Normal, branched, monounsaturated fatty acids were abundant in the samples from Dalangtan play, in the Qaidam Basin, and a strong microbial activity has existed in the hypersaline ecosystem over the

Fig. 9.7. Sampling sites and outcrops at Dalangtan Playa which mainly consist of the chloride and sulphate. Exposed layering evaporates by mining: (a) and (b), gypsum crystals rich layer (c) and chloride/halite rich layer (d).

geological history. For each outcrop, the concentration of variations of branched fatty acids is highly related to salinity. In DLT-P_1, the concentrations of branched fatty acids increased from the lowest hypersaline unit to the upper detrital unit (Fig. 9.8), the same trend was also observed along DLT-P_2 (Fig. 9.9). Combined with the unique geochemical and geological features of the high salt environment and the distribution of fatty acids from the profiles, we concluded that bacterial community was dominated by the sulfate-reducing bacteria and Bacillus (Cheng *et al.*, 2017).

Archaeol, GDGT-0 were also detected in both sections, whereas cyclic archaeal iGDGTs and bacterial bGDGTs were close to or below the detection limit. GDGT-0 was predominant among archaeal iGDGTs, representing more than ca. 93% of total iGDGTs. Detrital unit typically contained abundant archaeol and little iGDGTs. In contrast, hypersaline unit contained predominantly archaeol, with small amounts of GDGT-0. The concentrations of archaeol and GDGT-0 showed comparable patterns along the profile. For each outcrop, concentration variation is highly

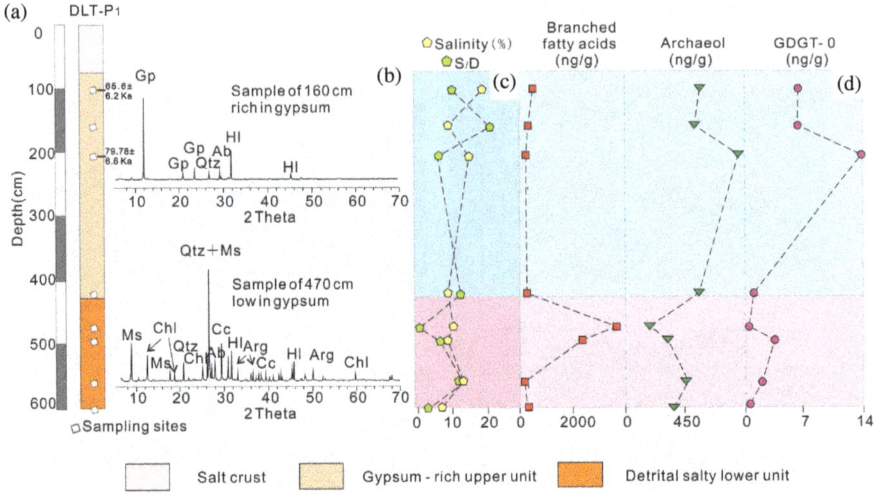

Fig. 9.8. (a) Stratigraphic sections and sampling sites of DLT-P_1. (b) Selected XRD analyses of the materials sampled at DLT-P_1. Samples come from depths of 160 and 470 cm. The mineral peaks show Ab (albite), Arg (aragonite), Cc (calcite), Chl (chlorite), Gp (gypsum), Hl (halite), Ms (muscovite), and Qtz (quartz). (c) Distribution of salinity and the S/D index that evaluates the intensity of main XRD peaks in salts (gypsum and halite) against detrital minerals (quartz, albite, muscovite, and chlorite) in the evaporites sediments of DLT-P_1. (d) Variation in branched fatty acids, archaeol, and GDGT-0 abundance with depth of DLT-P_1 (Cheng *et al.*, 2017).

related with salinity. In DLT-P_2, the concentrations of archaeol and GDGT-0 increased from the upper detrital unit to the lowest hypersaline unit, the same trend was also observed along DLT-P_1. The halophiles were the most important archaea in the hypersaline sediments, and they are treated as the contributor of archaeol. The lack of other GDGTs in the hypersaline sediments in most samples suggests a substantial contribution of methanogenic archaea to GDGT-0.

This systematic change in the relative abundance of archaea and bacteria with salinity was consistent with the fact that most obligate halophiles are archaea. Because of their different requirements for salt, halophilic bacteria and archaea tend to occupy different salinity niches, with the former being dominant at low salinity and the latter being dominant at high salinity. Although this generalization has been made previously (Jiang *et al.*, 2007), our study has definitively demonstrated this by lipids analysis.

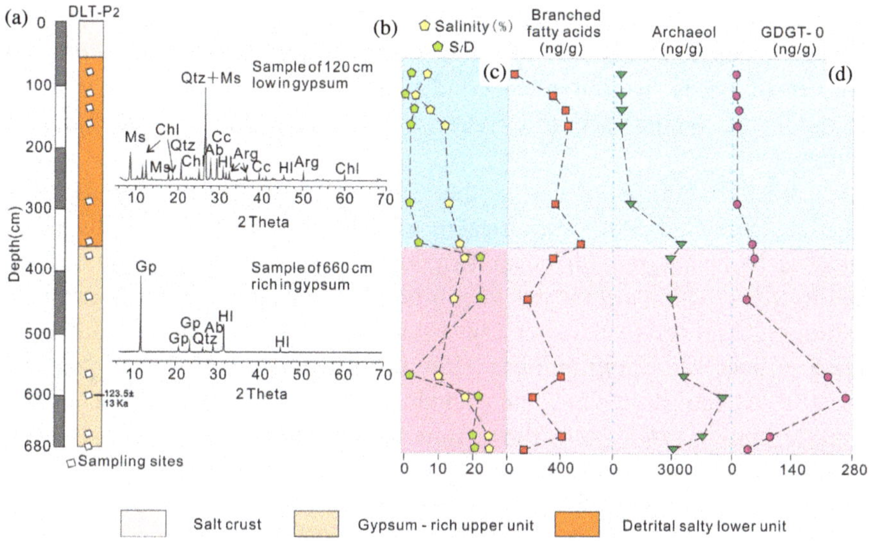

Fig. 9.9. (a) Stratigraphic sections and sampling sites of DLT-P_2. (b) Selected XRD analyses of the materials sampled at DLT-P_2. Samples come from depths of 120 and 660 cm. (c) Distribution of salinity and the S/D index that evaluates the intensity of main XRD peaks in salts (gypsum and halite) against detrital minerals (quartz, albite, muscovite, and chlorite) in the evaporite sediments of DLT-P_2. (d) Variation in branched fatty acids, archaeol, and GDGT-0 abundance with depth of DLT-P_2 (Cheng *et al.*, 2017).

9.6 The Astrobiological Implications from the Qaidam Basin

Global mineralogical mapping suggests that Mars experienced a wet period dominated by phyllosilicate formation in the Early Noachian period, followed by more arid conditions in the Late Noachian and Early Hesperian in which the evaporite deposits formed (Bibring *et al.*, 2006). This period is considered suitable for microbial survival, and there might be a biosphere dominated by chemoautotrophic microorganisms in the paleolake during the post-Noachian period (Grotzinger *et al.*, 2014). If they did exist, their remains would be preserved in sediments. The key point is the form of the remains of life on Mars if any, and the vehicle for preserving them.

Lipids are direct evidence that organisms existed in geological environments or participated in biogeochemical cycles. It has certain stability in the geological body. Obviously, it is more advantageous when restoring the appearance of biological communities as biomarkers were formed during geological history. Microbial cell membranes are important sources of lipids in geological conditions, which determines that these lipids can be synthesized and preserved in large quantities in geological bodies.

Taking Dalangtan Playa in the Qaidam Basin as an example, the physical and chemical properties of surface sediments and subsurface samples in Dalangtan Playa confirm the practicability of its analogous role to Mars. Abundant fatty acid compounds were detected in subsurface sediments, and relatively abundant branched fatty acids proved the existence of bacterial communities below the surface, mainly gram-positive bacteria. At the same time, the content of branched fatty acids changed with the salinity gradient of the sediments, and the traces of bacteria were more abundant in the low-salinity environment. The contents of archaeol in sediments were significantly higher than that of other compounds, followed by iGDGT-0 synthesized by archaea. While the contents of other iGDGT compounds and bGDGTs synthesized by bacteria were lower than the detection limit in most samples. The GDGT-0 compounds in the evaporative sediments may have originated from the contribution of methanogenic archaea, and halophilic archaea could have been the main contributor of archaeol compounds in the sediments. Moreover, since evaporites have been preserved over geological times where surface hydrological cycles are scarce, they are good at protecting traces of life from degrading (Cheng *et al.*, 2016). The preservation of lipids in the underground sediments of playas in the Qaidam Basin reveals the life activities in geological history and provides a paradigm for searching for the fossilized microbial life in evaporite deposits on Mars.

Modern Mars surface suffered from extreme drought, cold, and is exposed to a high dose of radiation. These characteristics present a huge challenge for any life to survive there. To avoid the effects of solar radiation on potential living matter on Mars, the depth of life-searching is at least 7.5 meters below the surface, according to the simulation (Parro *et al.*, 2011). However, evaporating regions, especially their subsurface, still have a certain potential for searching of life traces on Mars. The translucent salt crystals can filter part of the solar radiation, then less UV damage will be subjected to the possible organic life; most evaporites are deliquescent and can create a moist micro-environment. Life (if any exists) will make use of the trace amount of water to survive (Wierzchos *et al.*, 2006). What's more, briny water has a lower freezing point (as low as −60°C) that could provide available water activity for life's surviving under low-temperature conditions. Considering it, modern Mars under some certain situations could become habitable (Schuerger *et al.*, 2017).

The extreme situation of the very dry and saline playas in the Qaidam Basin is harsh for the surviving of microorganisms, while the watery lakes have more microbial diversity. Studies on microorganisms in watery lake have shown that halophiles, especially some firmicutes, are tolerant to low TOC and high salinity. Microorganisms with habitats in the sediments of modern saline lakes originate from lakes' water. This dynamic change is worthy of long-term monitoring to provide references for the search for life in ancient lacustrine sediments. The study of Dalangtan Playa's microbial diversity reveals the advantages of spore-producing microorganisms surviving in the environment characterized by super-salinity, desiccation, and elevated radiation. All this proves the superiority of *Bacillus* in this environment.

In fact, Most *Bacillus* can form endospores under harsh conditions, which are strongly resistant to UV radiation and desiccation. It has been reported that under extreme desiccation and high vacuum condition (10^{-6} Pa at 77 K for 24 h), 55% of *Bacillus subtilis* spores are still able to germinate. Dormant spores of various *Bacillus* species, including *B. subtilis,* are 5–50 times more resistant to UV radiation than the corresponding growing cells (Setlow, 2001). Due to their properties, members of *Bacillus* genus are considered as targeted microorganisms both for astrobiology and for planetary contamination preventive strategies. It has been reported that 66% of 358 strains of cultivable microorganisms collected from the surface of Mars Science Laboratory (NASA) were members of the *Bacillus* genus (Smith *et al.*, 2017). Several studies on *Bacillus* exposed to simulated Mars environmental conditions have been conducted. Under simulated Martian atmospheric pressure and composition condition for 19 days, spores of *B. subtilis* had impairment for germination but similar survival rates than those under Earth conditions (Nicholson and Schuerger, 2005). When exposed to simulated Mars Solar Radiation (254-nm UV light), *B. subtilis* had retained the possibility to initiate germination-associated metabolic processes and to produce biological signature molecules (Tauscher *et al.*, 2006). Moreover, under multiple simulated factors (high UV, low pressure, low temperature, and CO_2 atmosphere) for 24 h, 16.7% of *B. subtilis* still exhibited positive growth (Kerney and Schuerger, 2011). In addition, *B. pumilus* spores were able to grow in the dark after 18-month's exposure to both space and Mars simulated conditions (Vaishampayan *et al*., 2012). Due to the strong resistance to extreme conditions, spore-forming bacteria can withstand space environmental conditions. Liquid water is required for spore germination and subsequent growth. If present on extraterrestrial planets (e.g. from deliquescent evaporites or brines), spores of *Bacillus* can germinate despite the high salt concentrations (Nagler *et al*., 2016).

All these studies strongly support the interest of *Bacillus* in relation to life exploration on Mars particularly in evaporites which formed under saline conditions. According to the biomarkers found in the Qaidam Basin, the arid Martian environment is also conducive to the preservation of biomolecules. Further results will need to be revealed by further Mars exploration, particularly the sample return program and the deep Mars life exploration program. In the meanwhile, as studies on the deep deposits of the Qaidam Basin are still scarce, they should be continued to uncover the ancient biomarkers preserved in deeper depths. It could provide suggestions for finding traces of life in deep subsurface of Mars.

9.7 Summary

Mars is a cold, arid, seemingly lifeless planet. However, it has been proved to have a watery history. Even today, the planet's atmosphere contains as high as 100 pr-mm of water vapor. There may be temporary surface water on the surface (albeit for short durations and in little amounts). There is a lot of water ice underneath, and liquids thawing under pressure. These pieces of evidence suggest Mars may have existing or past traces of life considering the Earth's situation. The search for signs of life on Mars includes *in situ* investigation and study of meteorites. Evidence of S-bearing organics has been found in ancient Martian mudstone (~3.5 Ga). However, there are not enough clues to confirm whether it has a biological or abiotic origin. Methane in the Martian atmosphere has been one of the hot spots for life-searching because it can be produced on Earth by microbes, such as methanogenic archaea. Mars meteorites contain amino acids and complex carbonaceous organics. As meteorites are prone to contamination by Earth's materials, it makes it difficult to interpret the possible life material. In this case, taking the extreme conditions of the Earth as analogs of Mars, looking for traces of life in these environments, and testing the possibility of detecting life in the Martian environment with the simulated conditions of Mars is the current approach. The Qaidam Basin lake system (including watery saline lakes and arid salt flats) provides an excellent site for the study of extremophiles. Traces of life in this environment have been isolated or detected in both water bodies and arid sediments. The very dry salt flats in the Qaidam Basin are likely to represent the limits of life on Earth because there is not yet enough DNA that has been extracted successfully for metagenomic analysis, which is uncommon among Mars analogs. Thus, isolated life in the Qaidam Basin has a strong tolerance to extreme drought, high salinity, and high radiation. They are good candidates for studying whether life could exist in the Martian environment. Life signatures in the ancient salt rocks may provide indications for life in the Martian deep subsurface. As a good analog for Mars paleolakes, the study of life signatures in the Qaidam Basin will provide a steady stream of methodological support for the exploration of life on Mars.

Acknowledgments

The authors gratefully acknowledge support of the National Natural Science Foundation of China (No. 41830214) and the Science and Technology Development Fund project (121/2017/A3). We also thank Prof. Jiang Hongcheng and Dr. Liu Wen who have applied materials for this study.

References

Achille, G. D. and Hynek, B. M. (2010). 8-Deltas and valley networks on Mars: Implications for a global hydrosphere, In: Cabrol, N. A., Grin, E. A. (Eds.), *Lakes on Mars*, Elsevier, Amsterdam, pp. 223–248.

Amils, R., González-Toril, E., Fernández-Remolar, D., Gómez, F., Aguilera, Á., Rodríguez, N., Malki, M., García-Moyano, A., Fairén, A. G. and de la Fuente, V. (2007). Extreme environments as Mars terrestrial analogs: The Rio Tinto case. *Planet. Space Sci.*, **55**, 370–381.

Aoudjehane, H. C., Avice, G., Barrat, J.-A., Boudouma, O., Chen, G., Duke, M., Franchi, I., Gattacceca, J., Grady, M. and Greenwood, R. (2012). Tissint Martian meteorite: A fresh look at the interior, surface, and atmosphere of Mars. *Science*, **338**, 785–788.

Bibring, J.-P., Langevin, Y., Gendrin, A., Gondet, B., Poulet, F., Berthé, M., Soufflot, A., Arvidson, R., Mangold, N. and Mustard, J. (2005). Mars surface diversity as revealed by the OMEGA/Mars Express observations. *Science*, **307**, 1576–1581.

Bibring, J.-P., Langevin, Y., Mustard, J. F., Poulet, F., Arvidson, R., Gendrin, A., Gondet, B., Mangold, N., Pinet, P. and Forget, F. (2006). Global mineralogical and aqueous Mars history derived from OMEGA/Mars Express data. *Science*, **312**, 400–404.

Bibring, J.-P., Langevin, Y., Poulet, F., Gendrin, A., Gondet, B., Berthé, M., Soufflot, A., Drossart, P., Combes, M. and Bellucci, G. (2004). Perennial water ice identified in the south polar cap of Mars. *Nature*, **428**, 627–630.

Biemann, K. and Bada, J. L. (2011). Comment on "Reanalysis of the Viking results suggests perchlorate and organics at midlatitudes on Mars" by Rafael Navarro-González *et al.*, *J. Geophys. Res. Planets*, **116**.

Blaut, M. (1994). Metabolism of methanogens. *Antonie Van Leeuwenhoek*, **66**, 187–208.

Brennecka, G., Borg, L. and Wadhwa, M. (2014). Insights into the Martian mantle: The age and isotopics of the meteorite fall Tissint. *Meteori. & Planet. Sci.*, **49**, 412–418.

Brock, T.D. and Freeze, H. (1969). Thermus aquaticus gen. n. and sp. n., a non-sporulating extreme thermophile. *J. Bacterio.*, **98**, 289–297.

Brown, A. J., Hook, S. J., Baldridge, A. M., Crowley, J. K., Bridges, N. T., Thomson, B. J., Marion, G. M., de Souza Filho, C. R. and Bishop, J. L. (2010). Hydrothermal formation of clay-carbonate alteration assemblages in the Nili Fossae region of Mars. *Earth and Planet. Sci. Lett.*, **297**, 174–182.

Byrne, S., Dundas, C. M., Kennedy, M. R., Mellon, M. T., McEwen, A. S., Cull, S. C., Daubar, I. J., Shean, D. E., Seelos, K. D. and Murchie, S. L. (2009). Distribution of mid-latitude ground ice on Mars from new impact craters. *Science*, **325**, 1674–1676.

Cabrol, N. A., Grin, E. A. (1999). Distribution, classification, and ages of Martian impact crater lakes. *Icarus*, **142**, 160–172.

Cabrol, N. A. and Grin, E. A. (2010). *Lakes on Mars*. Elsevier.

Carr, M. H. (1986). Mars: A water-rich planet? *Icarus*, **68**, 187–216.

Cheng, Z., Fernández-Remolar, D., Izawa, M., Applin, D., Chong Díaz, M., Fernandez-Sampedro, M., García-Villadangos, M., Huang, T., Xiao, L. and Parro, V. (2016). Oxalate formation under the hyperarid conditions of the Atacama desert as a mineral marker to provide clues to the source of organic carbon on Mars. *J. Geophys. Res. Biogeosciences*, **121**, 1593–1604.

Cheng, Z., Xiao, L., Wang, H., Yang, H., Li, J., Huang, T., Xu, Y. and Ma, N. (2017). Bacterial and archaeal lipids recovered from subsurface evaporites of Dalangtan playa on the Tibetan plateau and their astrobiological implications. *Astrobiology*, **17**, 1112–1122.

Clancy, R., Grossman, A., Wolff, M., James, P., Rudy, D., Billawala, Y., Sandor, B., Lee, S., Muhleman, D. (1996). Water vapor saturation at low altitudes around Mars aphelion: A key to Mars climate? *Icarus*, **122**, 36–62.

Doran, P. T., Lyons, W. B. and McKnight, D. M. (2010). *Life in Antarctic Deserts and Other Cold Dry Environments: Astrobiological Analogs*. Cambridge University Press.

Dundas, C. M., Byrne, S., McEwen, A. S., Mellon, M. T., Kennedy, M. R., Daubar, I. J. and Saper, L. (2014). HiRISE observations of new impact craters exposing Martian ground ice. *J. Geophys. Res. Planets*, **119**, 109–127.

Ehlmann, B.L., Mustard, J. F., Murchie, S. L., Bibring, J.-P., Meunier, A., Fraeman, A. A., Langevin, Y. (2011). Subsurface water and clay mineral formation during the early history of Mars. *Nature*, **479**, 53.

Ehlmann, B. L., Mustard, J. F., Murchie, S. L., Poulet, F., Bishop, J. L., Brown, A. J., Calvin, W. M., Clark, R. N., Des Marais, D. J. and Milliken, R. E. (2008). Orbital identification of carbonate-bearing rocks on Mars. *Science*, **322**, 1828–1832.

Eigenbrode, J. L., Summons, R. E., Steele, A., Freissinet, C., Millan, M., Navarro-González, R., Sutter, B., McAdam, A. C., Franz, H. B. and Glavin, D. P. (2018). Organic matter preserved in 3-billion-year-old mudstones at Gale crater, Mars. *Science*, **360**, 1096–1101.

Etiope, G., Fridriksson, T., Italiano, F., Winiwarter, W. and Theloke, J. (2007). Natural emissions of methane from geothermal and volcanic sources in Europe. *J. Volcanol. Geoth. Res.*, **165**, 76–86.

Fassett, C. I. and Head III, J. W. (2008). Valley network-fed, open-basin lakes on Mars: Distribution and implications for Noachian surface and subsurface hydrology. *Icarus*, **198**, 37–56.

Feldman, W., Boynton, W. V., Tokar, R., Prettyman, T., Gasnault, O., Squyres, S., Elphic, R., Lawrence, D., Lawson, S. and Maurice, S. (2002). Global distribution of neutrons from Mars: Results from Mars Odyssey. *Science*, **297**, 75–78.

Fernández-Remolar, D. C., Morris, R. V., Gruener, J. E., Amils, R. and Knoll, A. H. (2005). The Río Tinto Basin, Spain: Mineralogy, sedimentary geobiology, and implications for interpretation of outcrop rocks at Meridiani Planum, Mars. *Earth Planet. Sci. Lett.*, **240**, 149–167.

Fischer, E., Martínez, G. M., Elliott, H. M. and Rennó, N. O. (2014). Experimental evidence for the formation of liquid saline water on Mars. *Geophys. Res. Lett.*, **41**, 4456–4462.

Formisano, V., Atreya, S., Encrenaz, T., Ignatiev, N. and Giuranna, M. (2004). Detection of methane in the atmosphere of Mars. *Science*, **306**, 1758–1761.

Gendrin, A., Mangold, N., Bibring, J.-P., Langevin, Y., Gondet, B., Poulet, F., Bonello, G., Quantin, C., Mustard, J., Arvidson, R. and LeMouélic, S. (2005). Sulfates in Martian layered terrains: the OMEGA/Mars express View. *Science*, **307**; 1587–1591.

Grant, W. (2004). Life at low water activity. *Philos. Trans. R. Soc. B: Biol. Sci.*, **359**, 1249–1267.

Grice, K., Schouten, S., Nissenbaum, A., Charrach, J. and Damsté, J. S. S. (1998). Isotopically heavy carbon in the C21 to C25 regular isoprenoids in halite-rich deposits from the Sdom Formation, Dead Sea Basin, Israel. *Org. Geochem.*, **28**, 349–359.

Grima, C., Kofman, W., Mouginot, J., Phillips, R. J., Hérique, A., Biccari, D., Seu, R. and Cutigni, M., (2009). North polar deposits of Mars: Extreme purity of the water ice. *Geophys. Res. Lett.*, **36**.

Grotzinger, J. P., Sumner, D., Kah, L., Stack, K., Gupta, S., Edgar, L., Rubin, D., Lewis, K., Schieber, J. an Mandgold, N., (2014). A habitable fluvio-lacustrine environment at Yellowknife Bay, Gale crater, Mars. *Science*, **343**, 1242777.

Guzman, M., McKay, C. P., Quinn, R. C., Szopa, C., Davila, A. F., Navarro-González, R. and Freissinet, C. (2018). Identification of Chlorobenzene in the Viking Gas Chromatograph-Mass Spectrometer Data Sets: Reanalysis of Viking Mission Data Consistent With Aromatic Organic Compounds on Mars. *J. Geophys. Res. Planets*, **123**, 1674–1683.

Hecht, M. H. (2002). Metastability of liquid water on Mars. *Icarus*, **156**, 373–386.

Holt, J. W., Safaeinili, A., Plaut, J. J., Head, J. W., Phillips, R. J., Seu, R., Kempf, S. D., Choudhary, P., Young, D. A. and Putzig, N. E. (2008). Radar sounding evidence for buried glaciers in the southern mid-latitudes of Mars. *Science*, **322**, 1235–1238.

Jakosky, B. M., Farmer, C. B. (1982). The seasonal and global behavior of water vapor in the Mars atmosphere: Complete global results of the Viking atmospheric water detector experiment. *J. Geophys. Res. Solid Earth*, **87**, 2999–3019.

Jaramillo, E. A., Royle, S. H., Claire, M. W., Kounaves, S. P. and Sephton, M. A. (2019). Indigenous organic-oxidized fluid interactions in the Tissint Mars meteorite. *Geophys. Res. Lett.*, **46**, 3090–3098.

Jiang, H., Dong, H., Yu, B., Liu, X., Li, Y., Ji, S. and Zhang, C. L. (2007). Microbial response to salinity change in Lake Chaka, a hypersaline lake on Tibetan plateau. *Environ. Microbio.*, **9**, 2603–2621.

Jiang, H., Huang, J. and Yang, J. (2018). Halotolerant and Halophilic Microbes and Their Environmental Implications in Saline and Hypersaline Lakes in Qinghai Province, China, Extremophiles in *Eurasian Ecosystems: Ecology, Diversity, and Applications*. Springer, pp. 299–316.

Kaneda, T. (1991). Iso-and anteiso-fatty acids in bacteria: biosynthesis, function, and taxonomic significance. *Microbio. Molecu. Bio. Rev.*, **55**, 288–302.

Kates, M. (1996). Structural analysis of phospholipids and glycolipids in extremely halophilic archaebacteria. *J. Microbiolog. Meth.*, **25**, 113–128.

Kerney, K. R. and Schuerger, A. C. (2011). Survival of Bacillus subtilis Endospores on Ultraviolet-Irradiated Rover Wheels and Mars Regolith under Simulated Martian Conditions. *Astrobiology*, **11**, 477–485.

Korablev, O., Vandaele, A. C., Montmessin, F., Fedorova, A. A., Trokhimovskiy, A., Forget, F., Lefèvre, F., Daerden, F., Thomas, I. R. and Trompet, L. (2019). No detection of methane on Mars from early ExoMars Trace Gas Orbiter observations. *Nature*, **568**, 517.

Kounaves, S. P., Carrier, B. L., O'Neil, G. D., Stroble, S. T. and Claire, M. W. (2014). Evidence of Martian perchlorate, chlorate, and nitrate in Mars meteorite EETA79001: Implications for oxidants and organics. *Icarus*, **229**, 206–213.

Kral, T., Goodhart, T., Howe, K. and Gavin, P. (2009). Can Methanogens Grow in a Perchlorate Environment on Mars? *Meteor. Planet. Sci. Supple.*, **72**, 5136.

Kwok, S. (2017). Abiotic synthesis of complex organics in the universe. *Nat. Astron.*, **1**, 642–643.

Lanza, N., Meyer, G., Okubo, C., Newsom, H. E. and Wiens, R. (2010). Evidence for debris flow gully formation initiated by shallow subsurface water on Mars. *Icarus*, **205**, 103–112.

Leshin, L., Mahaffy, P., Webster, C., Cabane, M., Coll, P., Conrad, P., Archer, P., Atreya, S., Brunner, A. and Buch, A. (2013). Volatile, isotope, and organic analysis of Martian fines with the Mars Curiosity rover. *Science*, **341**, 1238937.

Liu, X., Hou, W., Dong, H., Wang, S., Jiang, H., Wu, G., Yang, J. and Li, G. (2016). Distribution and diversity of cyanobacteria and eukaryotic algae in Qinghai–Tibetan lakes. *Geomicrobio. J.*, **33**, 860–869.

Malin, M. C., Edgett, K. S., Posiolova, L. V., McColley, S. M. and Dobrea, E. Z. N. (2006). Present-day impact cratering rate and contemporary gully activity on Mars. *Science*, **314**, 1573–1577.

Marlow, J.J., Martins, Z. and Sephton, M. A. (2011). Organic host analogues and the search for life on Mars. *Intern. J. Astrobio.*, **10**, 31–44.

Martín-Torres, F.J., Zorzano, M.-P., Valentín-Serrano, P., Harri, A.-M., Genzer, M., Kemppinen, O., Rivera-Valentin, E.G., Jun, I., Wray, J. and Madsen, M. B. (2015). Transient liquid water and water activity at Gale crater on Mars. *Nat. Geosci.*, **8**, 357–361.

Mellon, M. T., Feldman, W. C. and Prettyman, T. H. (2004). The presence and stability of ground ice in the southern hemisphere of Mars. *Icarus*, **169**, 324–340.

Mellon, M. T. and Jakosky, B. M. (1995). The distribution and behavior of Martian ground ice during past and present epochs. *J. Geophys. Res. Planets*, **100**, 11781–11799.

Ming, D., Archer, P., Glavin, D., Eigenbrode, J., Franz, H., Sutter, B., Brunner, A., Stern, J., Freissinet, C. and McAdam, A. (2014). Volatile and organic compositions of sedimentary rocks in Yellowknife Bay, Gale Crater, Mars. *Science*, **343**, 1245267.

Mumma, M. J., Villanueva, G. L., Novak, R. E., Hewagama, T., Bonev, B. P., DiSanti, M. A., Mandell, A. M. and Smith, M. D. (2009). Strong release of methane on Mars in northern summer 2003. *Science*, **323**, 1041–1045.

Nagler, K., Julius, C. and Moeller, R. (2016). Germination of spores of astrobiologically relevant Bacillus species in high-salinity environments. *Astrobiology*, **16**, 500–512.

Nagy, B., Meinschein, W. G. and Hennessy, D. J. (1961). Mass spectroscopic analysis of the Orgueil meteorite: Evidence for biogenic hydrocarbons. *Ann. N. Y. Acad. Sci.*, **93**, 27–35.

Navarro-González, R., Rainey, F. A., Molina, P., Bagaley, D. R., Hollen, B. J., de la Rosa, J., Small, A. M., Quinn, R. C., Grunthaner, F. J., Cáceres, L., Gomez-Silva, B. and McKay, C. P. (2003). Mars-Like Soils in the Atacama Desert, Chile, and the Dry Limit of Microbial Life. *Science*, **302**, 1018–1021.

Navarro-González, R. and McKay, C. P. (2011). Reply to comment by Biemann and Bada on "Reanalysis of the Viking results suggests perchlorate and organics at midlatitudes on Mars". *J. Geophys. Res. Planets*, **116**.

Navarro-González, R., Vargas, E., de La Rosa, J., Raga, A. C. and McKay, C. P. (2010). Reanalysis of the Viking results suggests perchlorate and organics at midlatitudes on Mars. *J. Geophys. Res. Planets*, **115**.

Nicholson, W. L. and Schuerger, A. C. (2005). Bacillus subtilis Spore Survival and Expression of Germination-Induced Bioluminescence After Prolonged Incubation Under Simulated Mars Atmospheric Pressure and Composition: Implications for Planetary Protection and Lithopanspermia. *Astrobiology*, **5**.

Orosei, R., Lauro, S. E., Pettinelli, E., Cicchetti, A., Coradini, M., Cosciotti, B., Di Paolo, F., Flamini, E., Mattei, E. and Pajola, M. (2018). Radar evidence of subglacial liquid water on Mars. *Science*, **361**, 490–493.

Osterloo, M., Hamilton, V., Bandfield, J., Glotch, T., Baldridge, A., Christensen, P., Tornabene, L., Anderson, F. (2008). Chloride-bearing materials in the southern highlands of Mars. *Science*, **319**, 1651–1654.

Osterloo, M. M., Anderson, F. S., Hamilton, V. E., Hynek, B. M. (2010). Geologic context of proposed chloride-bearing materials on Mars. *J. Geophys. Res.*, **115**.

Oze, C., Jones, L. C., Goldsmith, J. I. and Rosenbauer, R. J. (2012). Differentiating biotic from abiotic methane genesis in hydrothermally active planetary surfaces. *Proceed. Nat. Acad. Sci.*, **109**, 9750–9754.

Parro, V., de Diego-Castilla, G., Moreno-Paz, M., Blanco, Y., Cruz-Gil, P., Rodríguez-Manfredi, J. A., Fernández-Remolar, D., Gómez, F., Gómez, M. J. and Rivas, L. A. (2011). A microbial oasis in the hypersaline Atacama subsurface discovered by a life detector chip: Implications for the search for life on Mars. *Astrobiology*, **11**, 969–996.

Peters, K. E., Walters, C. C. and Moldowan, J. M. (2005). *The Biomarker Guide*. Cambridge University Press.

Pizzarello, S. and Shock, E. (2010). The organic composition of carbonaceous meteorites: the evolutionary story ahead of biochemistry. *Cold Spring Harb. Perspect. Med.*, **2**, a002105.

Poulet, F., Bibring, J., Mustard, J., Gendrin, A., Mangold, N., Langevin, Y., Arvidson, R., Gondet, B., Gomez, C. and Berthé, M. (2005). Phyllosilicates on Mars and implications for early Martian climate. *Nature*, **438**, 623–627.

Preston, L., Shuster, J., Fernández-Remolar, D., Banerjee, N., Osinski, G. and Southam, G. (2011). The preservation and degradation of filamentous bacteria and biomolecules within iron oxide deposits at Rio Tinto, Spain. *Geobiology*, **9**, 233–249.

Schmitt-Kopplin, P., Gabelica, Z., Gougeon, R. D., Fekete, A., Kanawati, B., Harir, M., Gebefuegi, I., Eckel, G. and Hertkorn, N. (2010). High molecular diversity of extraterrestrial organic matter in Murchison meteorite revealed 40 years after its fall. *Proceed. Nat. Aca. Sci.*, **107**, 2763–2768.

Schuerger, A. C., Ming, D.W. and Golden, D. C. (2017). Biotoxicity of Mars soils: 2. Survival of Bacillus subtilis and Enterococcus faecalis in aqueous extracts derived from six Mars analog soils. *Icarus*, **290**, 215–223.

Sephton, M., Wright, I., Gilmour, I., De Leeuw, J., Grady, M. and Pillinger, C. (2002). High molecular weight organic matter in Martian meteorites. *Planet. Space Sci.*, **50**, 711–716.

Setlow, P. (2001). Resistance of spores of Bacillus species to ultraviolet light. *Environ. Mol. Mutagen.*, **38**, 97–104.

Smith, M. D. (2002). The annual cycle of water vapor on Mars as observed by the Thermal Emission Spectrometer. *J. Geophys. Res. Planets*, **107**, 25-21–25-19.

Smith, P., Tamppari, L., Arvidson, R., Bass, D., Blaney, D., Boynton, W., Carswell, A., Catling, D., Clark, B. and Duck, T. (2009). H_2O at the Phoenix landing site. *Science*, **325**, 58–61.

Smith, S. A., Benardini III, J. N., Anderl, D., Ford, M., Wear, E., Schrader, M., Schubert, W., DeVeaux, L., Paszczynski, A. and Childers, S. E. (2017). Identification and Characterization of Early Mission Phase Microorganisms Residing on the Mars Science Laboratory and Assessment of Their Potential to Survive Mars-like Conditions. *Astrobiology*, **17**, 253–265.

Sobron, P., Lefebvre, C., Leveille, R., Koujelev, A., Haltigin, T., Du, H. W., Wang, A., Cabrol, N., Zacny, K., Craft, J. and Team, L. (2013). Geochemical profile of a layered outcrop in the Atacama analogue using laser-induced breakdown spectroscopy: Implications for Curiosity investigations in Gale. *Geophys. Res. Lett.*, **40**, 1965–1970.

Sobron, P., Wang, A., Mayer, D. P., Bentz, J., Kong, F. and Zheng, M. (2018). Dalangtan Saline Playa in a Hyperarid Region of Tibet Plateau: III. Correlated multiscale surface mineralogy and geochemistry survey. *Astrobiology*, **18**, 1277–1304.

Steele, A., McCubbin, F., Fries, M., Kater, L., Boctor, N., Fogel, M., Conrad, P., Glamoclija, M., Spencer, M. and Morrow, A. (2012). A reduced organic carbon component in Martian basalts. *Science*, **337**, 212–215.

Stevenson, A., Hamill, P. G., O'kane, C. J., Kminek, G., Rummel, J. D., Voytek, M. A., Dijksterhuis, J. and Hallsworth, J. E. (2017). Aspergillus penicillioides differentiation and cell division at 0.585 water activity. *Environ. Microbio.*, **19**, 687–697.

Stillman, D. E., Michaels, T. I., Grimm, R. E. and Hanley, J. (2016). Observations and modeling of northern mid-latitude recurring slope lineae (RSL) suggest recharge by a present-day Martian briny aquifer. *Icarus*, **265**, 125–138.

Summons, R. E., Amend, J. P., Bish, D., Buick, R., Cody, G. D., Des Marais, D. J., Dromart, G., Eigenbrode, J. L., Knoll, A. H. and Sumner, D. Y. (2011). Preservation of Martian organic and environmental records: Final report of the Mars Biosignature Working Group. Mary Ann Liebert, Inc. 140 Huguenot Street, 3rd Floor New Rochelle, NY 10801 USA.

Tauscher, C., Schuerger, A. C., Nicholson, W. L. (2006). Survival and germinability of Bacillus subtilis spores exposed to simulated Mars solar radiation: implications for life detection and planetary protection. *Astrobiology*, **6**, 592–605.

Vaishampayan, P. A., Rabbow, E., Horneck, G. and Venkateswaran, K. J. (2012). Survival of Bacillus pumilus spores for a prolonged period of time in real space conditions. *Astrobiology*, **12**, 487–497.

Webster, C. R., Mahaffy, P. R., Atreya, S. K., Flesch, G. J., Mischna, M. A., Meslin, P.-Y., Farley, K. A., Conrad, P. G., Christensen, L. E. and Pavlov, A. A. (2015). Mars methane detection and variability at Gale crater. *Science*, **347**, 415–417.

Webster, C. R., Mahaffy, P. R., Atreya, S. K., Moores, J. E., Flesch, G. J., Malespin, C., McKay, C. P., Martinez, G., Smith, C. L. and Martin-Torres, J. (2018). Background levels of methane in Mars' atmosphere show strong seasonal variations. *Science*, **360**, 1093–1096.

Weiss, D. K. and Head, J. W. (2014). Ejecta mobility of layered ejecta craters on Mars: Assessing the influence of snow and ice deposits. *Icarus*, **233**, 131–146.

White, D., Davis, W., Nickels, J., King, J. and Bobbie, R. (1979). Determination of the sedimentary microbial biomass by extractible lipid phosphate. *Oecologia*, **40**, 51–62.

Wierzchos, J., Ascaso, C. and McKay, C. P. (2006). Endolithic cyanobacteria in halite rocks from the hyperarid core of the Atacama Desert. *Astrobiology*, **6**, 415–422.

Wilson, J. T., Eke, V. R., Massey, R. J., Elphic, R. C., Feldman, W. C., Maurice, S., Teodoro, L. F. (2018). Equatorial locations of water on Mars: Improved resolution maps based on Mars Odyssey Neutron Spectrometer data. *Icarus*, **299**, 148–160.

Xiao, L., Wang, J., Dang, Y., Cheng, Z., Huang, T., Zhao, J., Xu, Y., Huang, J., Xiao, Z. and Komatsu, G. (2017). A new terrestrial analogue site for Mars research: The Qaidam Basin, Tibetan Plateau (NW China). *Earth-Science Rev.*, **164**, 84–101.

Zelles, L. (1999). Fatty acid patterns of phospholipids and lipopolysaccharides in the characterisation of microbial communities in soil: A review. *Biol. Fertil. Soils*, **29**, 111–129.

https://doi.org/10.1142/9789811224027_bmatter

Index

A
abiotic igneous processes, 298
aeolian erosion, 257, 267
Aeolian landform, 53, 157
Aeolis Dorsa, 158–159
Aeolis Planum, 161
Aeolis-Zephyria Planum (AZP), 163
Alar River, 130
albedo, 3
alcove, 278, 280–281
ALH84001 Martian meteorite, 21
alluvial-diluvial fans, 87, 95
alluvial fans, 6
Alpha Particle X-ray Spectrometry (APXS), 9
Altyn Mountain range, 84
Altyn strike-slip fault system, 78
Altyn-Tagh Mountains, 130
Amazonian paleolakes, 123
Amonick Mountain, 86
amplification of DNA, 308
analogies, 11
analog research, 14
analog study, 1, 11, 24
ancient Martian mudstone, 318
annual evaporation, 123
annual precipitation, 105, 123
anticline, 79
apron, 278, 280–281
aqueous minerals, 5, 145
aqueous processes, 206, 238
Arabia Terra, 259
archaeal, 312, 316
archaeol, 313
arid, 305
arid climate, 90
arid desert, 14
arid desert climate, 137
aromatic hydrocarbons, 298
aromatics, 298
Aspergillus, 306
astrobiological interest, 301
atmospheric pressure, 157
automatic identification, 258
average maximum temperature, 105
average minimum temperature, 105

B
Bacillus, 311
bacterial, 312
barchan dune, 102, 184–186, 188–189, 190–191
basaltic columns, 202
basaltic composition, 3
basin lake system, 213

bidirectional, 164, 166, 173, 178, 185, 187
binary structure, 266
biological diversity, 309
biological origin, 300
biomarkers, 302
biomass, 308
blunt heads, 159, 162, 170, 173
borehole, 140, 142
brackish lakes, 112
branching angle, 253, 261
brine evolution, 145
brine lake evolution, 137
brine lakes, 124
brine pool, 128
brine upward migration, 223
briny water, 296

C

canyons, 249
canyon systems, 294
cap layers, 161, 167–168, 175
carbonates, 4, 5
carbonate type, 134
carbon dioxide, 282–284
CARD-FISH (Catalyzed Reported Deposition-Fluorescence In Situ Hybridization), 306
Cenozoic sedimentation, 76
Cenozoic strata, 73
center coordinate, 126
central volcanoes, 43
Chahansilatu playa, 142, 144
Chaka, 135
channel, 6, 278, 280–281
chassignites, 20–21
chemical properties, 143
chemical sediment, 125
chemoautotrophic microorganisms, 315
China Space Agency, 7
chloride, 230, 233, 236, 295
chloride-bearing deposits, 207, 210, 226, 237
chloride-bearing sediments, 211
chloride-bearing terrain, 206, 208–209, 227, 230, 232, 235
chloride-containing salty sediments, 232
chloride deposits, 234
chloride type, 134
chlorinated hydrocarbons, 299
chlorobenzene, 299
classification of lakes, 114
clayey plain, 87
clay-rich deposits, 210
climate change, 63
climate evolution, 144, 147
climate modeling, 261
climate models, 23
climatic change, 260
climatic oscillation, 145
clinopyroxene, 3
closed-basin lakes, 114–115
closed basins, 207
closed depression, 51
columnar jointing, 202, 205
community composition, 304
compound polygons, 216
compressional structures, 38, 228
computer modeling, 23
computer simulations, 22
concentration, 313
cones, 45
conglomerate, 75
cracking, 199
cross beddings, 266
cross section, 267
cross-sectional shapes, 252
crust continues, 220
crystal flowers, 128
culturable bacteria, 304
curiosity (MSL), 9
cyanobacteria, 303

D
Dabuxun Formation, 78
Dalangtan Playa, 214, 225, 305
Dalangtan salt polygons, 232
Daqaidam and Xiaoqaidam Lakes, 131
dark dune fields, 183
dark streaks, 163
debris flow, 59
deep-water environment, 139
degree of mineralization (g/l), 126
dehydration, 226
delta, 181
delta front facies sediments, 127
deltaic and perennial lacustrine environments, 301
dendritic drainage, 253–254, 256
desiccation, 233
desiccation cracks, 201, 235
desiccation polygons, 205
Devonian–Cenozoic sediments, 73
discharge rates, 268
DNA extraction techniques, 307
dome dunes, 187
doming effect, 220
drainage area, 262, 264
drainage density, 258
drainage pattern, 268
drainage system, 48, 261
dry salt lakes, 82, 88
dry valleys, 264, 266
dunes, 54, 183
dust devils, 53
dust devil tracks, 53
dust storms, 52
dynamic changes of modern lake environments, 145
dynamic convection, 219

E
early Noachian, 61
Earth analogs, 15
Earth-like planets, 11
East and West Taijinaier Lakes, 124
East Asian Monsoons, 105
E'boliang Yardang Geopark, 169
Elashanstrike-slip fault system, 78
electromagnetic spectra, 2
elevated UV conditions, 305
Elysium, 41
endogenic, 61
endogenic geomorphology, 36
endogenous processes, 238
endospores, 317
epifluorescence micrographs, 309
Erhai, 135
erosional mechanisms, 97
Eurasian Hinterland Basin, 105
evaporate minerals, 236
evaporation, 88, 124, 210, 223
evaporation rate, 132
evaporative pumping mechanism, 218
evaporite, 142–143
evaporite deposits, 315
evaporite minerals, 90
exogenic, 61
exogenic forces, 62
explosive eruption, 44
exposed basin floor deposits, 119
extensional faults, 37
extensional rifting, 37
extensional structures, 37
extraterrestrial analogs, 12
extreme environments, 301
extremophile, 302

F
fan deposits, 117
fatty acid compounds, 316
fatty acids, 312
faults, 39, 78
fault system, 78
Firmicutes, 306
fixed dunes, 102

floor-fractured craters, 237
flow lobes, 44
fold, 79
formation mechanisms, 259
fossilized microbial life, 316
fractured polygon rims, 221
fractures, 41
fresh lake, 112
frosted dunes, 187–188
fungi, 303

G
Gahai, 135
Gale Crater, 118, 120, 158, 180
gamma-ray spectra, 2
Gasikule Lake, 128–129
GDGT-0, 313
genesis and sedimentary model, 139
geochemical parameters, 135
geological history, 59–60
geological phenomena, 12
geological processes, 111
geologic and climatic history, 145
geomorphical evidence, 294
geomorphological features, 12
geomorphology, 7
giant polygons, 205
glaciers, 84
global orbiting probes, 24
Gobi, 88
Golmud River, 264–265
gorges, 249
graben, 46
gram-positive bacteria, 316
granular disintegration, 98
gravity field, 7
groundwater, 276, 282
groundwater sapping, 256, 261
groundwater upwellings, 237
gullies, 6, 49–50, 59, 275–278, 280–285, 287–289, 294
Gusev Crater, 120
gypsum, 142, 165, 167–170, 175

H
habitable environment, 15–16, 24
habitable planet, 293
halite, 141–142, 218
halite crystallization, 219
halophiles, 314, 317
Hercynian movement, 81
Hercynian orogeny, 81
Heterotrophic microbes, 308
high-centered polygon, 201
high plateau playa, 305
high-relief centers, 209
High Resolution Imaging Science Experiment (HiRISE), 5
High Resolution Stereo Camera (HRSC), 6
high-throughput sequencing, 307
high ultraviolet radiation, 311
Himalayas, 82
HiRISE data, 208
honeycomb-like pattern, 226
honeycomb patterned polygons, 209
honeycomb structure, 217
hydrated minerals, 120
hydration, 98
hydrocarbons, 297
hydrochemical characteristic coefficient values, 138
hydrochemical characteristics, 132
hydrochemical properties, 133
hydrochemical type, 126
hydrological characteristics, 252
hydrological processes, 213
hypersaline, 305

I
Ice cap, 56
Ice-sublimation, 56
ice wedges, 205

impact basins, 210
impact craters, 57–58, 62, 113
impact hardening, 182
impairment for germination, 317
India and Eurasia plates, 73
inflow rivers, 87
inflow system, 87
infrared radiation, 2
inhomogeneous regional distribution, 262
inland environment, 139
inland rivers, 262, 269
inlet valley, 115–116, 123
InSight, 9
in situ exploration, 8–9, 24, 120–121
intermountain basins, 86, 89
inverted valleys, 257, 267
investigation of paleolakes, 113

J

Jezero crater, 10–11

K

Keluke and Tuosu Lakes, 131
Kumkuli–Qimantag–Burhanbuda–Dulan belt, 75
Kunlun Mountain, 125
Kunlun Mountain range, 82
Kurnakov–Valyashko classification, 133

L

labyrinthiform, 100
lacustrine, 125, 127
lacustrine deltas, 116
lacustrine deposits, 82, 112
lacustrine environments, 234
lacustrine mudstone, 299
lacustrine/playa deposits, 203
lacustrine sediments, 73, 164, 168, 170, 175, 178
lacustrine silt saline plain, 87
lake basins, 111
lake chains, 114
lake's sedimentary sequences, 306
landing sites, 8
landscapes on Earth, 261
large-sized salt polygons, 230
layered deposits, 117–119
layered structures, 162
LIBS, 9
linear dunes, 185–188, 190–191
lipids, 311
liquid water, 62, 293
lobate scarps, 40
longitudinal dunes, 101
longitudinal valleys, 254
long ridge yardangs, 158
long-term drought, 311
long-term drying of the Asian inland, 145
low-centered polygon, 201
Lulehe Formation, 76
Lunae Planitia, 42
Lvliangshan Mountain, 86

M

macromolecular carbon, 298
magnesium sulfate type, 133
magnetic field, 7
Mangala Vallis, 46, 47
Maoniu Mountain, 86
Marikh Vallis, 250
Mariner 9, 13
Mars analog site, 213
Mars exploration, 24
Mars Express, 4, 6
Mars Global Survey (MGS), 3, 6–7
marshes, 87
Mars Odyssey, 2, 4
Mars Pathfinder, 9
Mars simulation chamber, 23

Mars's past and present habitability, 301
Mars's wet past, 295
Martian atmospheric water vapor, 295
Martian craters, 6
Martian meteorites, 1, 15, 22–23, 297
Martian paleolakes, 112–113, 122
Martian polygons, 203
Martian salt polygons, 224, 233
Martian southern highland, 93
Martian valleys, 250
mass movements, 59
mass wasting, 57–58, 60
Mawrth Vallis, 54
medium-sized polygons, 228
Medusae Fossae Formation, 54
Meridiani Planum, 120
mesa yardangs, 158
Mesozoic Yanshanian movement, 83
metagenomic analysis, 319
meteorological stations, 105
methane on Mars, 300
methanogenic archaea, 314
methanogenic archaebacteria metabolism, 300
methanogens, 300
MFF, 182
mineral assemblage, 144
mineral composition, 119
mineralization, 133, 136
mineralogical diversity, 235
mineralogical precipitations, 211
mineral resources, 137
mineral types of salt playas, 141
Miocene, 127
mirabilite, 141–142
mobile dunes, 102
model ages, 122
moisture content, 310
MOLA, 36
Mössbauer spectrometry, 9
mud cracks, 201
mud volcanoes, 44–45
multidirectional wind dunes, 101

N

Nakhlites, 20–21
Nalenggele River, 262–264
neotectonic movement, 124, 141
Nili Fossa, 41
northern lowlands, 35
north pole, 35

O

open-basin lakes, 114–115
opportunity, 4
organic compounds, 297
orthogonal polygons, 90
outflow channels, 45–47, 250, 255

P

paleoclimatic characteristics, 143
paleocrater lake basins, 51
paleolake, 127, 234
paleolakes on Earth, 147
paleolakes on Mars, 147
paleo-yardangs, 158
parabolic dunes, 102, 190–191
perchlorate, 297, 299
perennial rivers, 88
periglacial regions, 200
periodic evaporation, 93, 217
permafrost, 83
permafrost region, 56
pH, 126
pH values, 134, 136
phyllosilicates, 5
physical simulation, 22
plagioclase feldspar, 3
planetary mapping, 13
planimetric patterns, 253–254, 269
playas, 62, 89, 180, 203
playa sediment cracks, 233

playa sediments, 206
Pleistocene, 165, 168, 170
Pliocene, 127
polar cap, 4
polar deposits, 55–56
polar layered deposits, 296
polygonal cracks, 206
polygonal features, 199, 207
polygonal fold systems, 222
polygonal fracture patterns, 217
polygonal fracturing, 221
polygonal landforms, 211
polygonal networks, 90, 205
polygonal patterned ground, 199
polygonal ridge networks, 236
polygonal ridges, 91, 229
polygonal structures, 238
polygonal surface structures, 90
polygonal terrain, 91, 199, 214, 226, 238
polygon-concentrated patches, 230
populus, 103
Populus diversifolia, 104
porous medium convection, 219
post-lacustrine modification events, 121
post-Noachian period, 315
precambrian metamorphic rocks, 73
precipitation of chlorides, 237
preservation of organic compounds, 302
pressure ridge-like features, 221
prevailing wind, 105
pyramid dunes, 102

Q

Qaidam Basin, 15, 71, 212, 214, 224, 234, 236, 238, 261
Qaidam salt polygons, 217
Qarhan Lake, 128
Qigequan Formation, 78
Qilian Mountain range, 84
Qilianshan-northern marginal fault system, 78
Qimantag Mountains, 129
Qinghai, 135
Qinghai-Tibet Plateau, 81, 111, 213
Quaternary sediments, 80

R

radial drainage, 255
raised ridge polygons, 231
raised rim polygons, 230
raised-rim salt polygons, 210, 236
random orthogonal polygons, 215
recurring seasonal slope, 294
recurring slope lineae (RSL), 50, 59, 275
remote sensing, 1–2
recurring slope lineae (RSL), 6
residual ice sheets, 296
rifts, 39
rift valleys, 249
rock record, 294
RSL (recurring slope lineae), 49, 276
Russell impact crater, 50

S

Saishiteng Mountain, 86
saline-alkali lands, 125
saline pan cycle, 217
salinity, 88, 132–133, 314
salinity gradient, 304
salt cracked polygons, 209
salt crust, 214, 220
salt crystal growth, 207
salt crystallization, 98, 218–219
salt depositions, 139
salt deposits, 73, 90, 140
salt-encrusted playas, 213
salt flats, 234, 302–303
salt-forming regions, 140, 147
salt forming sequence, 144
salt lake, 82, 88, 203

salt lake deposits, 137
salt layers, 143
salt minerals, 295
salt–mud mixture sediment, 218
salt playas, 124
salt polygons, 90, 203, 206–208, 214, 216, 219–220, 222, 224–227, 235
salt ponds, 233
salts, 97
salt solution tubes, 224
sample return, 10
sample return program, 318
sand dunes, 101
sand ripples, 163
S-bearing organics, 318
seasonality, 276
seasonal temperature variation, 200
sedimentary basin, 71
sedimentary center, 141
sedimentary landforms, 116, 118
sedimentary materials, 218
semi-fixed dunes, 102
Shanggaichaigou Formation, 77
shergottites, 20
shield volcanoes, 40, 43
Shizigou Formation, 77
silicic composition, 4
simulated Mars environmental conditions, 317
single valleys, 255
sinuous fold system, 221
sinuous ridges, 229, 257
skeleton tafoni, 99
smaller rifts/grabens, 37
small-sized polygons, 228
snow and ice meltwater, 136
sodium sulfate, 134
soil salinization, 103, 136
solubility of mirabilite, 144
solubility of salts, 137
soluble inorganics, 299
southern highlands, 35
south pole, 35
space environmental conditions, 318
spatial and temporal distribution of dry salt playas, 139
spatial distribution of salt playas, 140
spectral data, 119
Spider web patterns, 57
Spirit/Opportunity (MER), 9
spore-producing microorganisms, 317
stalactite-like salt flakes, 223
star dunes, 192
streamlined, 158–159, 162–164, 170–171, 173–175
stream order, 252
sublimation, 282–283
Sugan Lake, 130
summit fractures, 229–230
surface polygonally fractured deposits, 237
surface water activity, 294
survival rates, 317

T

Tafoni, 98
Taijinaier Lake, 262–263
tapered tail, 159, 162, 170
Tarim Basin, 83
Tarim Desert, 83
Tectonic features, 36
tectonic settlement, 141
telescope observations, 1
temporary liquids, 296
terraces, 117
terrestrial analogies, 14
terrestrial analogs, 12, 301
terrestrial paleolakes, 303
terrestrial playa conditions, 216
terrestrial playa environment, 211
Tertiary Period, 124
Tertiary strata, 73
Tharsis, 37, 40–41, 46
Tharsis rise, 39

The Amazonian Era, 61
theater-shaped head, 256
the availability of liquid water, 309
The Context Camera (CTX), 6
the diversity of culturable microorganisms, 306
The Hesperian Era, 61
the limits for life's surviving, 302
the limits of life, 319
THEMIS, 5
The Noachian Era, 61
The Qaidam Basin, 123
The recurring seasonal slope, 295
thermal contraction cracking, 201
thermal contraction polygons, 200
Thermal Emission Imaging System, 4
Thermal Emission Spectrometer, 3
three sediment zones, 139
Tianwen-1, 7, 8
Tibetan Plateau, 82
Tiemulik River, 130
Tissint, 298
total organic carbon, 304
translucent salt crystals, 316
transverse dunes, 101
trellis dunes, 102
Triassic strata, 75
T-shaped orthogonal system, 222

U
universal probe EUB388I-III mix, 309
U-shaped valleys, 249
Utopia Planitia, 8

V
Valles Marineris, 37
valley distribution, 258, 259
valley network, 46, 48–49, 51, 253, 294
ventifacts, 96
vermiculated surfaces, 100
vermiculation, 100
Visible and Infrared Mineralogical Mapping Spectrometer (OMEGA), 4
volcanic activity, 260
volcanic features, 40, 62
volcanic lava fields, 14
volcanic modification, 121
volcanic plains, 39, 43
volcanic weathering, 101

W
Warrego Valles, 254
Warrego Vallis, 48
water activity, 310
water ice, 4, 35, 55, 294, 296
water source, 123, 128–129, 131, 136, 260
water source of the paleolakes, 147
water sources, 122, 267–269
weathering, 182
wind, 62
wind-driven drops, 100
wind erosion, 97, 173–175, 178–179
wind tracks, 52
wind velocity, 157
wrinkle ridges, 38

X
Xiaganchaigou Formation, 76
Xitieshan Mountain, 86
X-ray diffractometry, 9
X-ray radiation, 2

Y
yardangs, 54, 79, 94–95, 125, 158, 160–165, 167–169, 171–175, 177–181
Youshashan Formation, 77

Z
Zephrria Planum, 158
Zephyria Planum, 163

Lightning Source UK Ltd.
Milton Keynes UK
UKHW020915311220
376190UK00001B/20

9 789811 224010